D1069072

THE STRUGGLE TO UNDERSTAND

"A summary of scientific discoveries that ran counter to superstitious, religious, and traditional beliefs, the consequent quenching of scientific thought, and the warnings that these struggles imply for the next century and the remainder of this one."

—Herbert C. Corben

THE STRUGGLE TO UNDERSTAND

A HISTORY OF HUMAN WONDER & DISCOVERY

HERBERT C. CORBEN

Prometheus Books
Buffalo, New York

Published by Prometheus Books

95 94 93 92 91 5 4 3 2 1

Library of Congress Cataloging-in-Publication Data

Corben, Herbert C. (Herbert Charles).
The struggle to understand : a history of human wonder and discovery / by Herbert C. Corben.
p. cm.
Includes bibliographical references and index.
ISBN 0-87975-683-7
1. Science—History. 2. Research—History. 3. Inventions—History. I. Title.
Q125.C64 1992
509—dc20 91-26727
CIP

Printed in the United States of America on acid-free paper.

To Beverly
and to
Deirdre, Sharon and Gregory,
their children and children's children . . .

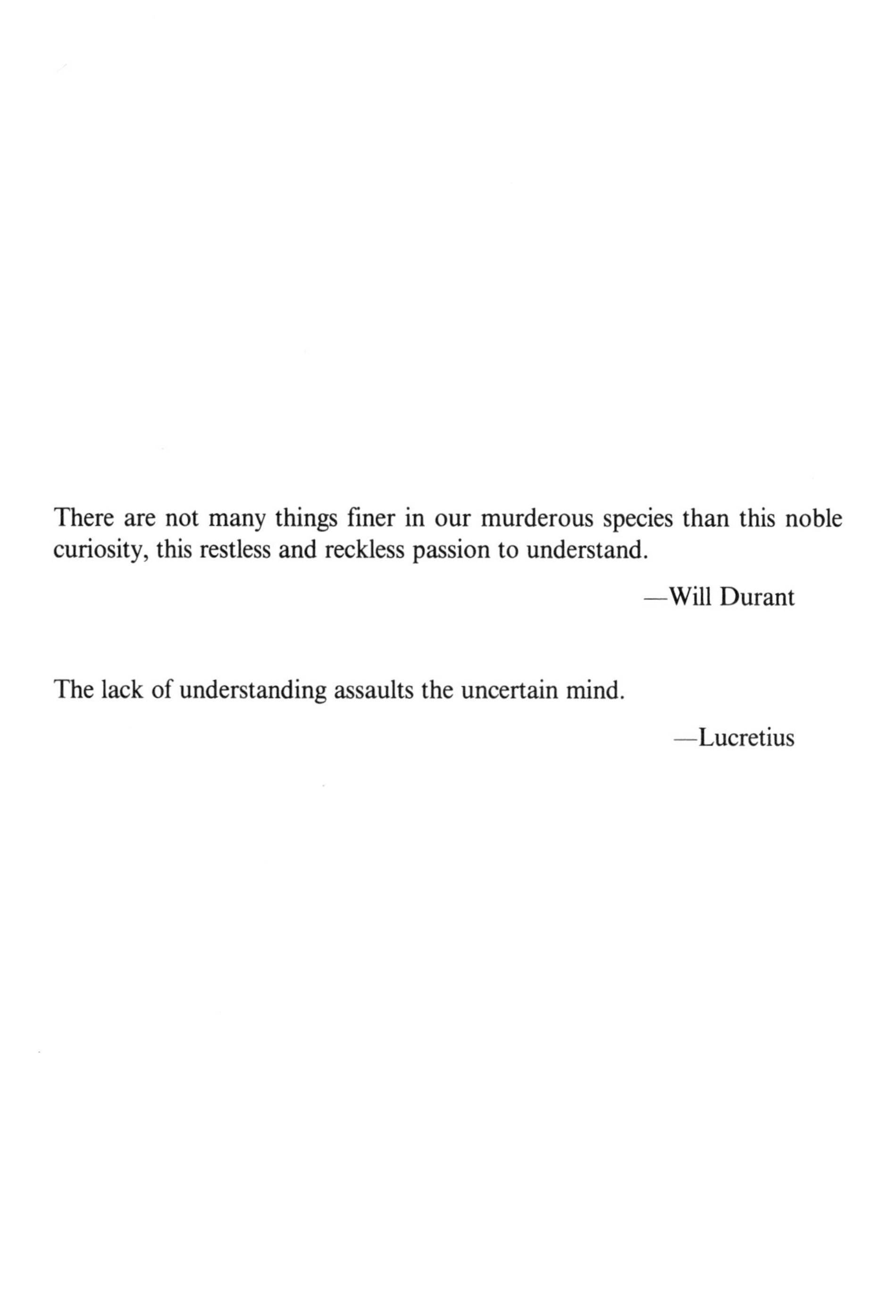

There are not many things finer in our murderous species than this noble curiosity, this restless and reckless passion to understand.

—Will Durant

The lack of understanding assaults the uncertain mind.

—Lucretius

Contents

Preface

For thousands of years, our ancestors pursued the spiritual and intellectual quests of trying to understand the world that surrounds us and the world that lies within. The desire to understand the environment was preceded by the desire—even the desperate need—to control it. The first things to do were to improve personal safety and the production of food, and for these purposes inventions and technology were developed from simple beginnings. In fact, technology preceded man, with the ape who used a stick to reach something that otherwise it could not get, or with countless other acts of animals that we ascribe to intelligence. By the time Stone Age civilizations were developing, many inventions had been made, with no idea of course of the scientific principles behind them—weapons such as boomerangs, bows, arrows, javelins, harpoons, spears and spear-throwers, along with more peaceful devices and techniques: the control of fire, pigments for painting, the lever, grindstone, wedge, hooks, sewing and, sometime before 3400 B.C., the wheel.

But these were not enough—it was necessary to form a theory that would provide both a guide to understanding and a prescription for control. More than that, it had to be an idea that not only accounted for the external phenomena, it had to satisfy an unexpressed urge to add meaning to what was a very precarious life. The first theory was that there was a spiritual being—a god—hidden in each part of nature, and that to control nature all that you had to do was to please the appropriate god. Then came the beginnings of careful science—the use of accurate observations of the stars and planets, and, in Egypt, of the symptoms of diseases. From these origins, we shall be concerned here with the laws of science, the revelations of religion, and the passion and controversy that have surrounded any change ever made in our understanding of either of them. We shall also examine some threads in the complex fabric of superstition that has guided so many in the past and continues to have an enormous impact today.

Our ancestors made serious mistakes in their efforts to understand natural phenomena; they got hung up for centuries on what we see as ridiculous ideas, and carried out research (as we do, even if very careful) with one hand tied by their own prejudices and the other bound by the beliefs of their contemporaries. But they were just as smart as people are today. They did not have the benefit of a modern education, and even now those who have had that opportunity and have not learned from it are back with them in the Dark Ages of superstition, peddling their beliefs in astrology, numerology, and the literal truth of every word in the book of Genesis.

Five hundred years ago, astrology seemed to be a reasonable hypothesis, since the suggestion from the pre-Christian era that the stars are at immense distances from us had been forgotten. Three hundred years ago, the idea that not only the earth but the stars, planets, Man—the whole works—were created in October 4004 B.C. wasn't such a stupid idea, if again you ignored some ancient writings and believed passionately in those that appear in the Bible. It is obvious to anyone with common sense that the earth is not spinning or moving around the sun—we would feel the motion if it did, wouldn't we? Clearly, heavier bodies fall faster than lighter bodies—just drop a penny and a feather and see for yourself. And those old stories of that man Ziusudra, from Sumer, building himself an ark to escape a tremendous flood, or of Utnapishtim or the Greek Deucalion doing the same thing, are just pagan myths. The only one that this really happened to was named Noah.

We shall trace these and other ideas through their successes and failures, through brilliant developments obstinately opposed or ignored, the lucky guesses, the incorrect hypotheses strongly clung to, and the personal dangers that were the rewards of many scholars who struggled to understand them. As far as possible, we restrict our attention to ideas that, through intuition, logical analysis, careful study, and accurate observation have turned out to be correct, and to those wrong ideas that have hindered the development of true understanding by being built also on intuition, logical analysis, careful study and even, in some cases, accurate observations.

The basic difference between the two methods does not lie so much in any of these procedures as much as in the assumptions that are made before beginning an investigation. One way is to choose from all of the writings and sayings of mankind a very small fraction and assume that bit to be absolutely true. It might be the Delphic Oracle, the Bible or the Koran, the pronouncements by the Pope, the works of Aristotle, Thomas

Aquinas, or even James Ussher (who set 4004 B.C. as the date of the Creation), or maybe the words which a charismatic speaker assures us come directly from God. There is so much material written and spoken, and it is absolutely impossible to comprehend more than a tiny fraction of it. The easy way out is to pick a few authorities and stay with them. It is the method of intellectual Fascism. In the other way of trying to understand some aspect of nature, you still have to choose a small fraction of the information and ideas of others, but you don't assert that these authorities are absolutely correct, and you use reason to sort them out, not to confirm them. This leads to uncertainty, which is troublesome to many people, but democracy always leads to uncertainties. Just as surely as democracy is wary of dictators, so intellectual democracy must be wary of intellectual dictators who destroy freedom of thought in themselves and their own children and would like to do the same thing to others.

Superstition also offers to some people the hope that it will help them to understand apparent mysteries. The meaning of the word depends on where and when you are defining it, and to whom you are talking. In 1531, when Henry VIII issued a law forbidding gifts and bequests for "superstitious uses" he meant for it to apply to anyone whose beliefs and rites were not strictly those of Henry's Catholics—Roman Catholics, Jews, and anyone else who might exalt the Pope or fail to exalt Christ. In the same spirit, but not the same letter, the word can now mean anything that we don't believe in. To an atheist, religion is a superstition; to an agnostic, it might be. However it is categorized, superstition has had a major influence on society and therefore on generations of philosophers and scientists who have tried to extend our knowledge of nature. Many of them were astrologers and alchemists themselves, with no idea that these aspects of their struggle to understand nature were on the wrong track. After they had died, their other research, combined with that of many with similar motives, revealed astrology and alchemy as forms of superstition. It was more serious when it was *required* to hold superstitious beliefs—devils inside the physically and mentally ill, witches controlling the weather, comets as portents and warnings from the Almighty, magicians working evil and becoming precursors to the image of the Mad Scientist.

A single document can dig deeply into only a microscopic section or, like this one, treat a large area more superficially. This book is intended for those who would like to think with me a little about "this restless and reckless passion to understand" and the excitement and pain that has come with it.

We omit so very much of the world's tradition by concentrating here only on our heritage from the Near East, from Jews, Muslims, and Christians, with particular emphasis on the religion of Christianity and the sciences and superstitions of Western Civilization.

The first three chapters of *The Struggle to Understand* tell of ancient gods and religions and the ways in which these beliefs and ceremonies became essential parts of Christianity. The next four chapters outline developments before the Christian era in physics, astronomy, biology/mineralogy, and medicine. We then study the growth of astrology, alchemy, and numerology up to more recent times. Chapter 9 discusses some of the scientific beliefs and theological conflicts of the first Christian millenium, and Chapter 10 outlines the magnificent contributions of Muslim scientists and physicians during the latter part of that period. We then go on to examine the impact on Christian thinking that occurred when these Muslim works were translated into Latin and Hebrew, and the struggles that followed in astronomy and physics, in the persecution of witches and the treatment of the insane, and in Biblical criticism and Christian methods of protection against weather and diseases. Chapter 15 centers on the development of the physical sciences of the eighteenth and nineteenth centuries and the ways in which erroneous scientific assumptions stood in the way of progress. This is followed by an outline of nineteenth-century discoveries in geology, paleontology, and biology, and their impact on the theory of evolution and the impact of evolution on religious dogmas. The final chapter, "Past, Present, and Future," is concerned with lessons from the past, with examples from the present, and fears and hopes for the future. The other chapters are examined individually for their modern relevance, with quotes and incidents taken from contemporary newspapers and periodicals. The struggle between democracy and dictatorship continues on the intellectual and religious level, for there remain brakes on free inquiry that have secular as well as religious origins.

Following the main text is a section *In Memoriam* providing short biographies of some who have attempted to use the methods of science and who, for a variety of reasons, have been harassed, accused, tortured, and even executed. The first time the names of these people appear in the text, an asterisk(*) is used to remind us of the hostile environment in which they worked. This is followed by a list of "Helpful Despots"—notes on emperors, caliphs, and other despots throughout history who have supported the scientific method of enquiry and those who practised it. A dagger(†) is used to denote the first time that their names appear.

I am very grateful to the magnificent and dedicated scholars whose works have inspired and enlightened me. The section titled "Select Bibliography" at the end of the book provides a list of some relevant works, with special note of those that I have found to be particularly valuable. However, this book is not in the same category as any of them—I am not a professional historian or encyclopedist; I am a scientist, rather, one who is deeply interested in the lessons we can learn from history. I write this about what historians of science and religion tell us, adding a few scientific details not usually presented and offering a number of personal judgments, leavened with a little levity.

Most of the difficulties that scholars have encountered and that are related in this book were of external origin—ridicule, lack of financial and moral support, persecution, torture. With these went internal, personal feelings, often hidden, some of them experienced by all scientists, past and present. While we tell of the external influences, you are invited to share these inner feelings—the excitement, even rapture, of discovery, the letdown on finding out that you were on the wrong track, the fear that someone else would scoop you, the anger at the theft of your ideas, and the temptation to give up the struggle to understand.

1

Gods, Goddesses, and Time

So many gods, so many creeds
So many paths that wind and wind
When just the art of being kind
Is all this sad world needs.

—Ella Wheeler Wilcox (1850–1919)

Time! It has been measured and mismeasured, used wisely and wasted; it has enslaved some and enriched others, and, except for those effects that are described by the theory of relativity, it has resisted all attempts to advance or delay it.

Time has been integrated mathematically with space, but nobody really knows why time and space are so different from each other. Each of us is imprisoned in a tiny increment of eternity, partly free to move in space but, in time, physically confined between birth and death. Slowly we become aware that there were those who preceded us and are gone, and those who will follow us when *we* are gone and whom we will never be able to see.

We resent this time-trap, although we are usually less upset about the time gone before than we are about missing out on future time. We try to escape the dilemma by espousing superstitious theories or religious beliefs, both of which are supported by formalized systems.

Realistically, the only way that we can control the flow of time is in our imagination. For example, imagine for a moment that you are visiting a time 20,000 years in the past, and you are in a winter-chilled cave carved by nature into the side of a hill. You have just finished dining hands-on from a slab of bison, and soon it will be time to lie down on the pallet of animal skins that your hostess has prepared for you.

Darkness fills every crevice of the cave except for the light of the small fire that serves both as kitchen stove and as heater. Your host and hostess would be amazed at what you know, were they able within their ancient experiences to understand even a small part of it. Your wrist-watch, for example—they would be very impressed to learn that you can tell the time just by consulting a number that flashes regularly on a small screen, provided that they knew the word "time" or what it stands for. They divide time up into day and night, not minutes and hours.

You try your best to fathom the world as they do—in simplistic terms like getting up, lighting a fire, deciding whether the food supply requires a day of hunting animals or of gathering plants. If there is to be a hunt, then there follows the preparation for it, testing the spear-thrower, perhaps, and sharpening the spears. After would come the triumphal return, preparation of the feast, a good meal, and so to bed, on the floor.

Tonight, the woman's attention is entirely centered on her child, a little girl of about three who has been ill for a week. You know that she is ill because you have looked at her and have felt her warm forehead. She has a high fever. You call the child Zoda, and you have named her parents, Xa and Yo. That is the closest your ear will come to making sense of what they say to each other in their primitive but flowing talk that we call "Early Nostratic." Now Xa pokes the fire, the sign that he is ready to sleep. Yo, who is already half asleep, gathers animal skins around Zoda's shoulders.

Suddenly you see that Yo is lightly shaking the child, trying to wake her. Xa stops attending the fire to watch. Yo gives a sudden fierce low cry. Zoda has gone to sleep quietly by the fire on this cold night, and now she will sleep forever.

You ache for this grief-stricken couple; the woman sobs and rocks her baby in her arms as if she could bring life back into the child just by holding her; the man stands at the entrance to the cave gazing blindly at the star-lit heavens. She wants to know why this has happened, who is responsible? You even see something in her face that surprises you—something more than fear and different from pain—it is awe and dread, mixed with acceptance, as she listens to her husband explain.

Mysteriously, or perhaps by his gestures, you are able to understand what he tells her. It is the star he has seen for several nights now that has come for their daughter and has taken her spirit away. Ever since seeing that star he has had the strange feeling that it has come into the sky especially because of Zoda, and that it is now content.

Yo believes that her husband tells the truth, and she carries her permanently sleeping daughter to the entrance of the cave and looks up at the heavens to where her husband points. Yes, she too sees the star, a very bright one, and she feels comforted.

Xa takes a piece of hollowed-out soap-stone that holds some animal fat. He lights the greasy fat with a stick at the fire and picks up a carved and sharpened antler. He moves back into the cave, out of sight, and starts digging. . . . Yo will soon dress the dead child in fine skins taken from a recently caught hind and will prepare some red ochre to pour over the baby to give her the blood of life. The child will stay close to them until they join her. It is their custom.

Xa, Yo, and Zoda are the hypothetical names of a family of Cro-Magnon Advanced Hunters, as we call them. Their ancestors arrived in southern Europe 15,000 years earlier, depending on their superior intelligence and inventiveness to displace the more primitive Neanderthals. The ridges of their brows were a little more prominent than ours tend to be, their faces perhaps a little longer. Otherwise, put a suit on Xa and walk him down Fifth Avenue, and nobody would notice his Age. His compatriots have given us the earliest statues that have been found so far, the first graven images of goddesses. From Willendorf, in Austria, comes a four-and-a-half inch stone statue of a very fat woman with enormous breasts—her width at the waist is half of her height. Other ancient fertility goddesses have been found in various parts of southern Europe, some in ivory, like the one found at Lespugne, southwest France, and others in bone and stone.

These people were struggling to understand birth and death, to make some sense of these boundaries to their lives—something that we are still trying to do. Along with their ancestors and descendants, they made up stories to explain death, to comfort each other in their losses, and stories about birth and sex and inheritance and taboos. From these came the myths—stories handed down in spoken tradition, stories that seemed to explain so many puzzles of the human predicament.

More tangible legacies from those times are the pieces of equipment for hunting and fishing that have been found—fish hooks, spear throwers, harpoons, and devices to catch birds, fashioned from bones, antlers, and stones. Five or six thousand years after the fictional Xa, Yo, and Zoda, their descendants, the advanced Cro-Magnon Magdalenians unintentionally left time capsules that would shock the late nineteenth century.

After buying a piece of property, it is always exciting to discover on it something that you didn't know was there—something positive, that is, like flowers sprouting from hidden bulbs and seeds, or maybe gold or oil, or even buried treasure. When Señor Marcelino Sanz de Sautuola acquired some land in the Cantabrian Mountains of northern Spain during the 1870s, he had no idea of the treasure that came with it, locked in a cave that had been sealed and camouflaged by natural debris for thousands of years.

Fortunately, he decided to build on this property. When inspecting the progress of the construction, he was intrigued to find that blasting for the foundations had opened up the entrance to a cave that had been tightly closed with rocks and stalagmites.

He brought his young daughter with him to explore it; the ceiling was low and he had to bend over, but the young girl was able to stand up straight and look around. Suddenly, "Papá, una vaca!" she may have said. No, it wasn't a "cow," but there was a painting of a large bison above them. Further exploration revealed paintings of hinds, horses, wild boars, and a whole herd of more than twenty male and female European bisons—sometimes called aurochs—standing, jumping, watching, foraging, wallowing. Separately, one wounded bison was dying, fallen but still defiant.

Mixed with the paintings are various engravings representing deer and what appear to be small huts—these cave dwellers did not spend all of their domestic time in caves. Also prominent are engravings of "antropomorfos," as described in the brochure of the *Centro de Investigaciones y Museo Altamira*—polytheistic deities in human form. Found on the floor there are shells and bones used as paint receptacles, together with some paints—naturally occurring carbon and manganese ochres, some mixed with grease, others in solid form for use as crayons.

We now know that these paintings at Altamira have been there for some fifteen thousand years, but when Sr. de Sautuola claimed that they were very ancient the leading archaeologists of the day did not believe him. Darwin's *Descent of Man* had appeared only eight years earlier, and biologists, geologists, and other experts were busy sorting out its implications. People generally had been in an uproar since 1859 over Darwin's contention in *The Origin of Species* that the various species were related to each other and were changing naturally, instead of staying fixed as, it was believed, God had created them nearly six thousand years earlier. Now comes this outrageous claim from Darwin that people themselves are part of the evolutionary chain.

In this intellectual environment, the experts were not about to suddenly accept evidence of the existence of man some thousands of years before the Christian date of creation. The leading scientific opponent of evolution, Louis Agassiz, professor of zoology at Harvard, died in 1873, but there were other scientists who had their doubts, particularly when it came to the evolution of man. At the 1880 Congress of Lisbon, the professionals in the various related fields rejected, sight unseen, the idea that the Altamira paintings were the works of late ice-age artists.

Scientists are forever steering between the Scyllan monster of skepticism and the Charybdian whirlpool of gullibility. This time they stayed on the side of the monster until 1910; by then similar caves had been discovered in the nearby southwestern part of France, and thin stalactites that covered some of the paintings had testified to their age. However, Sr. de Sautuola had died, unable during his lifetime to convince the experts of the significance of his discovery. (Strangely enough, two years later, anthropologists foundered in the gullibility whirlpool, judging and believing for the next forty years that the fake skull that had been left to be discovered in Piltdown, Surrey, England, was a genuine link, "Piltdown Man," in the chain of evolution.)

What motivated these early painters to portray this assortment of animals so vividly and dynamically? Sheer enjoyment? Certainly. In a similar cave at Pech-Merle, France, artists left their marks, each placing his hand against the wall and blowing paint over and around it, so that paintings of horses are interspersed with accurate images of human hands. There must be other reasons for the animal paintings, apart from the joy of self-expression—to remind all comers where the food comes from?—to help the hunters "psych up" for the chase? Did they wish to portray animals as so important to the community that they were elevated to the status of gods?

We do not know, because it was a long time before written records, but we do know that four thousand years or so later, with the techniques of agriculture and the domestication of animals spreading from centers such as Jericho, gods were believed to dwell not only in animals, but in all life. The religious cult of the goddess of Mother Earth continued as a basic belief for thousands of years as she became the goddess of sex, of fertility, and of the miracles of harvest and birth.

More than two million years earlier in the evolution of mankind, there began to emerge a new method by which evolution could occur. Under the general heading of "survival of the fittest" came a specialty that we could call "survival of the smartest." The slow change in the position of the thumb allowed our ancestors to work more easily with their hands, and the equally slow change in the foot allowed them to walk on two legs, leaving the arms free for other activities. With these changes came a slowly increasing brain size and a corresponding increase in intelligence.

About one and a half million years ago, the first tool makers, developing in Africa, were about four feet tall. Their brains were about the size of a fully grown grapefruit. More precisely, the volume of the brain of one of these very early people was, on the average, that of a sphere with a 11 cm (4 1/3″) diameter. That gave them the advantage over a more primitive group of the time, who had brains about the size of a 10½ cm (4⅛″)-diameter ball, like the brains of large apes.

Over the next *million* years, these tool-makers spread from Africa through Europe and Asia, growing to five feet in height. If one of their brains was moulded into a ball, it would be approximately 12.4 cm (4⅞″) across. That is a 65 percent increase in volume over a diameter of 4⅛″. They had learned to control fire and they invented the handaxe, a pointed and sharpened stone that was so useful that specimens have been found in many of the places where these people lived.

They had survived two long ice-ages and were about to enter a third. Their descendants looked more and more like us, not every day, perhaps, but the change could be noticed at least every hundred thousand years or so. When they encountered the last ice age 70,000 years ago, we call them "Neanderthals," the name of the valley near Düsseldorf in Germany where their remains were first uncovered.

The Neanderthals had a whole array of weapons, tools, and warm clothing. There was a thick bony ridge above their eyes—that and their receding chins would make them stand out in a modern crowd. However, these people had made a leap of consciousness and abstract thought. They buried their dead with ceremony, marked their graves, and they appear to have believed in an afterlife. The expanding intelligence allowed our early ancestors to survive, but it was a double-edged sword—it also allowed them to think and to wonder.

As time moved on, in order to bring order out of the chaos of their imaginations, the sages of antiquity looked up from the dust at their feet and thought about the world. They looked at the immense complexities of nature that surrounded them, sensed the vagaries of their own human nature and animal instincts, and decided that there must be a number of supernatural beings responsible for it all.

Some of these gods and goddesses were animal-like, others more human, with a number of combinations of half-and-half—mermaids, centaurs, the god Pan, and several Egyptian gods, for instance. Nearly all were endowed with human fears and hopes, whims and strivings, vengeances and lusts, together with superhuman powers, like the power of immortality, or the power to transcend time.

Unaware of their evolution from animals—*The Ascent of Man* as Dr. Jacob Bronowski called it—people tried to understand where they and everyone and everything else came from, and postulated a *Descent of Man* from the gods. While the morality of some of these gods can certainly be questioned, there began a struggle between man's animal nature and his perceived divine essence, a struggle to understand man's place in the world that has echoed through the centuries.

Ultimately animals, plants, and all other parts of creation were perceived to be powered by a god—the setting sun, the rising moon, the circling stars, the wandering planets, the roaring ocean. . . . All of nature was a congregation of supernatural beings, some of them very dangerous and distinctly evil.

Identification with the "Earth Mother" and fear of punishment combined to mitigate against the senseless destruction of nature. Would that same respect for nature were always with us today. But there are mortal dangers—lightning, earthquakes, volcanoes, tornadoes, droughts, floods, and the rumble of thunder. In the absence of any understanding of the physical nature of these perils, how else would you explain them except as manifestations of angry gods or evil demons, magnified images of man? In the struggle to understand, the belief that natural phenomena arose from the free will of supernatural beings was the first step. Even today, a natural disaster in which human agency has played no part is called an "act of God."

How could men best protect themselves from these great powers? Appeasement through human sacrifice might work and get rid of some enemies at the same time. They could also use a magic ritual to endow their victim with supernatural qualities, and then acquire some of his power for themselves by eating his body and drinking his blood. Maybe sacrificing someone who was very dear to them would bring even better results.

We all know of the command of God to Abraham (Genesis 22:2) "Take your only son Isaac, whom you love, and go to the land of the Amorites [Palestinian people who lived in the mountainous regions] and offer him there for a burnt offering upon one of the mountains which I will tell you," and the happy outcome of the replacement of Isaac by a ram for the sacrifice. In the same theme but in a different culture, Greek mythology tells how Agamemnon offended Diana, who responded by becalming his fleet in Aulis some forty miles north of Athens. Here he was, twelve hundred and fifty years before the birth of Christ, all set to lay siege to Troy, and the goddess had decreed that she wouldn't let the winds fill his sails until he sacrificed a virgin! He chose his daughter Iphigenia, and Tennyson has her speak at the moment of sacrifice:

"My voice was thick with sighs/ As in a dream. Dimly I could descry/ The stern black-bearded kings, with wolfish eyes/ Waiting to see me die." Euripides reports that, fortunately, Diana relented at this last moment and carried Iphigenia away, leaving a hind to be sacrificed in her place.

In II Kings 3:27, we are told that, after being badly defeated by the combined armies of Joram and Jehoshophat, King Mesha of Moab "took his eldest son who was to reign in his stead and offered him as a burnt offering." The Moabite Stone discovered in 1868 confirms this conflict of the ninth century B.C. Mesha and his armies apparently gathered strength from the sacrifice and wreaked some vengeance on the Israelites. Two hundred years later, Josiah "destroyed the altar of Topheth in the Valley of the Sons of Hinnon, so that no one could ever again use it to burn his son and daughter to death as a sacrifice to Moloch (II Kings 23:10,

Living Bible Translation). This was uncomfortably close to more modern times—the first philosopher of Greece, Thales of Miletus, was born while Josiah was King of Judah. Unfortunately, Josiah was unable to stamp out completely the practice of sacrifice by fire; in the Judeo-Christian tradition, but not its spirit, Catholics and Protestants alike have burned uncounted numbers of "heretics" and "witches." Even today ritualistic murders continue to be performed by followers of Satanic cults.

Slowly, however, the practice was generally replaced by the sacrifice of animals, but it still wasn't all that easy to figure out what the gods really wanted. Early Babylonians, for example, examined the livers of sacrificed animals. A Babylonian clay model of an animal liver is in the British Museum, and a later bronze model of a sheep's liver was found in Italy about the same time that Sr. Sautuola found the Altamira paintings. Some of the Etruscans, who preceded the Roman Empire in Italy, must have brought this model with them from Mesopotamia, so that Roman priests also learned the fine art of interpreting animal livers. The theory is simple: the liver holds a great quantity of blood—it is therefore the seat of the soul. The god who accepts the sacrifice identifies with the animal itself; therefore, the soul of the god can be read in the liver.

Other techniques were developed, all aimed at determining the wills of the gods and therefore the course of human events. Roman priests (*augurari*) supplemented the "knowledge" that they gained from entrails by observing "omens" such as the flights of birds. We still recognize this in our language with words like "augury," "ominous," and "divination." In Judea, the throwing of dice was a common method of achieving the same results. Other methods included the examination of the shapes, and presumably the colors, of oil drops on water and, above all, the use of dreams to predict the future. (Now that dreams are known to reveal the past rather than the future, people are supposed to consult their psychiatrists rather than their astrologers.) The other techniques for prying into the future have their echo today in the interpretation of sequences of playing cards or configurations of tea leaves, or that offshoot of astrology, the chiromantic art, palmistry, of interpreting the creases and bumps on the hand, sometimes justified by the passage from Exodus (13:9): "It shall be to you for a sign upon your hand."

As the laws of chance allow some predictions to be fulfilled, the methods are seen to be valid, especially if the cases where they don't work are ignored. The laws of chance are then seen to be really the laws of the hidden order that, in our various ways, we are all seeking. This search, and the belief in the existence of order, form the fundamental creed of science, religion, and even superstition. It was strongly reinforced thousands of years ago by observing the regular motions of the stars and other heavenly bodies.

Clearly, above all this local chaos of the weather and of subterranean fires there was perfect, repetitive order, waiting to be properly interpreted. Thus was born belief in a god of rule and order in the heavens. Also born were astrology, concerned with the impact of the sun, moon, planets, and stars on human affairs, and astronomy, more concerned with establishing a calendar and using it for predicting floods, planning for planting and harvesting, and fixing holy days. We return to these in chapters 4 and 5.

The gods of early civilizations began to acquire specialties and receive names. Our records begin when the essence of writing—cuneiform, that is, wedge-shaped marks on clay tablets, but not yet the alphabet—began to appear more than five thousand years ago in the civilization of Sumer, in Mesopotamia, near the southern part of the front line of the recent Iran-Iraq war. Some two hundred years later, the Egyptians began to use hieroglyphic writing on papyrus, made by weaving strips of the pith of a sedge of the same name. Very soon they caught up with and passed the Sumerians in their level of sophistication.

Gods and goddesses were arranged in hierarchies, the chief god representing a source of great power: Enlil of Sumer, Marduk of Babylon, Ahura ("azure sky")-Mazda, or Osmazd, of Persia and his son Mithras, Amon of Egypt with Seb and Nut (earth and heaven) and possibly (there are conflicting stories) their son and daughter, Osiris and Isis, Zeus of Greece who, along with Amon, became Jupiter of Rome, Odin and his wife Frigga and son Balder from the North, Yahweh, or Jehovah, of Judea and, later, Gautama Buddha of India, Jesus of Nazareth, Allah of Islam.

The cross was a sacred symbol of the god Tum—or Atum, or Amon—the chief god of Thebes who became the supreme god of Egypt. He carried this cross, called an ankh, that signified life—it was like the much later Christian cross, except that the short vertical part above the cross-bar was replaced by a loop. This god created the first couple, Shu the god of sunlight, and his wife Tefnut, the lion-headed goddess of rain. Shu was sometimes identified with the cat-god Bes; in any case he and Tefnut produced the earth god Seb (or Geb) and his wife and sister Nut, the heavens—and then pulled them apart.

They apparently got together again, at least from time to time, as Nut gave birth to the powerful god Ra, from whom the pharaohs claimed descendancy—despite the fact that Ra was represented with the head of a lion, a cat, or a hawk. Osiris and Isis also came from the marriage of Seb and Nut and we return to them and their son Horus in chapter 2. We note here that they had another son Anubis who had the head of a jackal. Anubis worked with Thoth, the god of time and of knowledge, the god with the head of an ibis, the god sometimes called "cynocephalic," a fancy way of saying that he was often depicted with the head of a dog.

Another couple born to Seb and Nut was the husband-wife, brother-sister pair Seth and Nephthys. The question equivalent to "Where did Cain get his wife?" was not raised here—there was no taboo against brother-sister marriages among the Egyptian or Greek gods and goddesses. However, like Cain, Seth slew his brother Osiris. Thereafter Seth was represented as a beast with a pointed snout. If the early Israelites were enslaved in Egypt, as the book of Exodus states (but this is disputed by many modern scholars), then these myths must have been known to Moses and the Israelites during their captivity.

A true story that also must have been known to them was that of a pharaoh of Egypt who believed that there was just one god and who tried to force that belief on his people. In 1380 B.C., Amenhotep IV became that pharaoh and he changed his name to Ikhnaton ("Aton is pleased") because of his faith in one and only one god, the sun, Aton. It was not just the sun, but all of the bounty that the sun brings, the heat of the sun, the universal god of love, the god of all life, and the god of all nations. His beautiful foreign wife, Nefertiti, influenced him in this faith.

Interpreting the wills of the gods became very complicated, and there was no way for ordinary people to understand how to do it. Priests had therefore set themselves up as the only ones capable of "divining" divine wills, and it soon became apparent that the kings and pharaohs of Babylon and Egypt were also subject to the gods, effectively giving full power, and the corruption it brings, to the priests. Almost one thousand years before Ikhnaton, King Urukagina of Lagash, in Mesopotamia, had issued edicts forbidding the priests from taking for themselves the items that the people had offered to the gods, and from charging excessive burial fees. Perhaps the Babylonian use of animal entrails for divination arose because that was all that was left of the sacrifices after the priests were through with them! Ikhnaton was equally incensed with the corruption of the priests of Amon. "Amon" needed lavish gifts, "Amon" needed sacrifices of fine animals, "Amon" needed a large harem, . . .

Ikhnaton's simple conception of one universal God of Life and Love, named Aton, turned the powerful priests into his deadly enemies, and understandably so, for Ikhnaton was very dogmatic. Aton was his god, and he was determined that Aton should be everyone else's god too. When he went further and ordered the removal of the names of all other gods from all tombs and monuments, the people also turned against him. He was a poet and a visionary, not a warrior, and his empire collapsed around him. He died at the age of thirty, a broken man, a century or so before Moses led the Israelites out of Egypt.

Under different circumstances of history, some of his poetry might have become part of our legacy of Christian hymns. Its more immediate legacy was Psalm 104, the twenty-fourth verse of which reads: "O Lord, how manifold are thy works, in wisdom thou hast made them all; the earth is full of thy riches." Ikhnaton wrote it this way:

> How manifold are thy works
> They are hidden from before us
> O sole God, Whose powers no other possesseth
> Thou didst create the earth according to thy heart
> While thou wast alone:
> Men, all cattle large and small,
> All that are upon the earth,
> That go about upon their feet;
> All that are on high
> That fly with their wings
> The foreign countries, Syria and Kush
> The land of Egypt
> Thou settest every man into his place,
> Thou suppliest their necessities . . .
>
> (Du 1, 208)*

At that time the Jewish people worshipped a number of gods of the Near East. In Ezekiel 8:14 we read, "there sat women weeping for Tammuz"—the Babylonian god was mourned and resurrected annually, leaving his permanent imprint as the name of the tenth month of the Jewish calendar. The brazen serpent made by Moses was worshipped until the time of Isaiah, around 720 B.C. So also were Moloch, with his thirst for sacrifice of the first born, and Baal, the Golden Calf, derived from more ancient worship of the Bull, the principle of fertilization by the male; in fact, "every form of creeping things and abominable beasts" (Ezekiel 8:10) was worshipped. The first commandment recognizes the early Jewish polytheism, "Thou shalt not have any *other* gods *before me*." It was King Josiah, again, whose high priest "found" the forgotten copies of the Law and used them to justify purging Judea of all other gods, including those that had been installed in Solomon's Temple three centuries earlier. Thereafter, there was, at least officially, only one Hebrew God, Jehovah, his personality made more loving through the writings of the Isaiahs and, later, the teachings of Jesus.

Civilizations as far apart as the islands of the Mediterranean and the islands of the Pacific independently came up with the idea that there is a supreme god surrounded by a number of assistants, enemies, and specialist sub-gods. Many gods such as Inti of the Incas, Rangi of Polynesia, Mani-

*For abbreviations of citations, see Select Bibliography.

bozho of the Algonquins, Kombu of the Bantus, Jumelo of the Finns, Indra of India were the chief gods of the universe, each believed in by his own people. Throughout the world there were, and in some places still are, gods and goddesses of the sun, the moon, the earth, and the sea, gods of love, sex, and fertility, gods of the dead, gods of lightning and thunder—even the Mayan god Huracan, the god of hurricanes.

Early Greek civilization developed an interlocking hierarchy of gods and goddesses that led people to believe that there were explanations for many physical phenomena, while at the same time it projected the foibles of mankind onto a large screen, where they were characterized by divine behavior. The lightning and thunder were evidence of the power of Zeus—Odysseus tells in Book VII of *The Odyssey* how "with one of his blinding bolts Zeus had smashed my good ship to pieces out in the wine-dark sea." The waves of the sea were subject to the will of Poseidon. The sounds of voices bouncing back from the hills came from the nymph Echo, who talked too much and thereafter was condemned to say only the last word of anything that was said to her. (Worse than that, she fell in love with Narcissus, who fell in love with himself, looking incessantly at his reflection until he turned into a flower.)

There were Aphrodite, goddess of love; and Mneme, goddess of memory; Clotho, Lachesis, and Atropos, the three fates; Dionysos, the god of wine; Asclepios, the god of medicine; and many more. There was even the lady Leda, who apparently in one evening was impregnated with two pairs of twins, one from her husband and the other from the ever-philandering Zeus, dressed as a swan, maybe to fool the husband Tyndareus. The love affairs of Zeus with his aunts, sisters, and female cousins, as well as with an unending stream of lovely mortals like Leda, outdo Rabelais' *Gargantua* and *Pantagruel,* Balzac's *Droll Stories,* and a dozen popular romances.

The trunk of the genealogical tree of the Greek gods and goddesses, along with some of its branches, is shown in the diagram.

The monsters that emerged from the union of Tartarus with his mother Gaea—the earth mother—show an early recognition of the perils of incest. Tartarus and his mother produced the fire-breathing giant with a hundred heads, Typhon (from which our word "typhoon" is derived), and Echidna, half woman and half serpent. Brother and sister then got together to produce a whole range of horrors: the fire-breathing, lion-dragon-goat monster Chimaera, a personification of the volcano of the same name; the woman-lion sphinx; Cerberus, the dog-serpent who guarded the entrance of Hades; Scylla, already referred to; the Nemean lion, slain by Hercules; the eagle that ate the liver of Prometheus; and other personifications of evil.

We shall not dwell here on the internecine feuds, the other incestuous relationships and the subtle symbolism of these gods, so thoroughly explored

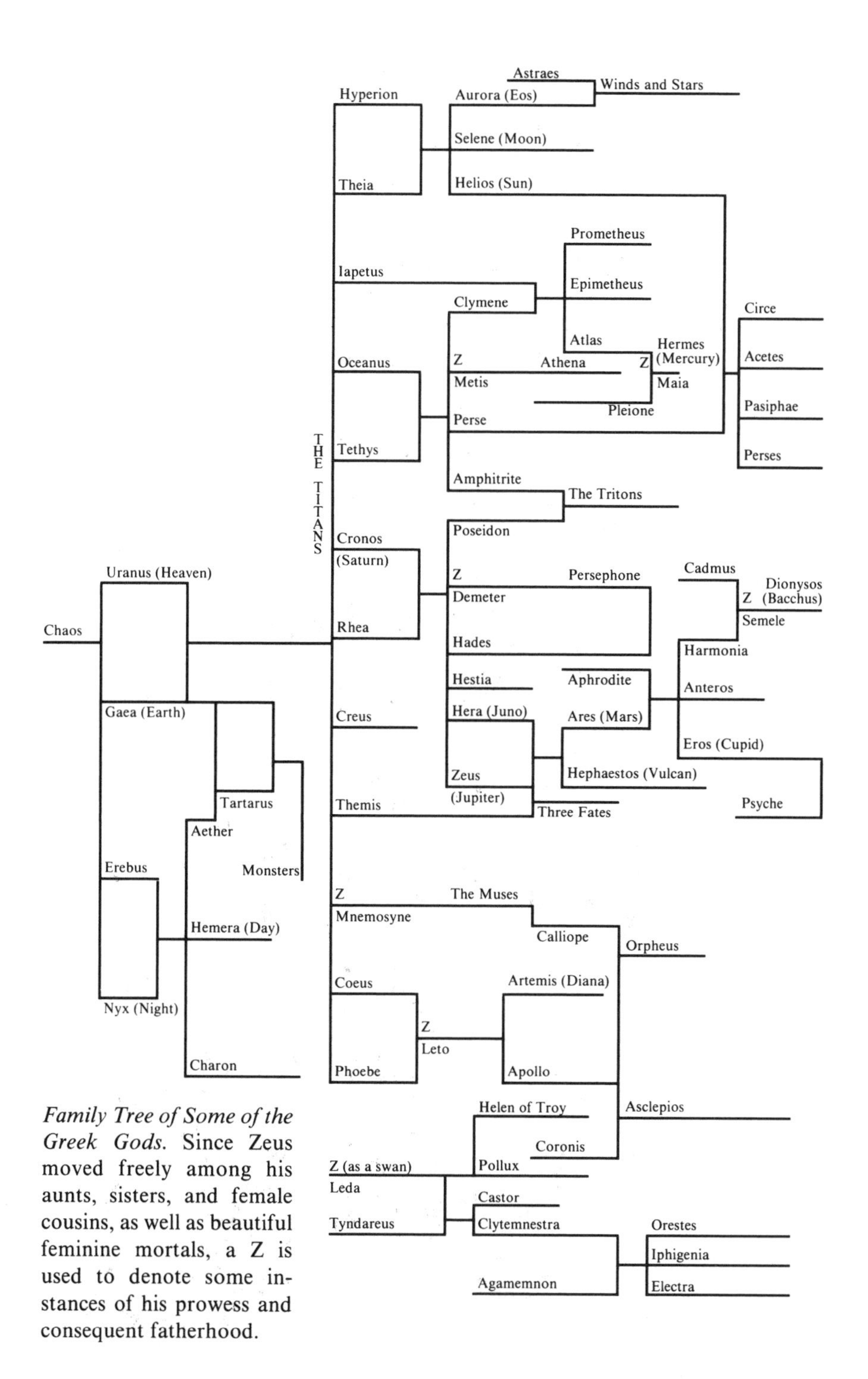

Family Tree of Some of the Greek Gods. Since Zeus moved freely among his aunts, sisters, and female cousins, as well as beautiful feminine mortals, a Z is used to denote some instances of his prowess and consequent fatherhood.

and described by Joseph Campbell (see Select Bibliography). We merely emphasize that this was an early attempt to construct a religious model that would explain many widely different effects—from physical phenomena to mental quirks—by attributing them to gods.

There was a criticism of this religious approach to understanding as early as the end of the fifth century B.C. Before he was killed in the year 404 B.C., the Greek tragic poet Critias, a pupil of Socrates, had this to say about the gods:

> I fancy then some subtle sage conceived
> What mortals needed was to find out how
> Fear might be laid on evil-doers, if aught
> They do or speak or think in secret wise:
> That then he introduced the Being Divine,
> As spirit blooming in perpetual life,
> Hearing and seeing and thinking with the mind,
> Forever keeping watch on those misdeeds,
> And as a god, with power to see and hear
> Whate'er was done or said among mankind;
> Even if in silence you frame evil wishes,
> You shall not hide it from the gods, for thought
> Is the gods' essence.
> Speaking in such words,
> He must have introduced grand moral teachings,
> Concealing truth with mask of lying phrase;
> Asserted that the gods dwelt here on earth,
> To strike dismay to men and lead them on.
> He noted, too, that fears came on them thence
> Adding new hardships to their wretched life:
> The motions of the sky that brought about
> The lightning's glare, the fearful thunder crash,
> The starry host—resplendent broidery
> Of time, sage artificer; thence beside
> The dazzling meteor shot the heavenly way,
> The laden storm-cloud moved along the land.
> These all about them pierced their souls with fear;
> Thereby his speech gained credit, when a place
> He chose as fit to build the god a home,
> And crushed the headstrong by the laws he made
> Thus first, methinks, men must have been persuaded
> By some man to obey the spirit's law.
>
> (UA 4, 281)

The Romans took count of the balance of trade in deities, dividing them carefully into "domestic" and "imported"—to translate freely their *di indigetes* and *di novensides.* The principal indigenous gods were Jupiter, Vulcan, Ceres, Vesta, Quirinus, Mars, and Janus. There was Februus too, and of course Juno, wife of Jupiter, remembered with the others who gave their names to our months, as special parts of the year became associated with them. As early as 217 B.C., however, the influence of Greece was so great that the Romans set up the *di consentes*—twelve Graeco-Roman gods whose statues were erected in the Roman Forum. Zeus and Hera became Jupiter and Juno, Poseidon and Athena became Neptune and Minerva, Ares and Aphrodite became Mars and Venus, Apollo and Artemis became Apollo and Diana, Hephaestus and Hestia became Vulcan and Vesta, and Hermes and Demeter became Mercury and Ceres.

The goddess Rhea, the mother of Zeus and other gods and goddesses, became the earth-mother, also known as Cybele. We return briefly to the religion of Cybele and Attis in chapter 2; here we note only that her followers found a meteorite. Some of them must have seen it fall from heaven—proof-positive that some gods were up there throwing stones at each other or, more likely, warning their terrified earthlings. The holy meteorite was triumphantly installed in Rome as a focus for the worship of Cybele.

Five hundred and fifty years elapsed between the birth of Buddha and the birth of Jesus of Nazareth, and another five hundred and fifty years from his crucifixion to the birth in Mecca of Muhammad, the Prophet of Islam. Like the prophets of Judea so many years earlier, Muhammad was appalled by the immorality of his people as contrasted to the moral behavior of the Hebrews who shared with them a common heritage through father Abraham. Yet "there is one God," say the Arabs, "and His name is Allah"; "there is one God," say the Christians, "and Jesus Christ is his Son"; "there is one God," say the Jews, "and his name is not to be spoken." It isn't just the names, of course, it is all that has been contributed throughout history—psychologically, economically, and politically—to the meanings of the names. And it isn't as if there were three or more choices and you choose one after careful consideration of each of them. You are a Jew because your parents were Jews, a Muslim because your parents were Muslim, a Christian because your parents were Christian, or you were brought up in a Jewish, Muslim, or Christian society. The splits continue indefinitely. You and your parents follow Muhammad's son-in-law Ali, and you are a Shiite, separated from the Sunnis and their subgroups the Hanbalites, Hanafites, Malikites and Shafiites, or you follow Martin Luther, or you split from other Protestants and follow John Wesley . . . and so it goes on. It is one of the great tragedies of civilization that the members of these religious movements, united in banning idolatry and in requiring high moral standards, are so often united in mutual hate.

In the struggle to understand, it was a giant step forward to recognize that nature does not present a different and independent god wherever you look, suggesting that there is an underlying unity in all of the things that we see. Scientific thought is founded on the same idea. But it was a giant step backwards for each religion and each subsection of each religion to proclaim that its particular view of the world is the only correct one—for members of each generation to assert that the beliefs, fears, and hates of their fathers are automatically their beliefs, their fears, and their hates.

We try to balance at the intersection of three levels of life. Animal level number one is the 99 percent or so of our genetic makeup that is identical with that of some animals. It includes all of our bestial instincts plus, let us hope, the love and care that animals usually show for their offspring—and sometimes for people. Then there is the 1 percent or so that is human in our genes, a very small percentage that can take over much of the other 99, at least in the living conditions for many, though not for our homeless, or for the countless poor and persecuted throughout the world. There are two levels of that 1 percent: the animal level number two, and the spiritual level.

Our animal level number two is the struggle to survive in the world of money, a world often referred to as a "jungle," or as a "dog eat dog" environment, with a political philosophy called "social Darwinism." It is a struggle for survival on a different level—to get food, clothing, shelter, enjoyment—preferably more and better than the others get. It takes up a lot of our time, often consuming it all. The question "How can I find the time?" reverberates through our lives. Happy are they whose spiritual and second animal levels overlap a lot.

The spiritual level is the sense of wonder at the magnificence of so many facets of the universe—the intriguing world of animals and plants, the breathtaking geological structures, the views through microscope and telescope, the beauty of another human being, the excitement and joy of human accomplishment, from cathedrals to space probes, from science to poetry, from superb athletic performances to courageous deeds, and from honesty to care for others.

Some see a divine hand behind all of this and feel closer to God; others find this even more wonderful because they see it as happening without the help of any god at all. In either case, it is a series of beautiful and often isolated moments, times of awareness of wonder. It is what true religion is really about, rather than the pettifoggery of doctrinal disputes.

2

Birth, Death, and Resurrection

Spring

The Cosmos is more than ten billion years old,
But its stars are too hot and its space is too cold
For life to get even a tenuous hold
Except on a planet that circles not far
From the right sort of long-living average type star
Like this earth, for example, right here where we are.
As it follows its predestined circular run,
North and South take their turns at facing the sun.
When it's our turn, we say that our Spring has begun.
With the showers of April, the flowers of May
Are encouraged to grow by the sun's warming ray.
It's a time to be grateful and, some say, to pray.
As we count up the billions of stars that we see,
The chances are good that there really could be
Other people out there who are like you and me,
People who work hard for what life can bring,
People who cry, but can dance and can sing,
People who also are warmed by their Spring.

—Author

Conditions have to be just about right for life to evolve, and there are plenty of places in the universe where it might occur, where people "who are like you and me," or who are very different from us may be living. All this could happen, but maybe some other civilizations would not experience a spring or for that matter a summer, fall, or winter. We are fortunate that the axis about which the earth spins each day is not parallel to the axis of the earth's yearly near-circular motion around the sun and that the angle between these axes, approximately 23.5°, is enough to give much of our land significant annual changes in the weather, without wild extremes. Each year the cycle continues: the birth of nature in the spring, the apparent death of nature in the fall and winter, followed by the resurrection of nature in the spring. How to understand it? There must be a god who is responsible, a deity so identified with nature that when he is alive and well on earth it is spring and summer, when he weakens in the autumn and dies in the winter, the earth dies with him, and when he is resurrected it is spring again. This must be a half-man, half-god, with the powers of the gods but the mortality of men, or maybe it is a goddess who controls the seasons by going to the underworld every winter, leaving the earth desolate without her presence.

Clay tablets from about 2000 B.C., found in Nippur, Ur, and many other ancient sites on the lower Euphrates, tell of the legendary ruler Dumuzi. The original Sumerian myth of Dumuzi and his divine consort Inanna was adapted by the later Babylonians and Assyrians, who called the two Tammuz and Ishtar, respectively. The myths and religious cult surrounding this pair were passed on through many generations and cultures and influenced even Hebrew and Greek thinking. As noted above, the cult of Tammuz is mentioned in the Bible. In Greek myth, Tammuz became Adonis, the lover of Aphrodite (the Greek version of Ishtar), and was killed by a wild boar.

In Mesopotamian religion, Inanna-Ishtar represented the supreme mother goddess and symbolized fertility and the life force in nature. Dumuzi/Tammuz was also a fertility god, originally revered by shepherds and herdsmen as the god of the flocks. The Babylonian epic "The Descent of Ishtar" (based on an earlier Sumerian original) was an attempt by this ancient people to explain the waning and waxing fertility of nature during the different seasons of the year. The epic describes the goddess's descent into the realm of the dead, known as Aralu, where she confronts her sister, Ereshkigal, who is the ruler of this lower region. Ereshkigal gains the upper hand and strips Ishtar of each of her powers, represented by seven articles of clothing which she must forfeit as she passes through the seven gates of the underworld. When she is finally rendered powerless, all life on earth begins to perish until the god Ea (the god of wisdom and fresh water) is able to persuade Ereshkigal to let her sister go. Ereshkigal relents but only on the condition that Ishtar provide a substitute for her in the underworld. This substitute turns out to be Tammuz who is hunted down by Ereshkigal's ministers of death who drag him down into the realm of the dead. Ceremonies of mourning and wailing marked Tammuz's abduction to the underworld. In the Sumerian version of the epic, Dumuzi is ultimately released from the underworld when his sister, Belili, agrees to spend half the year in Aralu in his place. A lament recovered from the ancient site of Nippur, the religious center of the world's oldest civilization, recounts the effect of Tammuz's absence from the earth.

> The wailing is for the plants; the first lament is that "they grow not"
> The wailing is for the barley; the ears grow not.
> For the habitations and flocks it is; they produce not
> For the perishing wedded ones, for perishing children it is;
> The dark-headed people create not.

and then

> Where grass was not, there grass is eaten
> Where water was not, water is drunk
> Where the cattle sheds were not, cattle sheds are built.
>
> (We, p. 39)

Tammuz was hailed in Syria with the cry "The Lord is risen" and his ascension was celebrated in the closing scenes of his festival. We have seen that he was accepted by the early Jews. Ishtar assumed many forms and many names, each civilization, sometimes each city, having its own special name for the earth mother who was both the goddess of fertility and the goddess of love, and therefore of women and, in some cultures, of the moon. She was Inanna in parts of Sumer and later her fertility personality became Demeter among the Greeks and Ceres among the Romans. For many years the festival of Ceres was held on the nineteenth of April as a salute to spring. More well known is Ishtar's role as the goddess of love, becoming Aphrodite to the Greeks and Venus to the Romans—the deification of the beauty of women. The story of Aphrodite and Adonis is really that of Ishtar and Tammuz. Adonis was killed by a wild boar and Aphrodite, who loved him, prayed to Zeus to return him to life. He was allowed to return for part of each year. He also was mourned with consuming sorrow and his resurrection in spring was marked by the wildest joy, particularly by the women. Priests also mourned, some mutilating themselves to imitate the mortal injury in the thigh that Adonis had suffered.

At different times and places, religious cults developed around this same theme, helping to satisfy the urgent needs of the people to understand the seasons, and ensure a good food supply and, on a different level, to identify with a universal spirit and look forward to a happy life after death. One such spirit was Orpheus of Thrace, who descended to Hades in order to beg from Pluto the return of his beloved wife Eurydice, who had died from a snake bite. (In a painting by Nicholas Poussin [1594–1665] a passerby obscures Orpheus's vision just as she is being bitten.) He played the lyre and sang so beautifully that Pluto relented and let her follow Orpheus out of the underworld provided that he did not look back, which unfortunately, like Lot's wife, he did. Eurydice disappeared into the darkness of Hades, and as Alexander Pope describes it in "Ode to St. Cecilia's Day":

> But soon, too soon, the lover turns his eyes;
> Again she falls, again she dies, she dies.
> . . . .
> Now under hanging mountains
> Beside the fall of fountains
> All alone He makes his moan
> And calls her ghost
> For ever ever ever lost!

The distraught Orpheus was murdered and his head was thrown into the Mediterranean, where it floated, *still singing,* to the Island of Lesbos. The cult of Orpheus, developing from this scrap of "history," attracted followers with mystical hymns and sacred rites, which included sacramental feasts of raw flesh. With time, however, some very current ideas emerged from it—man's immortality and the promise of escape from pain in a future life, the necessity of purification from evil, the idea of a man-god incarnate, suffering, and dying. In view of Orpheus's lapse, instructions were also provided for conduct of the soul in the underworld.

Orpheus's father, Apollo, the god of archery, prophecy, and music, also disappeared during the winter, but he went, not to the underworld, but to the land up north, inhabited by the fabled Hyperboreans "beyond the northern winds." Each spring, the return of Apollo was celebrated with the Festival of Theophany, which means the physical manifestation of a god by his incarnation in a human body. Apollo eventually became the sun-god, chief god of the Delphic Oracle.

The great earth mother flourished from early times as the goddess Cybele of Anatolia, in Asia Minor, where she and her lover, the god Attis, were also worshiped in rites that symbolized the dying and rejuvenation of the earth. Legend says that Attis died and went to Hades, and then rose from the dead. Cybele's original shrine was centered on a sacred stone that fell from heaven—a meteorite—which in 204 B.C. was brought to Rome. There her cult expanded, requiring worshipers to sacrifice a bull together, and be bathed in its blood. The sacrifice of animals led to their deification, particularly in the forms of the golden calf and the sacred bull, whose death brings forth life and who in Spain is still a very special animal in the bull rings and in the Pamplona stampede every July.

From Persia came the worship of Mithras, son of the god Ahura-Mazda, or Ormazd, before whom after death, all men must appear to be judged, unclean souls being handed over to Ahriman for everlasting torture. The eternal battle between Ormazd, the embodiment of light and goodness, and Ahriman, the power of evil, is basic to the religion of Zarathustra, which dominated Persia until the time of Muhammad. The birthday of the son Mithras was celebrated in December with the consumption of consecrated bread and wine and of a mystic meal, served in sacred vessels, a bell being sounded at special moments. Mithras had killed the divine bull, and the beneficial plants and animals had emerged from its body. Mithras had then been transported to heaven to watch over and help the faithful until the day of his second coming. This religion also came to Rome, becoming very popular, especially with the Roman soldiers during the reigns of Trajan (98–117 A.D.) and Commodus (180–192 A.D.); Mithraism merged partially with the religion of Cybele and Attis, by that time well-established in Rome. Writing in 346 A.D., Firmicus Maternus, who had been an initiate of this religion and then attacked it, described part of the ritual in his book, the title of which is easily translated: *De Erroribus Profanarum Religionum:*

> On a certain night the image (of the dead man-god) is placed, lying on its back on a litter, and the devotees mourn it with rhymic lamentations. At length, when they have satisfied themselves with this pretended lamentation, a light is brought in. Then the throats of all those who were weeping are anointed by the priest, and after they have been anointed, he slowly murmurs these words to them:
>
> "Have courage, O initiates of the Savior-God,
> For there will be salvation for us from our toils."

The Magi were members of an offshoot of this religion. They believed in the advent of a Saviour, but later they became so much involved with sorcery and, yes, *magic,* that they fell into disrepute. However, we still remember three of them every Christmas. Dionysos, called Bacchus by the Romans, was another supernatural being who started off as a god of all vegetation, and the Dionysian cult included mystic rites that revealed the resurrected deity and promised worshipers eternal life. Unfortunately, Dionysos/Bacchus narrowed his interests, concentrated on the grape and its products, and became an early alcoholic. However, he shared a lot of power with Apollo.

In Egypt, as noted in chapter 1, the earth goddess was Seb, who married Nut (heaven) and was the mother of Osiris. The latter was the god of male fertility and the Nile River, and his wife and sister, Isis, became an earth mother in her own right. Isis, miraculously according to some, or with the help of Osiris to others, conceived a son, Horus, the god of day. Each winter, near the shortest day of the year, the temples of Horus showed sacred statues of Isis, nursing the infant Harpocrates, which was the name of the young Horus, god of the morning sun. At the spring festival of Isis, initiates mourned with her the confinement of her son, and sometimes her daughter, to Hades, and on the third day a play was given representing the resurrection. The birth of a god representing the sun or the day was celebrated in late December, as the days begin again to get longer —a time of great joy, because if the day continued to shorten it would be the death of all. The equinox in March was a time of triumph of day over night, the resurrection of life, particularly when the moon also became full. If our ancestors had lived in the Southern Hemisphere, Christmas Day would be June 25, and Easter Sunday would be the first Sunday after the first full moon after the equinox, September 21.

We have mentioned the murder of Osiris by his jealous brother Typhon, who cut up his body into fourteen pieces, which became sacred relics. Wherever one was "found" a temple was erected. Osiris sat in judgment on all and insisted on a very high level of moral behavior. After dying, a person was required to swear, among other things, that he had not repulsed or scorned God, that he had not stolen the food of the gods, and that he had not oppressed members of his family. He then went before forty-two judges in the Hall of Osiris, each judge specializing in a particular transgression. The Ten Commandments, coming more than a millennium later, are hard enough to obey at all times, but it was beyond human ability always to avoid every one of these forty-two possible violations. They emphasized in some detail what were to become the third commandment (no false oaths) and the last five (murder, adultery, robbery, false witness, and covetousness). Other rules decried insolence, a haughty voice, anger, hasty judgments, cheating, and stirring up conflict. In principle, these people were impossibly virtuous, but they quickly learned from the priests how to bribe the gods (and assist the priests), and how to beat the system by buying a copy of *The Book of the Dead* (written by the priests) that would help them fool even Osiris. One had to be careful, though, because Osiris had the first lie detector—a feather, against which the heart of the supplicant was weighed.

Osiris was related spiritually to the sacred bull Apis, whose death brings forth life, as we have seen. Whenever Apis died, another Apis would arise. Apis was embalmed and placed in a tomb in the rock-cut Serapeum in the cemetery at Memphis, and his soul passed on to the world beyond in the form of Osiris-Apis. The Greeks dropped the "Osiris" and called this spirit "Serapis," the spirit of the underworld and rebirth from it.

According to James Hastings (1852–1922), author of *Dictionary of the Bible,* Osiris was the "prototype of man who after a virtuous life must die, but who afterwards rose again to live forever."

During the first century B.C. the Roman poet Lucretius, to be quoted later for his ideas about motion and the atomic theory, did not subscribe to this belief in resurrection:

> What has this bugbear death to frighten man,
> If souls can die as well as bodies can?
> For, as before our birth we felt no pain,
> When Punic arms infested land and main,
> When heaven and earth were in confusion hurled
> For the debated empire of the world,—
> So when our mortal frame shall be disjoined
> The lifeless lump uncoupled from the mind,
> From sense of grief and pain we shall be free;
> We shall not *feel,* because we shall not *be.*
>
> (*De Rerum Natura,* Book III, lines 830–844;
> translated by Palmer Bovie, UA 5, 281)

In India, the Vedic hymns tell of Dyaus, the sky; of Dyaush-pitar, the "Heavenly Father"; and of Aditi, the mother of all gods. The departed souls inhabited three levels—the upper sky, the middle area, and the earth's atmosphere, all presided over by their god Yama. The *Hymn to Death* from the tenth Mandala of the *Rig-Veda* tells us that Yama died and led the way for others to follow him to heaven:

> To Yama, mighty-king, be gifts and homage paid,
> He was the first of men that died, the first to brave
> Death's rapid, rushing stream, the first to point the road
> To heaven, and welcome others to that bright abode
> No power can rob us of the home thus won by thee . . .
> Return unto thy home, O soul! Thy sin and shame
> Leave them behind on earth; assume a shining form—
> Thy ancient shape—refined and from all taint set free.
>
> (translated by Sir Monier Monier-Williams)

Here we are exposed to very early ideas—written three thousand or more years ago—about resurrection, eternal life, and forgiveness of sins, all arising from the death and resurrection of the god Yama.

In England, we have the authority of the Venerable Bede* (673–735) that the pre-Christian Anglo-Saxon goddess of the Spring, Eostre, was worshiped in a festival held each year in April after the vernal equinox of March 21. We derive our word "Easter" from her name, and we have taken the pagan symbol of resurrection—the egg—and associated it with the Easter celebrations. Eostre is alleged to have amused the children by turning her pet bird into a rabbit that managed to retain an egg-laying ability despite the metamorphosis. Others see the role of rabbits at Easter as symbolizing the fertility of spring-time.

There are many more similar myths from northern Europe, Asia, the other continents, and the islands of the Pacific. It is enough to have briefly summarized here these pre-Christian myths of the seasons that, by incorporation into Christian beliefs and practices, have so deeply affected western civilization. They are centered on the fruitfulness of Mother Earth, and her fertilization by a god, the death and resurrection of a man/god whom she loves, with associated ceremonies of mourning, purification and joy, and promises of personal immortality and eternal happiness in an after-life. Ceremonies included the rituals of eating the flesh and drinking the blood of the one who was sacrificed, also of fasting, of practising continence, praying, confessing, and undergoing immersion in holy water. The god is born at the winter solstice, the shortest day of the year in northern latitudes. It was even said that Isis nursed the baby god Horus in a stable. Later, the god is killed, descends into the land of the dead, and is resurrected on a date when day begins to be longer than night, and the light of the sun and moon overcome darkness. And the Day of Judgment loomed ahead, to assign everyone to the everlasting extremes of bliss or torment.

Each of these pre-Christian religious theories of the nature of the seasons was believed by a large fraction of the population of its time and place to be literally true. For many people, the adherents of other religions were following myths, worshiping idols, but theirs was the true religion based on the death and resurrection of their particular god, who really *did* die, and who really *was* resurrected.

Changing views of the gods throughout history are reflected in statues and paintings of them that are now in our museums and cathedrals. The colossal statue of Ikhnaton found in his capital Tel-el-Amarna is that of a demi-god with an unrealistically elongated smiling face—nothing at all like his death mask. Statues of Isis evolved from that of a mystical symbol of the fertility of the earth and all women, with the face of a cow to match the bull Osiris-Apis, into that of a woman/goddess holding her baby god. As Isis became transformed into the Mother of God, some early Christians worshiped her. Statues of Isis and the young Horus were renamed Mary and Jesus. The feast of purification of Isis became the Feast of the Nativity, the resurrection of Attis was replaced by the resurrection of Christ, magic incantations to drive out devils began to include Latin prayers and the sign of the cross. Images change, and by the first part of the thirteenth century Mary was represented in Notre Dame Cathedral as the Queen of Heaven, being crowned at the side of an adult Christ. Two hundred years later, she was again a mother with a holy child. At the same time, statues of Christ evolved from that of an awesome God to that of a loving Jesus. As Andre' Malraux wrote in his *Metamorphosis of the Gods,* "These gods and ancestors, heroes and priest-kings, immortals and obscure dead—art had the mission to express their deliverance from the human condition, their release from time."

The sacking of Jerusalem by the superstitious, idolatrous, free-loving Babylonians under Nebuchadnezzar II reaffirmed the faith of the Jews that their one stern God was superior in every way to Marduk and all of the other Babylonian deities. It also left an indelible impression on the minds of Jews and, later, Christians that all other religions are pagan, immoral, superstitious, and therefore evil. With that attitude, the pagan nature of many Christian beliefs, ceremonies, and hopes, and even some of the high moral values, tended to be forgotten—even deliberately ignored or denied.

It all started out with the struggle to understand the seasons and with the solution in terms of an attractive religious theory. Each spring, in different places and eras, when Tammuz, Adonis, Attis, Apollo, Dionysos, and others were resurrected from the dead, the earth blossomed and the people rejoiced. The early Church Fathers adapted and adopted not only the pagan ceremonies but the pagan myths as well. Perhaps they were wise. It is doubtful if Christianity would have survived if they hadn't. On the other hand, in this process, Jesus became clothed in a pagan costume and surrounded by pagan superstitions and cruelties, the very opposite of his message.

3

Creation, Fall, and Flood

Heaven and Earth, Center and Circumference, were created all together, in the same instant. This work took place and man was created by the Trinity on October 23, 4004 B.C., at 9 o'clock in the morning.

—John Lightfoot 1602–1675
Vice Chancellor
The University of Cambridge
Hebrew Scholar

There, you have it all—the prestige of a great scholar of a great institution, the meticulous detail that had obviously attended his studies, his knowledge of ancient languages, and his evident command of geometry combined with the certainty and precision of the statement to make it irrefutable. He was simply quoting James Ussher (1581–1656), archbishop of Armagh in Northern Ireland, who had set up this unfortunate chronology that was to appear in the margin of the Authorized Version of the Bible. Printed in that way, it seemed to be part of the text and therefore part of God's Word. Ironically, the good archbishop successfully opposed the attempt by English clerics to make the doctrinal standards of the Irish church conform exactly to those of the English! Perhaps both he and Dr. Lightfoot were inspired by the accuracy with which, a century earlier, Martin Luther had described the hour of the Fall: "They entered into the garden about noon, and, having a desire to eat, she took the apple, then came the Fall—according to our account about two o'clock." (ADW 1, 288) On the other hand, Geoffrey Chaucer (1340?–1400) had referred to "the month in which the world began / That hight *March,* when God first maked man." (Nun's Priest's Tale, lines 367–8) Earlier in that fourteenth century, Dante, in his *Divine Comedy,* told of meeting Adam in heaven and having him tell that from his death he had to wait 4302 years until the death of Christ allowed him to get out of limbo and into paradise. Adam also says that he had lived on earth for 930 years, thus setting creation at about 5200 B.C.:

> This wouldst thou hear, how long since, God
> Placed me in that high garden, from whose bounds
> She led me up this ladder, steep and long;
> . . . still was I debarr'd
> This council, till the sun had made complete,
> Four thousand and three hundred rounds and twice
> His annual journey; and through every light
> In his broad pathway, saw I him return,
> Thousand save seventy times, the whilst I dwelt
> Upon the earth.
>
> (*Paradise* Canto XXVI; HCl 20, 398)

Despite these words of Adam, the later date was settled on. The myth that the earth and the life on it were created in the year 4004 B.C. offered a certainty that captured people's minds. Even today many want to teach it as fact to our children, not always realizing that they are abandoning their God-given ability to think, in favor of the guesswork of a seventeenth-century Irish prelate!

History has moved in such a way that Jewish versions of the stories of very early times were preserved and, later, transferred unchanged to the Christian Church. They remained as dogmas rather than beautiful myths. They had been pieced together by earlier civilizations to account in a simple way for the wonderful phenomena of nature and to allay the fears and bolster the hopes of man.

Despite the absence of any evidence that Christ himself believed in, or advocated the adoption of, the story of Genesis as a fundamental creed, the early Christian Fathers were well aware of the hold that an absolutely certain doctrine can have on the minds of men and women. It took a little time to get organized, to decide which ancient documents were "The Word of God" and which were not. Around 300 A.D. the Christian writer Firminanus Lactantius made it clear that all other thoughts on the study of the creation must be subordinated to Scripture, a point of view echoed later in the century by Saint Ambrose, Bishop of Milan, and above all by Saint Augustine (354–430):

"Nothing is to be accepted save on the authority of Scripture, since greater is that authority than all the powers of the human mind," and that includes an unflinching belief in every word from Genesis to Revelation.

Over the centuries, this belief dominated Christendom, offering to the masses of people a strong measure of stability and of faith in an afterlife that would make up for all of the miseries of the present one. It was a foundation stone of the Catholic Church and, despite their differences with the Church on other matters, Martin Luther, John Calvin, and John Wesley adopted it unchanged. On the other hand, centuries before them, the great twelfth century Jewish scholar Maimonides in the introduction to his *Guide to the Perplexed* had this to say about the story of Creation as told in the Book of Genesis: "Creation has been treated in metaphors in order that the uneducated may comprehend it according to the measure of their faculties and the feebleness of their apprehension, while educated persons may take it in a different sense." But patronizing tones alienate the opposition rather than convert it.

In the middle of the last century, excavations began at the ancient city of Nineveh, on the east bank of the river Tigris. One series of seven cuneiform tablets discovered there provides us with a very early epic poem, called the *Enuma Elish,* which describes how the world was created. At first the world consisted of nothing but a watery chaos, personified by two deities named Tiamat and Apsu. Tiamat was a kind of sea monster and she was represented as a dragonlike creature with claws, wings, horns, and a tail. Apsu represented primeval fresh water, the source of springs and rivers. From the union of this original divine pair, a second and third generation of deities were created, who soon rivaled Tiamat and Apsu in authority and power. This rivalry ultimately led to a battle for survival, and Marduk (an astral god called the son of the sun-god) took up arms against Tiamut, Apsu, and their cohorts. The cataclysmic conflict reminds us of John's description of the Apocalypse (Revelation 12: 7–9); "And there was war in heaven: Michael and his angels fought against the dragon; and the dragon fought and his angels and prevailed not; neither was their place found any more in heaven. And the great dragon was cast out, that old serpent called the Devil, and Satan, which deceiveth the whole world." The original composition of "Enuma Elish" probably took place during the first Babylonian dynasty, some 1800 years before St. John the Divine. In the end, Marduk defeated Tiamat—as a matter of fact, he killed her and made the firmament of heaven out of her skin. The poem tells that the celestial bodies were then created, followed by the animals. Ultimately came the creation of mankind.

Alternatively, the ancient creation-myths described God as making a Siamese twin man-and-woman joined together, the creation of woman following from their later surgical separation. This and other myths were adopted in modified forms by the Jews, and we read in Genesis (5:2), "Male and female created he them and blessed them and called *their* name Adam."

In describing a symposium—a banquet with philosophical dialogue—held at the house of the poet Agathon in 416 B.C., Plato has Aristophanes speak about love: "At first there were three sexes, not two as at present, male and female, but also a third having both together." Each person had four arms and four legs, two heads and two genitals—a man-man, a woman-woman, and a man-woman. These early people were so strong that they threatened the gods, so Zeus had a thought, "I will slice each of them down through the middle, and if they choose to go on with their wild doings, I will do it again." Aristophanes graphically continues, "He sliced men through the middle, as you slice hard-boiled eggs with a hair." Thereafter two-legged creatures, separately male and female, inhabited the earth and "all men who are a cutting of man-woman are fond of women and all the women are mad for men . . . the women who are a cutting of the ancient women do not care much about men . . . those who are a cutting of the male pursue the male and while they are boys they are fond of men"—the first theory of sexual preference.

In Teutonic myths it is the Aesir (the good gods) against the powers of Hel, goddess of the dead, queen of the underworld and daughter of the evil Loki. The chief god Odin (or Woden) and his wife Frigga, goddess of the sky—our Wednesday and Friday—had a son named Balder, "god of light and peace, and of the good, beautiful, eloquent, and wise." Loki arranged to have him shot by the only thing that could kill him—a piece of mistletoe. Teutonic myth predicted that there would be a tremendous war between the good and evil gods—the *Götterdämmerung*. The main gods would be killed off, but Balder would rise from the dead and a man and woman—Lif and Lifthrasir, "life" and "desiring life," would be preserved to start re-peopling the earth. As with the fight between Marduk and Tiamat, or the Armageddon of the Apocalypse, this "Twilight of the Gods" was inspired by stories of past battles and of earthquakes and volcanoes.

There are myths about the early history of man and the world from all societies, at all times. In Egypt, Amon created Kneph and his wife Hathor, the goddess of love. By some accounts they were the parents of Osiris and Isis. In Egyptian temples at Philae and Denderah excavations have revealed representations of Nile gods, modeling lumps of clay into men. In India, one legend asserts that there was only water, so Bhagavan made a crow and sent it off to find some earth, reminiscent of Noah's sending out a dove for the same purpose. The crow learned from a tortoise that the land had been swallowed by an enormous worm, which was then forced to regurgitate twenty-one different varieties of earth. Alternatively, two huge eggs collided and broke, making the earth and her husband, the sky, and they became the mother and father of all living things. From the Rig-Veda, the most ancient writings of the Hindus, we learn that the first human couple, Yami and Yama, were twin brother and sister, hesitant to procreate because of the taboo against incest. Apparently they overcame their reluctance. Another hymn from the Rig-Veda describes the creation in magnificent verse:

> Nor Aught nor Nought existed, yon bright sky
> Was not, nor heaven's broad woof outstretched above
>
> . . . .
>
> The Only One breathed breathless by itself;
> Other than It, there nothing since has been.
>
> . . . .
>
> Who knows the secret? Who proclaimed it here?
> Whence, whence this manifold creation sprang?
>
> . . . .
>
> He from whom all this great creation came,
> Whether his will created or was mute,
> The Most High Seer that is in highest heaven,
> He knows it—or perchance even He knows not.
>
> (Du 1, 409)

In many legends in various parts of the world there is a story of paradise, a golden age of time past. The Greek historian Hesiod, writing about 750 B.C., affirmed the belief that in the early days men were very happy but that things weren't what they used to be. "Men lived like gods, without vices or passions, vexations or toil. In happy companionship with divine beings they passed their days in tranquility and joy." What happened? Regrettably, in most cases, a woman was blamed. We are familiar with the story of Eve, less perhaps with the Chinese legend of Poo See. One of the three hundred and five odes in the ancient Shi-Ching puts it this way: "All things were at first subject to man, but a woman throws us into slavery. Our misery came not from heaven but from a woman; she lost the human race. Ah, unhappy Poo See. Thou kindlest the fire that consumes us and which is every day increasing. . . . The world is lost. Vice overthrows all things." (Du 1, 330)

Closer to home, the Greeks believed that the giant Epimetheus, under the supervision of his brother Prometheus, was assigned the task of creating man and the animals and giving them the means for self-preservation. By the time he came to create man, however, he had bestowed on the animals all of the protective devices that he had—claws, shells, wings, strength, speed, etc. So Prometheus, aided by Minerva, went up into the heavens, lit his torch at the sun, and brought back and presented to man the superb gift of fire. (Perhaps Prometheus knew that the sun burns with a nuclear fire, which he left for others to bring later). Then Jupiter created a perfect woman, Pandora, and offered her to Epimetheus, some said to *punish* him for being an accessory in the theft of fire—history and mythology are filled with strange theories. Apparently Epimetheus was delighted with the idea and Pandora moved in. In his house there was a box that contained many evil things that he had explicitly not used in creating man. Pandora, curious, opened the box, releasing all of the mental and physical sicknesses that remain with us to this day.

We don't believe that the story of Pandora really happened, although her "Oops!" has echoed down the centuries, but many believe that the story of Eve actually occurred as written. The magnificent overpowering attraction that a woman has for a man appeared to reduce his power, destroy his will, and make him feel both vulnerable and guilty about his most normal reaction. She it was who tempted him with the symbolic or real apple, or was it the snake (his) that tempted her? Anyhow, Eve got the blame.

The "original" of the story of Adam, Eve, and Noah, that is, the earliest version so far discovered, comes from the writings of the Sumerian culture that existed 4300 years ago. Men and women were created, they sinned, and they were punished by a flood. According to one story, only one man, by name Ziusudra, survived. He was granted immortality and perfect health as a reward for having survived the ordeal of the flood. By another account, the god of wisdom, Ea, decided to save one couple—Utnapishtim and his wife. They built an ark, survived the flood, landed on Mount Nissir, and sent out a dove to reconnoiter the wet landscape. This story was handed down through the Sumerian and Babylonian civilizations and spread to the Israelites during the centuries of war and occupation, ultimately to appear in the Book of Genesis with very little changed except the names.

According to Greek mythology, Deucalion and Pyrrha, king and queen of Thessaly, were a pious couple, and because of this Zeus spared them when he sent a flood to punish the rest of the world. (The Greeks were not afraid to name wives—"Pyrrha" is so much more personal than "Mrs. Noah," "Mrs. Ziusudra," or even the impressive "Mrs. Utnapishtim.") Deucalion and Pyrrha floated in a boat for nine days and then came to rest on a mountain, where they were commanded to restore the race of men by "throwing the bones of their mother behind them." This puzzled and disturbed them until they realized that "mother" meant "Mother Earth," whose "bones" are the stones and rocks. So they picked up a few and threw them behind, and, sure enough, every one that they threw turned into a man or a woman. As the waters subsided an enormous serpent, the Python (like the worm in the story of Bhagavan), emerged from the ooze. Apollo killed it and celebrated by initiating the Pythian Games, precursors of our many musical festivals and our summer Olympic games.

The flood which gave rise to all of these myths was probably an immense overflow of the Euphrates. The Sumerian city of Ur, right on the river, bore the brunt of it. Excavations there have uncovered an eight-foot layer of clay and silt—below the remains of civilization. No wonder that the word went out that the whole world had been destroyed by an angry god. From an early Sumerian poem we get a feeling for the deep reverence that was held for the Creator:

> Who shall escape from before thy power?
> Thy will is an eternal mystery!
> Thou makest it plain in heaven and in the earth.
> Command the sea and the sea obeyeth thee.
> Command the tempest and the tempest becometh a calm.
> Command the winding course of the Euphrates
> And the will of Marduk shall arrest the floods.
> Lord thou art holy! Who is like unto thee,
> Marduk? Thou art honored among the gods that bear a name.
>
> (UA 1, 34)

The original version of this was written nearly five thousand years ago. It wasn't until the seventh century A.D. that the first poem to the Creator and His Creation was written in English.

In 657, the English Abbess St. Hild founded a monastery at Whitby in Yorkshire. A few years later, the organization of the Church in England was perfected there at the Council of Whitby. At this monastery there was a stableman named Caedmon, and the English scholar Saint (the Venerable) Bede, writing in Latin early in the next century, tells us this about him: "There was in this abbess's monastery a certain brother, particularly remarkable for the grace of God, who was wont to make pious and religious verses, so that whatever was interpreted to him out of Scripture, he soon after put the same into poetic expressions of much sweetness and humility, in English, which was his native language. By his verses the minds of many were often excited to despise the world, and to aspire to heaven." One night when he was asleep in the stable, an apparently divine being told him in a dream to "sing the beginning of created beings," and he replied, still asleep, by composing a song addressed to "the almighty preserver of the human race who created heaven for the sons of men as the roof of the house, and next the earth." Next day he sang it to his boss, the chief steward, who was so impressed that he arranged to have Caedmon sing before the Abbess herself. She in her turn appointed him as the resident poet of the monastery. We are not really sure which songs were composed by him and which by his successors, but through them the stories of Genesis, Exodus, and Daniel became epic poems, along with the story of Christ and Satan. As songs in English, they supplemented and confirmed the teachings of the Church. The story of Genesis came to the commoners of England, and the poem may have helped to inspire John Milton a thousand years later. In *Paradise Lost* the angel Raphael tells Adam about the Creation:

> Heaven opened wide
> Her ever-during gates, harmonious sound
> On golden hinges moving, to let forth
> The King of Glory, in his powerful Word
> And spirit coming to create new worlds.
>
> . . . .
>
> Chaos heard his voice. Him all his train
> Followed in bright procession, to behold
> Creation, and the wonders of his might.
> Then stayed the fervid wheels, and in his hand
> He took the golden compasses, prepared
> In God's eternal store, to circumscribe
> This Universe and all created things.
> One foot he centered, and the other turned
> Round through the vast profundity obscure,
> And said, "Thus far extend, thus far thy bounds
> This be thy just circumference, O World!"
> Thus God the Heaven created, thus the Earth,
> Matter unformed and void.
>
> (Book VII, 205–209; 221–233)

Perhaps Milton's contemporary, John Lightfoot, got the idea of "Center and Circumference," quoted earlier, from this part of *Paradise Lost,* but it is more likely that they both got it from the reference to "the circle of the earth" in Isaiah (40:22). The lines of Milton that follow, as the days of the Creation unfold, provide a detailed image of the whole process, interpreted by many as a factual account.

Similar images of the Creator had literally taken shape in the statues, stained-glass windows, mosaics, paintings, and illustrated missals of the Middle Ages, culminating in the magnificent work of Michelangelo. The Almighty assumed the form of a superpowerful large man, working with His hands to make the world. A human touch was added in some portrayals of Him on the seventh day, as an absolutely exhausted giant. It was then found that it was enough for Him to speak, to say, "Let there be light," and there was light. Milton noted this in the speech of Raphael.

The idea that the world was formed by the word of God is not confined to the third verse in Genesis or the first in Saint John's Gospel; it was postulated, apparently independently, by civilizations as far apart as the Mayan civilization of Central America, the Maoris of New Zealand, and the Sapangwa tribe of Tanzania, Africa. If not universal, the belief is widespread that the creation of the world was due to the word of God. Did some people confuse the spoken word of God with the written word of man? Before geologists and paleontologists began to understand the evidence for the enormous age of the earth and the life on it, there was nothing to guide people besides the written word. It is true that there were a few written words from Xenophanes, dating from the sixth century B.C., telling of marine shells found in the mountains, and imprints of various fish found in stones, but if anyone thought of them they were dismissed as imperfect copies that the Creator had discarded. The written word that guided all thinking throughout Christendom was the Bible. The theologian Francisco Suarez (1548–1617) insisted on a very literal interpretation of the words "day" and "year" in the Old Testament. People counted the "begats" in the Pentateuch back to the beginning of the world, as they saw it, but all that it was possible to know at the time could not be earlier than 3500 B.C., when the most primitive form of writing began. Perhaps oral history could extend the date beyond this, particularly for cataclysmic events.

It is a coincidence, perhaps, that the beginning of written records should occur so close to the dogmatically asserted date for the creation of the world (4004 B.C.), quoted at the beginning of this chapter. Chinese scholars and even Sumerian priests had placed creation as occurring much earlier than this. Until two hundred years ago, however, nobody had any solid, convincing evidence that anything at all happened before this date. Since then, it is a very different story, which we will discuss in chapter 16. Perhaps present-day creationists could recognize that there is no evidence for any written words before the date of 4004 B.C., and that there was no way for a writer as recent as the good Bishop Ussher to know for sure that anything had happened before then. It was a time when, in their thinking, God gave the Word of Creation, but it was the creation of the *written word,* without which we would not even *have* the Bible.

4

Matter, Atoms, and Vacuum

> From the beginning until now men have begun to philosophize because of wondering. At first, they wondered about the extraordinary things near at hand; then they moved on from little things to raise questions regarding greater happenings such as the phases of the moon and of the sun and the stars and concerning the origin of the universe. The man who is puzzled and is wondering is aware of his ignorance (wherefore the myth-lover is a philosopher after a fashion, for the myth is compounded of marvels).
>
> —Aristotle (384–322 B.C.), *Metaphysica*

The early Greek people had no sacred book of their own—no Hebrew *Talmud* or *Gemara,* no Persian *Avesta,* no Indian *Rig-Veda,* "only" the *Iliad* and the *Odyssey,* probably written during the ninth and eighth centuries B.C. Its authors—at least two of them—we call Homer. No magic was offered, no superstitions, and no priests, but many legends and myths, some of them based on oral history. These stories continued to evolve, partly under the influence of other cultures. However, religious beliefs did flourish in early Greece as the writings of Homer were supplemented by various religious cults, including Orphism, cult of the worshippers of Orpheus. Particularly in its finer aspects, this had a considerable influence on Pythagoras* and some of the other early philosophers.

In 596 B.C., the poet and prophet Epimenides went from Crete to Eleusis, a town near Athens, bringing with him religious mysteries and rituals that ultimately became the official Athenian religion. Temples to various gods proliferated and the Delphic oracle yielded the secrets of the earth goddess and, later, of Apollo and Dionysos. In describing how to set up an ideal city-state, Plato suggested "the founding of temples and sacrifices and worship of gods and spirits and heroes besides, tombs of the departed and whatever service is due to those who are in the next world, to keep them gracious. We will use no interpreter except the god (Apollo) of our fathers, for this god is the ancestral interpreter on such matters for all mankind, and he sits in the middle of the earth upon the navel, and interprets." (*Republic,* Book N k27C) With it all, the Greek civilization in the first six centuries before Christ raised the first questions of philosophy and provided a number of possible answers. We still refer to some of these thoughts as "philosophy"—much of the rest has evolved into our present physical and life sciences.

One of the first questions asked was, "What is everything made of?" Following the earlier idea of a primeval watery chaos, from which everything was constructed by the gods, and with the knowledge that water caused plants to appear and to grow, Thales of Miletos postulated that everything was made of water. The town of Miletos was on the mouth of the river Maeander, on the west coast of Ionia (presently Turkey) opposite the island of Samos. At the time of Thales, it had been for several hundred years one of the most important of the Greek cities and a center of trade with Egypt. The interval of time between the civilization of Sumer, or that of the Egyptian Old Kingdom, to the time of Thales was about as long as that between Thales and the present, so that a long list of theories, gods, and observations preceded him. Thales visited Egypt, where he learned practical knowledge and taught the elements of geometry, with its emphasis on logical deduction. Other philosophers from Miletos had different views of the nature of the fundamental *physis* out of which everything is made: Anaximandros (610–547 B.C.) suggested, somewhat mystically, that it was the infinite; Anaximenes, who died around 526 B.C., that it was air; and Heracleitos, of Ephesus (540–475 B.C.), that it was fire.

The idea that nature consists of just one substance led Parmenides, by metaphysical arguments that we will not repeat, to the conclusion that there is no empty space in the world. The vacuum does not exist because one cannot think of nothing. This denial of the existence of a vacuum impressed Aristotle, whose ideas eventually took a firm grip on the world of thought. Thus another wrong idea was born and lived through the centuries, and it still echoed in 1648 when Blaise Pascal discovered the principle of the mercury barometer and was severely criticized with Aristotelian arguments by the Jesuit Père Noel. The "vacuum" at the top of the tube supposedly could not exist. To affirm that it did would be to undermine Aristotle and the Church, and pave the way for the evils of atheism. It is hard for us to believe that a simple barometer could have such deep moral consequences.

The various ideas of Thales and his successors about the nature of matter culminated with Empedocles of Agrigentum, in Sicily (492?–430 B.C.), who added earth to the list. He pronounced that matter consists of various mixtures of four basic elements—earth, fire, water, and air—and that these enter into combinations to form the world. He believed that nothing new comes or can come into existence, that changes are due to attractive and repulsive types of forces—love and strife—between the elements. Empedocles was a poet, a philosopher, a statesman, a scientist, and, according to some, a worker of miracles. He was a rich man, and it is told that he walked the streets in purple robes and a golden girdle, with long hair and a retinue of slaves. He believed in the transmigration of the soul, and he shared this belief with followers of the Orphic mysteries and of the Hindu religion, as well as with at least one modern film star. The force of his personality caused the earth-air-fire-water theory to take hold. He soon became a legend, or at least legends developed about him. Anyone standing on the top of Mount Etna can be readily convinced that everything consists of earth, air, fire, and water, because that is all you can see or feel. It is therefore told that Empedocles threw himself into the crater to convince people by his disappearance that he was a god. Another story has it that he died and in the evening his friends gathered around him for a last supper before burial. A loud voice called him, the heavens were illuminated, and he disappeared.

The distinction between air and the heavens was recognized, and the word "aether" used for the fifth element, the "quintessence" that flew upwards during creation, the matter out of which the heavens are made. (Some later alchemists identified the quintessence with alcohol, and we still say that it makes people "high.") The theory of earth, air, fire, water, and aether became yet another one of those paradigms that were left unchallenged for two thousand years by all but a few wise men.

Thales and his earlier followers must have visited Ephesos, some forty miles north of their home city of Miletos, if only to see what was to be called one of the Seven Wonders of the World—the Temple of Diana—being built there at the time. In fact, Heracleitos, coming later, described his philosophy of fire and motion in a book which he deposited in that temple. (The original building was destroyed in 356 B.C. A magnificent rebuilt version was there when, centuries later, Saint Paul preached in the synagogue at Ephesos [Acts 19:24–41] and the silversmiths who made idols wanted to run him out of town, because he was bad for business.)

Another forty miles north of Ephesos is the birthplace of the last of the Ionian philosophers, Anaxagoras* of Clazomenae (500?–428 B.C.). He was a mathematician, an anatomist, and a philosopher who taught that Mind (*nous*) stirred up the chaos of seeds that existed in the beginning, organizing them into the objects that are seen everywhere. He moved to Athens and became the first teacher of natural philosophy there. Four hundred years later, Lucretius unsympathetically described his ideas about nature:

Let us now look at Anaxagoras
And his theory of "equalpartedness,"
Or "Homoeomeria,' as Greeks call it,
A term our native speech cannot provide
Although the thing itself we can express,
Easily enough, in our words. His theory is
That all things have similitude of parts,
That bones are formed of small and tiny parts
Of bone, for instance, or that flesh is made
From small and minute elements of flesh;
Blood is formed of particles of blood
When many drops have brought themselves together;
He thinks that gold consists of grains of gold,
That earth is a compound of bits of earth,
He pictures other things, and thinks of them,
In the same way. He does not yield the void
A place in objects, or establish limits
To the divisibility of bodies:
On both accounts he seems to me to err.

(L, Book I, 830–846)

However, Lucretius would have agreed with Anaxagoras that the sun is not a god. In 432 B.C., Anaxagoras was charged with impiety and indicted for rationalism, saved by the intervention of Pericles, but exiled for the rest of his life.

At the northern end of the Aegean Sea there was a town called Abdera, which prospered for nearly a thousand years until the middle of the fourth century A.D. Perhaps each civilization must have its laughing-stock, and in those days the air of Abdera was proverbial for causing stupidity. Jokes were told to illustrate how stupid the people of Abdera really were (like modern Canadian Newfie "jokes"), but Abdera could afford to laugh back, because it was there that the great philosopher and mathematician Democritos was raised, and possibly there that he had the idea that the universe is made of atoms. These atoms were eternal, invisible, and indivisible. They differed in figure, arrangement, position, and possibly weight, he thought. Some cling to each other, while others are smooth and roll over each other. Properties like hot or cold, sweet or bitter, hard or soft can be traced to the constituent atoms, because that is all there is—atoms and emptiness. Born around 460 B.C., Democritos was an older contemporary of Socrates.* He inherited some money and traveled to Egypt, and *en route* he may well have stopped at Miletos where Leucippos, some years his senior, developed essentially the same theory. It may in fact have been Leucippos who first had the concept of the atom and transmitted the idea to Democritos. Certainly Leucippos wrote that a vacuum makes motion possible and that space is infinite, and the number of atoms likewise.

During the height and subsequent decline of Athens, its greatest philosophers lived—Socrates (470–399), his pupil Plato (427–347), and his pupil Aristotle (384–322). Socrates took the view that all of these speculations on the nature of matter and the cosmos were completely fruitless. He believed, correctly, that earlier philosophers speculated on the nature of the universe without sufficient knowledge or clear definitions of what they were talking about, a habit from which some modern cosmologists are not completely free. As Xenophon reported, Socrates preferred to discuss "what is godly, what is ungodly, what is beautiful, what is ugly, what is just, what is unjust, what is prudence, what is madness, what is courage, what is cowardice, what is a state, what is a statesman."

In emphasizing clarity of argument and the necessity of being free of prejudice and superstition when trying to find truth, Socrates made a fundamental contribution to all knowledge. He too was accused of impiety and also of corrupting youth. He was arrested; he issued a defiant defense and was condemned to take poison, which he did in the presence of some of his disciples.

We have nothing that Socrates may have written, but his student Plato was one of those with him during those final hours, and his magnificent prose has been preserved. After Socrates' death, Plato traveled to Egypt, meeting expenses partly by selling olive oil, we are told. He then moved on to southern Italy to learn more about the ideas of Pythagoras, and then to Sicily where he ran into trouble with the imaginative tyrant of Syracuse Dionysius the Elder (430–367).

Like some other despots throughout history, Dionysius fancied himself as a poet and dramatist, even sending one of his works to Athens to compete for the prize for the best tragedy. He read his poems to those around him, and, scared to say anything else, they praised his dubious talents. When the poet Philoxenes (436–380) came to the court, he was less than appreciative of the despot's poetry, and was jailed for his opinion. Dionysius persisted, getting praise from all the others, until, very proud of a new creation, he told the jailer to bring Philoxenes to hear it. After reading the poem, Dionysius waited anxiously for the famous poet's comments, but Philoxenes turned not to him, but to the jailer, saying, with a sigh, "Take me back to prison." Apparently Dionysius appreciated his honesty and forgave him, but he certainly didn't appreciate Plato's honesty. He sold him into slavery.

Fortunately, Plato had real friends who raised funds to release him and to purchase for his use a gymnasium surrounded by a grove of olive and plane trees. The grove was a very special place on the river Cephisos that flows past the west side of Athens, dedicated to the god-hero Academos. During the previous century the trees had been carefully planted and pruned, and statues had been erected at the command of Cimon (510–449). The Academy was a beautiful place. Many of its students were rich, or became so, and as alumni or alumnae they were generous in their support. Theory of numbers, geometry, astronomy, law, philosophy, and "music," the inspiration of all of the muses, formed the curriculum.

Plato has been praised, blamed, and made fun of, perhaps more than any other man. "Plato is philosophy, and philosophy Plato" (Ralph Waldo Emerson); "His *Dialogues* are still the noblest body of philosophical thought in existence, and of matchless literary beauty" (Richard Garnett); "Sith of all Philosophers, he is the most poeticall" (Sir Philip Sidney); "It seems to us that Plato has wandered from the truth" (St. Thomas Aquinas); "If Jupiter were to speak it would be in a language like this" (the *Dialogues*: Cicero); "What advantage did Plato think to procure us in saying that man was a two-legged animal without feathers?" (Blaise Pascal). (The statement to which Pascal refers had led Diogenes to pluck the feathers off of a chicken and bring it to the Academy. "This," he said, "is one of Plato's men.")

One of the contemporary comic poets of Greece, Epicrates, lampooned Plato this way, in part:

> But it so happened
> that these most sage academicians sate
> In solemn consultation—on a cabbage.
> A cabbage! what did they discover there?
> Oh sir! your cabbage hath its sex and gender,
> Its provinces, prerogatives and ranks
> And, nicely handled, breeds as many questions
> As it does maggots. All the younger fry
> Stood dumb with expectation and respect,
> Wond'ring what this same cabbage should bring forth;
> The lecturer eyed them round. . . .

Suggestions are made that the cabbage is round, that it is an herb, or a plant, until

> . . . stepping forward, a Sicilian quack
> Told them their question was an abuse of time—
> It was a cabbage, neither more nor less,
> And they were fools to prate so much about it
> Insolent wretch! Amazement seized the troop
> Clamor and wrath and tumult raged amain

> Til Plato, trembling for his own philosophy,
> And calmly praying patience of the court,
> Took up the cabbage, and adjourned the cause.
>
> (UA 4, 310)

It is hardly a surprise that in setting up his theoretically perfect society, Plato excluded poets.

Plato wrote compelling dialogues on the ideal state, on ideal love, on rhetoric, and a host of other topics. He had some ideas about motion, too, whether of the souls of men or of the planets. His thoughts about men's souls have been taken seriously by many theologians; his thoughts on the motions of the planets were dead wrong, and adherence to them held up the development of physics and astronomy in Christendom for many centuries. We return to these unfortunate ideas in the next chapter.

It saved a lot of confusion that Plato was broad-shouldered—that was why he was called "Plato," from the Greek word for "broad" or "flat," from which our words "plate" and "place" are derived. His real name was Aristocles, easily confused with Aristotle, the name of his star student. (One letter can make a big difference, as we see in chapter 9, when the homoöusians and the homoioussians were arguing with each other at the first Council of Nicene, six hundred and seventy-two years after Plato's death.)

Aristotle performed research on essentially all areas of knowledge that were known at the time. His lectures on the thinking of his predecessors and on his own thinking were so comprehensive that they overwhelmed all opposition. Right or wrong, always inspired, sometimes sheer folly, Aristotle's notebooks constituted an enormous encyclopaedia of fact, logic, metaphysics, and science. His philosophy was deep, his biology pioneering, but his physics also was terrible, because he relied, quite naturally, on his common sense, an attitude that does not always work with physics. He had great respect for Democritos as a mathematician, but rejected his atomic hypothesis, piling argument upon argument against it. Aristotle had a point, of course; there really wasn't much evidence to support an atomic hypothesis. Philosophically, also, the idea that objects could consist of tiny atoms moving around in a vacuum was repugnant to him, and later it became a veritable official anathema, as we have seen. You couldn't see the atoms, and a vacuum didn't exist anyhow. His denial of the existence of a vacuum also had major consequences for his incorrect ideas about motion, discussed in the next chapter. However, Aristotle was a great man, with great ideas. He would have been horrified at the way these ideas became codified and distorted, just as surely as Jesus, coming nearly four hundred years later, would have been angered by the distortion of his message, and the ultimate incorporation of it into a phony synthesis with Aristotle's philosophy.

This powerful criticism of the atomic hypothesis caused the idea to be dropped for half a century until Epicuros revived and extended it. He supposed that more nebulous items, like gods, minds, and souls, were also made of atoms, although he assigned to them atoms that were smaller and more subtle than those that made up crass matter. Nevertheless, in Epicuros's philosophy, *everything* was made of atoms—real, solid, tiny atoms—making his the first philosophy of materialism, but not "materialistic" in the modern sense. Displaced from the Island of Samos, where he was born a few years after the death of Plato, Epicuros eventually bought a house and garden in Melita, a suburb of Athens. There he lived until his death, thirty-seven years later, surrounded by family, male and female disciples, and students. It was a remarkable society of people living together, dedicated to knowledge, and what we now call the Epicurean philosophy. Partly cut off from the world, these people sought happiness through pleasure, but not always the pleasures of the flesh so much as the pleasures of morality, prudence, justice, and temperance. In these ideas they were also indebted to Democritos. Epicuros summarized it this way: "When we say that pleasure is the chief good . . . we mean the freedom of the body from pain, and of the soul from disturbance . . . sober contemplation as examines the reasons for choice and avoidance, and puts to flight the vain opinions from which arises most of the confusion that troubles the soul." That, of course, is not the way that the world outside of the garden saw it—the very presence of women had tongues wagging, and the success of the school also caused resentment. The philosophy of atomic materialism didn't help either, among those who understood it. Epicuros was forgotten for two centuries, and with him the whole atomic theory.

During the first half of the first century B.C., there lived in Rome a man who was truly an anomaly. He was a poet, although along with Catullus, Virgil and, a few years later, Horace, and Ovid, he was not alone in that. He was a philosopher-poet with some very negative ideas about love. He was also a scientist, and although Rome had engineers, it had never produced a scientist concerned with fundamental ideas about nature. He was surrounded in Rome by superstitions and power struggles that disgusted him. Titus Lucretius Carus left us a magnificent 7500-line poem that throughout history has never received the attention it deserves. We have seen a quote from it about Anaxagoras, but this poem, *De Rerum Natura—On the Nature of Things,* covers much more. Lucretius believed that freedom came from knowledge rather than superstition, from understanding rather than belief in gods:

Let the first principle of Nature's Law
Start the thread of argument, as follows:
Nothing is ever born from nothing, as if
By act of the gods. Of course, the fear of this
Constrains mankind because they see so much
Proceeding in the sky and on the earth
Whose causes they do not know, and assign
To divine intervention. When we see
That nothing can be born from utter nothing,
We shall see what correctly follows that:
What everything is fashioned from, and how
These things proceed without the help of gods.

(L, Book I, 148–157)

Make this last word singular, and use a capital G, and it is easy to see why his writing was ignored. Even as recently as the first part of this century when British schoolboys were required to study Latin, discussion of Lucretius's work was carefully avoided. Now, of course, everything in the Latin language is avoided in many schools.

The quote at the beginning of this book summarizes the attitude of Lucretius: "The lack of understanding assaults the uncertain mind." Our task, as he saw it, is to improve understanding, replace inferior guesses and theories with better ones that describe things more clearly and more accurately. The soul also is mortal, and knowledge of this releases men from concern about whether eternal punishment or eternal bliss awaits them hereafter, leaving them free to enjoy life and learning while alive. He thought that the world was so much more exciting because it did not have gods manipulating it from behind the scenes. The wonders of nature were the source of his religion, the logic of nature a source of certainty.

Some of Lucretius's advanced ideas on motion and the cosmos are quoted in the next chapter. We conclude this one with quotes from some parts of his work in which he staunchly supported the ideas of Democritos and Epicuros, and in which he developed from common observation a number of arguments in favor of the atomic hypothesis, as translated by Palmer Bovie:

We smell various odors, but never see them
Approach our nostrils. We do not see heat waves;
Our eyes cannot discern the chilling cold;
We never see a voice. Yet all these things
Consist of bodies that touch on our senses:
Only a body touches and is touched.

Clothes hung out beside the wavebreaking shore
Get damp, and the same clothes dry out again
When spread in the sun; but one can never see
The moisture settle, or how it flies off
Because of the heat, so it must be dispersed
Into tiny particles too small for eyes
To observe. Just so, when, over many years,
The sun goes and returns on his cycle,
The inside surface of your ring grows thinner
By being worn. Water dripping from eaves
Will hollow out a stone; the curved iron blade
Of ploughshares dwindles down invisibly
In fertile fields. We see paving stones
Of streets worn down by the footsteps of the crowd;
At city gates bronze statues point out how
Their right hands have been thinned down by the touch
Of travellers greeting them as they pass by.
Such things are lessened, worn down, as we see,
But just what bodies disappear, and when,
Jealous Nature shields our sight from glimpsing.

. . . .

The atoms streak across the void, impelled
By weight or random blows against each other
And, striking, clash and leap apart again
As solid bodies do, compact in weight
With nothing pushing them on from behind.

. . . .

The nature of the atoms as a whole
Lies far below our threshold of perception
So where you cannot see them they withhold
Their movements from you, all the more so since
Things we can see often mask their motions
By being set off at tremendous distance.

As I will demonstrate
In these verses, the small atomic bodies
Sustain the entire universe themselves,
Unlimited in number, and unbounded,
By means of their eternal rain of blows
From every side.

But since the multitude
Of atoms, in their multitude of ways,
Having long since been used to fling, impelled
By blows and their own weights, and to meeting
In all manners, and trying out all ways
By which they might be able to form unions
It happens that in this great length of time,
Though widely scattered, they might still, at last,
By trying out all kinds of combination
And movement, have arrived at those which suddenly
Combined to form the first stage of great things,
Of earth, of sea, of sky, of living creatures.

(L, Book I, 298–321; Book II 83–89)

This is the last contribution of Greco-Roman philosophers before the birth of Jesus. After Lucretius, we hear of nobody for the next nine hundred years who was even aware of, much less supportive of, the hypothesis that all matter is composed of atoms.

1. HOMER 2. VIRGIL 3. JUPITER 4. DANTE 5. PLATO 6. LUCRETIUS

Humans first understood the universe in mythopoetic terms. Homer and Virgil (1, 2) crystallized the Greco-Roman mythic vision in great works of classical literature (the *Illiad, Odyssey,* and *Aeneid*). They depicted a world where gods and goddesses intermingled with human beings, and Zeus/Jupiter held sway over all destinies (3). Similarly, in the Middle Ages, Dante (4) expressed the Church's mythological conception of the universe as a hierarchy of heaven, purgatory, and hell in his *Divine Comedy.*

Rational criticisms of mythology and religion were first raised by ancient Greek and Roman philosophers. Plato (5) typified the idealist school that stressed the primacy of thought over matter. By contrast, Lucretius (6) interpreted the world in strictly naturalistic terms, much as modern science does today. This conflict between religion and reason and between idealism and materialism has been an ongoing theme of history up to modern times.

5

Sun, Planets, and Motion

I think of science as the well-ordered gathering of knowledge concerning the universe and ourselves; I think of it as a sacred contemplation and a spiritual fulfillment.

—George Sarton

It wasn't until Isaac Newton tied the whole thing together three hundred years ago that people began to realize there was any connection at all between the motion of a planet in the sky and the motion of an arrow through the air. After all, the planet was moving in the pure "aether," according to the eternal will of heaven, whereas the arrow was moving in the region below, which was controlled by death, and corrupted by sin. Because of this we will first examine some of the efforts that were made to understand the heavenly motions and then return to ideas about the movements of objects that are closer to home.

As early as 2700 B.C., using observations of the bright star Sirius, Egyptian priests had developed a calendar based on the year, while the Sumerians based their calendar on the moon, with corrections for an occasional extra month to bring it into line with the year. Their careful observations of Venus and Mercury as early as the twentieth century B.C., their early listing of many stars, and their interest in the periodic motions of stars and planets founded the science of observational astronomy, and bequeathed information to the Greek philosophers who tried to understand these phenomena.

At the time of Thales's death, Pherekydes had come from his native island, Syros, a hundred miles away, to teach philosophy at Samos. Undoubtedly he had studied at Miletos; we don't know much about him, but we do know that he believed in metempsychosis, the passing of the soul at death into another body. We also know that he had a star pupil, by name Pythagoras (580?–500?), born right on the island of Samos, the son of wealthy Mnesarchos. This very prosperous island, an important center of trade, with its own red naval ships to protect its merchant marine and expand its influence, was under the dictatorship of Polycrates (?–522).

Like a surprisingly large number of despots who followed him, Polycrates was very supportive of art and poetry. The lyric poet Anacrēōn (560?–475?) found refuge at his court when the part of the mainland where he lived was invaded by the Persians. The great artist-architect-engineer-engraver-sculptor Theodorus also graced the court and spread his influence across to Ephesus, on the mainland, in helping to design the second temple of Diana of the Ephesians.

Pythagoras traveled to Egypt and beyond; his belief in metempsychosis, learned from Pherekydes, may have been reinforced by a trip to India. He could not accept the dictatorship of Polycrates, having ideas of his own about how a country should be run. According to tradition he was fifty years old when, in 530, he left Samos to live in the health resort, Croton, on the southeast coast of Italy. There he married Theano; they had three children, their daughter Damo being later entrusted with his writings. Although left in dire poverty, she refused to sell them after he died.

Pythagoras set up a school of literature and philosophy for men and women; the women were also taught domestic and maternal arts. But it wasn't only a school; it was a religious community, a "club of three hundred," with members bound by vows of trust and support to each other and to the master, and with private goods shared communally. Dressed in his white robe, always freshly laundered, barefoot, sometimes playing the lyre, Pythagoras presided over the classes and dictated the community's diet. Wine was allowed in moderation, but water was preferred. No meat of any kind was permitted, partly for fear of consuming part of a human who had qualified to come back only as an animal, and who might otherwise come back as a future Pythagorean. Vegetarianism was the strict rule, but, possibly from some unfortunate experiences, Pythagoras banned the consumption of beans.

At least this is the tradition. It is not the way the comic poet Aristophon described it in the fourth century B.C., but by then the condition of the followers of Pythagoras had deteriorated:

> So gaunt they seem, that famine never made
> Of lank Philippides so mere a shade;
> Of salted tunny fish their scanty dole
> Their beverage, like frogs, a standing pool,
> With now and then a cabbage, at the best

The leavings of the caterpillar's feast.
No comb approaches their disheveled hair,
To rout the long established myriads there;
On the bare ground their bed, nor do they know
A warmer coverlet than serves the crow.

(Philippides was the famous long-distance runner who ran from Athens to Sparta to ask for aid in the battle of Marathon—a run of a hundred miles. The run from Marathon to Athens, bringing news of victory, was much shorter, only twenty-six miles. We don't know who ran the first marathon.) Aristophon didn't like the Pythagoreans, it is clear, and about their master, dead for a century, he adds:

I've heard this arrogant impostor tell
Amongst the wonders which he saw in hell,
That Pluto with his scholars sat and fed,
Singling them out from the inferior dead;
Good faith; the monarch was not overnice
Thus to take up with beggary and lice.

(UA 4, 309)

Pythagoras had really attempted to formulate and practise a complete synthesis of science and religion. The aim of life on earth was to make the soul perfect, and the brotherhood and sisterhood practised together a spiritual as well as an intellectual life. In fact, the two were indistinguishable. Following scientific studies was seen as a religious experience, and as the quote by George Sarton at the beginning of this chapter indicates, it can be such an experience today.

For the Pythagoreans, then as now, scientific study purifies the mind. The soul was seen as part of the divine soul of the world, leaving one body at death to assume another. The master himself remembered being at the siege of Troy some seven hundred years earlier, and he recalled to his disciples other lives that he believed that he had led. Ovid described Pythagoras's ideas about the soul:

Souls cannot die. They leave a former home
And in new bodies dwell, and from them roam.
Nothing can perish, all things change below
For spirits through all forms may come and go
Good beasts shall rise to human forms, and men
If bad, shall backward turn to beasts again.

> Thus through a thousand shapes the soul shall go
> And thus fulfill its testing below.
>
> (translated by John Dryden; UA 26, 274)

The scientific studies were centered on mathematics, the basic nature of anything being seen as the numerical relationships between its parts. Harmony was seen to lie in wisdom, including the numerical harmony of music. The planets and the gods, goddesses, and (later) the angels that accompanied them were believed to be incessantly singing, or at least humming, so that the theory of their heavenly journeys become part of the earliest theory of music. The fact that nobody heard the music was attributed to human mortality and, later, to sin (to which human mortality ultimately became attributed).

Pythagoras and his followers discovered a fundamental relation between mathematics and music when they noticed that harmonious notes are obtained by clamping the string of a monochord so that the lengths of the two parts are in the ratio of small integers, two to one for an octave, three to two for a fifth, four to three for a fourth, etc. This was the first time that numerical quantities had been used to express a general rule about an earth-bound phenomenon, and it was in fact the beginning of experimental physical science. Unfortunately, somewhat later, this discovery was extended to include the orbits of the moon, planets, and stars. Each of these was supposed to move in a circle, and the circumference of the circle was likened to the length of a string. Thus each heavenly object was thought to emit its own note. The idea of the circular motion as being a perfect motion and therefore the only one permitted by nature, coupled with the idea of the music of the spheres along which these celestial objects moved, was another erroneous idea that remained for centuries—as noted in Lorenzo's speech to Jessica in Shakespeare's *The Merchant of Venice* (1596):

> Sit, Jessica. Look how the floor of heaven
> Is thick inlaid with patines of bright gold;
> There's not the smallest orb which thou behold'st
> But in his motion like an angel sings,
> Still quiring to the young-eyed cherubins.
> Such harmony is in immortal souls;
> But whilst this muddy vesture of decay
> Doth grossly close it in, we cannot hear it.
>
> (Act V, Sc. i, 58–65)

or, as Milton wrote in "Arcades" (1633):

> Then listen I
> To the celestial Sirens' harmony . . .
> Such sweet compulsion doth in music ly
> To lull the daughters of Necessity,
> And keep unsteddy Nature to her law,
> And the low world in measured motion draw
> After the heavenly tune, which none can hear
> Of human mold with gross unpurged ear.
>
> (lines 62–3, 68–73)

Poor science, it would seem, but magnificent poetry. As science, though, it was right up to date. *The Merchant of Venice* appeared in the same year as Johannes Kepler's *Mysterium Cosmographicum,* in the twelfth chapter of which the young astronomer described the music of the spheres in considerable mathematical detail.

For a while the society of Pythagoras thrived, and he saw it as a model for other societies, starting with the town of Croton itself. The local people didn't like this at all, and they felt that this band of philosophers in their midst was a cause for suspicion anyhow. There began a "town versus gown" conflict that has surfaced from time to time throughout the centuries. Seventeen hundred years later, three murders that occurred in a like conflict at Oxford would lead some frightened students and faculty to leave and help to establish a college at Cambridge. In the case of the school of Pythagoras it was even more serious. The townspeople raided the place, setting the building on fire. Some of the disciples, and possibly Pythagoras himself, died in the flames.

Before we leave this remarkable man we may pause to think of some of his maxims, little parables we might call them, because nearly all of them have hidden meanings:

> Do not help men to lay down burdens, but to bear heavier ones.
>
> Do not stand upon your nail pairings or hair cuttings.
>
> Avoid a sharp sword.
>
> Keep your bed packed up.
>
> Efface the traces of a pot in the ashes.
>
> Do not walk in the main street. . . .

Perhaps the last five could be read this way:

Make each advance in character permanent.

Swords (read "guns") can be as dangerous to the owner as to another person.

Be ready for misfortune.

Keep your private affairs to yourself.

Make your own judgments; do not be persuaded by the crowd.

The Pythagoreans argued that nature is perfect, that the sphere is a perfect figure and therefore the earth, being made by nature, must be a sphere. They may also have noted the curve of the horizon or later associated the shape of the earth with its shadow on the moon during an eclipse. One of them, Philolaos of Croton, was apparently the first man to suggest that the earth was just another planet, and that it was not at the center of the universe. As he saw it, every day the earth completed a circle around a central fire that was always hidden from Greece, in the same way that anyone living on the other side of the moon would never see the earth. It would heat up the antipodes too much, perhaps, so he imagined that there was a "counter-earth" that shielded the antipodes from the central fire. Along with earth and counter-earth, the sun, moon, and the planets all moved in circles around the hidden central fire, with the fixed stars located on an outer sphere. Philolaos had the stars *fixed* and the earth *moving.* We refer to him as Philolaos of Croton, because he was born there, and because that is where Pythagoras had established his school. By the time Philolaos came along, however, the Pythagoreans had long been expelled from Croton and had moved up the coast to Metapontion. Even that wasn't safe, and they were harassed and persecuted there, so Philolaos moved inland and ultimately traveled to Egypt, settling in Thebes, which is why his idea was later referred to as the "Egyptian theory." He was the first to write down some of the thoughts of the school of Pythagoras, and for that we are deeply grateful to him.

A century later, Hicetas, probably from Syracuse, imagined that the earth was rotating, not in orbit around an invisible fire but around its axis! To quote Cicero, who wrote another three hundred years later, but himself quoted a much earlier author, the Pythagoreans believed that as the earth "turns and revolves around its axis with great velocity, all the phenomena come into view which would be produced if the earth were at rest and the heavens in motion."

Pretty advanced thinking, but Plato did not accept it. Over the centuries, so much has been written about Plato's philosophy that it should be unnecessary to add more here except for the fact that it started a continuing conflict about the best way to understand nature. In Plato's view, the idea of something is more fundamental than the thing itself—a man is mortal but the idea "Man" applies to all men and it lives forever. One should therefore start with the idea and deduce its consequences, rather than start with the individual objects themselves. (Nothing wrong with working out the consequences of an idea, of course, but if the conclusions you reach are inconsistent with the facts, you had better get another idea.) Plato also believed that the mathematics that described the motion of the divine planets was itself divine, and therefore perfect, and that the perfect figure is a circle. Here are some of his comments on the subject:

> "Each planet moves in the same path, not in many paths, but in one only, which is circular, and the varieties are only apparent."
>
> "The movements can be apprehended only by reason and thought, not by sight."
>
> "What is seen in the heavens must be ignored, if we truly want to have our share in astronomy."
>
> (GS 1, 449)

With this attitude, astronomy could not go anywhere. Plato decided that the distances to the moon, sun, and planets were in the ratios of integers, in accordance with the perfect harmony that Pythagoras had discovered about the lengths of strings that vibrated harmoniously. But Plato was a great philosopher, and his wrong ideas, not just about astronomy but how to go about the process of *learning* about astronomy, were believed by some scholars for two thousand years. In our struggle to understand something about nature, we have allowed too many authority figures to stand in the way.

Plato's student Eudoxos (408–355), however, moved away from his master's edicts and developed an intricate theory of twenty-seven rotating spheres. Eudoxos actually worked at an Egyptian observatory for a while and used his enormous mathematical skills to describe the apparently complicated motions of the planets in terms of the rotations of spheres. Each planet was stuck on a sphere that rotated, but the axis about which the sphere rotated was stuck on another sphere that rotated about a different axis, itself stuck on another sphere. To get agreement with the observations, the number of spheres was increased to fifty-six by Aristotle. With enough spheres rotating about enough arbitrarily chosen directions you can reproduce any bounded motion that you like, but this model suffered from a fatal flaw, first pointed out later in the century by Autolycos of Pitane: All of the spheres had their centers at the earth, so that, according to the theory, the planets would stay at constant distances from us, but in fact the planets appear to vary in brightness, so that they must be moving towards us and receding. Venus in particular showed significant periodic changes in brightness and seemed always to be close to the sun, sometimes rising with the dawn and at others setting in the evening. Heraclides of Pontas (c. 388–315) was another student of Plato who departed from the ways of his teacher. He conceived the idea that Venus was really revolving around the sun, not the earth, and that Mercury was doing the same thing. He accepted the earlier thought that the stars appeared to move because the earth was rotating on its axis but he saw the other planets and the sun as moving around the earth, with Venus and Mercury circling the moving sun like sea gulls around a ship.

Apollonios of Pergē (c. 262–190) came up with a different model. (Pergē is a town on the south coast of Asia Minor which Paul and Barnabas were to visit later, as recorded in Acts 13:13.) Imagine a point moving around the earth in a circle (the "deferent"), and then draw another circle (the "epicycle") with that point at its center. Finally, imagine a planet rotating around *this* circle. It worked! It described pretty well the motions of the planets relative to the stellar background. The reason it worked was that, relative to the stars, the sun appears to move in a near-circle around us, and the planets move in near-circles around it. The fact that the orbits are ellipses rather than circles meant that another circle or so had to be added to the model for each planet—you can always make an ellipse by superimposing two circular motions. But circles it had to be, according to the thinking of the time, although Apollonios himself wrote a detailed treatise on the mathematical properties of the hyperbola, parabola, and ellipse. His writings on that subject were very thoroughly studied by Johannes Kepler more than eighteen hundred years later as he showed that the orbits are indeed ellipses. Until the simplicity of this motion was seen, however, the old ideas became more complicated. Not only did you have to describe the planetary orbits correctly, you had to get the sun and each planet to the right place at the right time. To do this for the sun, it occurred to Apollonios or, if not, certainly to his successor Hipparchos, to introduce an "eccentric"—i.e., a circle with its center not at the earth, but displaced from it by a distance and direction chosen to make the model work. Ultimately, Ptolemy extended this idea to the planets. In chapter 12 we describe why it worked fairly well, but to agree with the observations it had to be refined with more revolving wheels until there were forty in all. The more assumptions that you make, the more accurate you can make any theory become.

According to Archimedes, another model of the solar system had been proposed by a man who had died when Apollonios was a boy:

> Aristarchos* of Samos brought out a book consisting of some hypotheses wherein it appears, as a consequence of assumptions made, that the real universe is many times greater. . . . His hypotheses are that the fixed stars and Sun remain unmoved, that the Earth revolves about the Sun in the circumference of a circle, the Sun lying in the middle of the orbit, and that the sphere of the fixed stars, situated about the same center as the center of the Sun, is so great that the circle in which he supposes the Earth to revolve bears such a proportion to the distance of the fixed stars as the center of the sphere bears to its surface.
>
> (GS 2, 57)

This remarkable statement, along with others by Archimedes, implies that around 270 B.C. Aristarchos had the whole picture—the sun at the center, the planets including the earth going around the sun, the moon going around the earth, the earth rotating on its axis (giving the impression that the stars were moving), and the stars so far away that they appeared to be in the same direction as the earth moved around the sun. Aristarchos was accused of impiety by Cleanthes, who became leader of the Stoics in 263 B.C. "for moving the hearth of the universe and trying to save the phenomena by the assumption that the heaven is at rest but that the Earth revolves in an oblique orbit while rotating around its own axis." No great calamity occurred to him because of this accusation, however. Aristarchos was the first person to estimate from measurements the distances and sizes of the sun and moon. His geometrical calculations were correct but based on inaccurate data; some errors that he made partially cancelled each other. He noted the size of the earth's shadow on the moon during an eclipse and measured the near-right angle between the directions of the sun and moon when one half of the latter was illuminated. He then estimated the angle subtended at the earth by the diameter of the moon and, knowing his geometry, figured out these various ratios.

It was left to Eratosthenes (273–192) to provide an accurate measurement of the circumference of the earth, although Eudoxos may have tried something similar years earlier. The town of Syene (modern Aswan) on the river Nile was measured to be 5000 stadia south of Alexandria. The best estimate of the length of a stadium from early documents is 158 meters (0.098 miles), although the Olympic stadium was longer. If we adopt the approximate value of one tenth of a mile, the distance Eratosthenes measured would be 500 miles, which is about right. Syene is a little north of the Tropic of Cancer, so that a vertical well allowed sunlight to penetrate to the bottom at noon on the summer solstice and a vertical stick cast almost no shadow. (The use of a "gnomon" like this had been known to Anaximandros.) At Alexandria, however, the shadow made an angle with the vertical that Eratosthenes measured as one fiftieth of a complete circle (i.e., 7.2°). Thus, he figured there are 50×5000 = 250,000 stadia, or 25,000 miles in the circumference of the earth, measured in the north-south direction. Rounded off to the nearest thousand miles, he was right on target, closer than the accuracy of his measurements warranted, and indeed more accurate than the assumption "ten stadia are equal to one mile" warrants. He extended his estimates of distances on the earth, mostly by studying tales that travelers brought back to Alexandria. Continuing in the north-south direction, his estimates of the distances between the latitudes were: Alexandria to Hellespont (8100 stadia), there to the Dnieper (5000), and from there to the latitude of Iceland, 11,500 stadia—a total of 24,600 stadia or, at ten stadia to the mile, about 35°. Not bad; Alexandria is at approximately 32° and most of Iceland is between 64° and 66°. Here also were the concepts of latitude and longitude, developed in detail by his successors Hipparchos, Marinus, and Ptolemy.

In the east-west direction, Eratosthenes judged the total length of the latitude through Athens to be 200,000 stadia—it is of course less than the length of the equator by a factor equal to the cosine of the latitude (38°), i.e., approximately 80 percent. Indeed, knowing that the earth is close to a sphere in shape, he used his north-south measurements to obtain the result. When he relied on travelers' tales to estimate east-west distances, however, he wasn't as accurate: east coast of India to the extreme west coast, 19,000 stadia; from there to the Caspian Gates (narrow pass on the west side of the Caspian Sea, used as a trade route for centuries), 14,000 stadia; on to the Euphrates, 10,000; to the Canopic mouth of the Nile, near Alexandria, 6,300; to Carthage, 13,500; and from there to the Pillars of Hercules, "at least 8000"—a total of at least 70,800 stadia, or 7080 miles, at the assumed ratio of ten stadia to the mile. Without any assumption about that ratio, it appears that he thought that the east side of India was about one third of the way around the world from Gibraltar. It is closer to one quarter, but then his informants couldn't stay on a particular latitude, or even great circle, and had to travel much farther as they followed the paths and made their ways through the mountains. Perhaps they also exaggerated a little.

In the following century there lived a man who was to have an enormous influence on the beliefs of the astronomers and geographers who followed him—Hipparchos (190?–125) of Nicaia, today northwest Turkey. Hipparchos was not only a great mathematician (he invented trigonometry), but also a very accurate observational astronomer and a major contributor to the theory of deferents and epicycles. He examined his own records and those of Aristarchos and other observers and assessed the length of the tropical year (the time from one spring or fall equinox to the next) as one three-hundredths of a day (4.8 minutes) shorter than the approximation 365¼ days. It is actually 11.5 minutes shorter. He then examined records that indicated the moon had gone around the earth 3760 times in 304 years so that the synodic month, the period from one new moon to the next, was 0.080851 years or 29.531 days, accurate to within better than one second! However, that would imply an overall accuracy of one part in three million. In fact, we have seen that his estimate of the length of the year was off by more than six minutes, so that the length of the month, derived from it, would be thirty seconds too much. For the numbers quoted, this small error is compensated by a small error in the number 3760 (it is closer to 3760.03). This is not to detract from Hipparchos's accurate pioneering work, but rather to show the level of accuracy at which he was calculating, so good that it doesn't need the exaggerated claims that some writers have made.

Hipparchos also observed a new star—a nova—in 134 B.C. and was probably the first person to discover the precession of the equinoxes, certainly the first to accurately measure this effect, which is due to the fact that the axis of the earth does not point in exactly the same direction at all times. Due to the earth's bulge around the equator, other bodies, the sun and moon in particular, can apply a torque to it, causing the earth's axis to "precess" slowly and come back to its original direction after 26,000 years. In addition to all of this, he developed the system of latitudes and longitudes for charting the earth, and he followed the mathematician Hypsicles in dividing the zodiac circle into 360 degrees. Unfortunately, Hipparchos rejected the idea of Aristarchos that the earth moves in a circle around the sun. The remarkable aspect of this rejection of such an important discovery is that he was right—the earth does not move in a *circle.* The accuracy of Hipparchos's observations told him that circular motions of the planets did not agree with the data, but at that time there was still no idea that anything but circular motion would be allowed in the aether. Of course, the model could have been patched up with epicycles and eccentrics, but nobody at the time seems to have thought it to be worthwhile. Hipparchos also rejected the measurements of Eratosthenes, so that these too fell into disrepute, leaving the field open for the belief of the Greek Stoic philosopher Poseidonios and others that the circumference of the earth was only eighteen thousand miles, and that a ship sailing west from Gibraltar would reach India after traveling only about seven thousand miles.

Poseidonios of Apamea, Syria (135?–51) taught Stoic philosophy to some of the famous men of Rome, including Cicero, the Roman orator, statesman, and philosopher, and Pompey the Great (called that because he annexed Syria and Palestine and precipitated a civil war).

The philosophy that Poseidonios taught was one of unity, and with that as a basis he hit on some very good ideas, and some that were less so. The grand concept that all parts of the solar system are related to one another, for instance, led him to see a vital force emanating from the sun, permeating the world, influencing all things, alive or inanimate, and, with the moon, causing tides, which he saw were clearly related to the phases of the moon. In addition to coming up with the low estimate for the circumference of the earth, however, Poseidonios espoused the popular belief in the validity of astrology, so that we return to him briefly in chapter 7. We note here, though, that if, as he believed, all parts of the solar system are related to each other, it follows logically that planets are related to everything else, including people. That they affect people was to him just as logical as the idea that the moon affects the tides.

In a monumental work on geography, Strabo (63 B.C.–21 A.D.?) argued that there is but one ocean, so that it would be possible to sail from Spain to the East Indies, but not directly. He thought that Eratosthenes had overestimated his distances and quotes him as saying, "if the extent of the Atlantic Ocean were not an obstacle, we might easily pass by sea from Spain to India, *still keeping in the same parallel.*" Here the practical geographer looked down on the mathematician, for he guessed that there would be other lands in the way, even populated areas. The Christian scholastics who would be appearing later on the scene probably never heard of Strabo's remarkable prediction: "It is quite possible that in the temperate zone there may be two or even more habitable earths, especially near the circle of latitude which is drawn through Athens and the Atlantic Ocean," (UA 5, 404) a circle at approximately 38° N, passing between Washington and Richmond and between Tokyo and Beijing. Thirteen hundred years later, Strabo could have been burned at the stake for saying a thing like that.

Strabo's *Geography* described many details of the structure of the land, all the way from Spain to the River Ganges, for Strabo traveled widely. For some reason that is not understood, his remarkable book had only an indirect effect on the succeeding generations of scholars, although as the Karl Baedeker of the early Christian era, Strabo must have provided invaluable information for travelers and merchants.

Another component of the solar system—comets—had been studied by Chaldean astrologers, and Greek scholars had differing concepts of the Chaldean theories that had resulted. Epigenes of Byzantion had presumably read or heard reports of the burning-air nature of comets, whereas Apollonios of Myndos (not the Apollonios of conic sections and epicycles) heard that comets were objects whose paths could be calculated like those of planets. This idea was supported by Seneca, a man born in or about the same year as Jesus: "One day a man will be born who will discover the orbits of comets and the reason why their paths are so different from those of other planets. Let us be satisfied with the discoveries already made, so that future generations may also add their mite to the truth." (GS 2, 168)

Lucius Annaeus Seneca (4 B.C.–65 A.D.) was the leading Stoic philosopher of Rome. As advisor to Nero he contributed to good government, or at least to the minimization of evil government, but he fell into disfavor. Before Nero commanded him to commit suicide, Seneca had a short time in his retirement to write *Questiones Naturales* in which he described his efforts to understand, not only comets, but his immediate physical surroundings, from rain, rainbows, hail, and snow to earthquakes, volcanoes, rivers, and springs. Like Strabo, he suggested that there might be another continent beyond the Atlantic Ocean, and he speculated "how many an orb moving in the depths of space has never yet reached the eyes of man." He explicitly anticipated Hamlet's

> There are more things in heaven and earth, Horatio,
> Than are dreamt of in your philosophy.
>
> (Act I, sc. v, 166–7)

As a Stoic, Seneca believed in the integration of natural phenomena, but he did not despise the slandered Epicuros. As adviser to Nero he had made an enormous fortune, much of which he donated to rebuild Rome after the great fire of 64 A.D. When Nero sent his command, Seneca allowed himself to bleed to death in a warm bath. He had used information about natural phenomena to inspire morality in his readers, and for this the fathers of the early church, notably Tertullian of Carthage (160?–230), encouraged the reading of some of his works, while condemning other Roman pagans like Lucretius. However, comets were too scary for anyone to believe that they were anything but divine warnings sent down into the lower regions below the moon, rather than planet-like objects as Seneca believed. A sesquimillennium after Seneca's death, Michael Maestlin recognized that the comet of 1577 lay above the moon, but he lived in fear of this discovery, which by then would dangerously violate orthodox Christian beliefs.

In the second century A.D., Marinus of Tyre followed Hipparchos in developing the latitude and longitude system, establishing the origin of longitude in the "Fortunate Islands" (Canary and Madeira) about 17° west of its present origin in Greenwich, England. He thought that Eratosthenes had underestimated the direct distance from Gibraltar to India, so that, according to Marinus, the longer voyage around the other way was considerably less than some others had thought. Still in the second century A.D., Ptolemy (Claudius Ptolemaeus) put together the whole picture of geography and astronomy as he saw it, quoting, modifying, and extending the works of others, and expressing them in mathematical terms. He was another great man who unfortunately became an Authority. He was the first to perform experiments on refraction and reflection of light, he followed Poseidonios in thinking that the circumference of the earth was only eighteen thousand miles (again assuming that a stadium is one tenth of a mile), and above all he presented in exquisite detail the earth-centered model of the solar system introduced by Apollonios and developed by Hipparchos and himself. Like his predecessors, he also endorsed astrology, even writing four books on the control of human affairs by the stars. He regarded the theory of planets fixed to circles which rode on other circles as a mathematical model, its only function, an essential one, being to predict accurately, for the years to come, the motions of the planets relative to the fixed earth. He would have been horrified to learn that it was to become a dogma and that some who challenged it would be made to suffer and even die for their skepticism. The theory that the earth and other planets move around the sun was no longer in competition, because it had no superior predictive power and brought with it serious difficulties in its conflict with common sense. Ptolemy's model worked, patched up from time to time, but dominating the thinking of Christian, Jewish, and Muslim astronomers for fifteen hundred years. We shall return to it in chapter 12.

Years later, if people had believed Eratosthenes's more accurate estimate of the size of the earth, rather than the smaller estimates made by Poseidonius, Marinus, and Ptolemy, Christopher Columbus (1446?–1506) would have had much more difficulty in obtaining the funding to sail westward to "India." However, the crowning argument that was to persuade Columbus and his sponsors was theological, rather than scientific. Cardinal Pierre d'Ailly (1350–1430) was called the "Hammer of Heretics"—he played a prominent part in the execution of many of them, including Jan Hus (1369?–1415) who, like Ikhnaton so many years earlier, and Martin Luther a hundred years later, spoke against the simony of the priests and uttered other heresies. The Cardinal read these words in the Apocryphal Second Book of Esdras (6:42): "Upon the third day Thou didst command that the waters should be gathered in the seventh part of the earth; six parts hast Thou dried up and kept them to the intent that of these some, being planted of God and tilled, might serve Thee." This was a divine message, d'Ailly saw, telling that water occupied only one seventh of the area of the earth. The distance westward from Spain to the lands Marco Polo had visited could not be very great. Also, Roger Bacon (1214?–1294) had written, "the sea between the end of Spain on the west and the beginning of India on the east is navigable in a very few days if the wind is favorable." Columbus obtained a copy of d'Ailly's *Image of the World* which contained these revelations, so that by writing the book the Cardinal had made it possible to prove that they were wrong.

Down here on earth, motion appeared to be nothing at all like the movement of the planets. Aristotle believed that nature has an overall design in which everything has its appointed level. If displaced from where it belongs, an object strives to return to its "natural" place by moving up, down, or sideways as the case may be, and once there it comes to rest. Stars are light and fiery, so that it is their nature to stay above us. Stones fall to the earth, the place nature designed for them; flames leap upwards, air rises through water but moves horizontally across the earth, water rests on earth. "We may distinguish the absolutely heavy, as that which sinks to the bottom of all things, from the absolutely light, which is that which rises to the surface of all things." As it turned out much later, this approach leads to confusion, but at the time it represented a very reasonable attempt to understand vertical motion. It was "common sense."

The concepts of force and weight were recognized very early, and it was also noticed that an object moving through a medium such as water would suffer a force that resisted its motion. Unfortunately, it was also thought that nothing could move without the continuous action of a force that pushed it forward. Aristotle used this point of view to show that a vacuum could not exist "as air resists motion; if the air were withdrawn, a body would either stay still because there was nowhere for it to go, or if it moved it would go on moving at the same speed forever. As this is absurd, there can be no vacuum." Although the phrase "go on moving forever" reminds us of Newton's law of inertia, Aristotle clearly rejected it—he could not imagine that an object could move under no forces. It would be many centuries before it was realized that force and weight are related to the acceleration, not the speed, because it is not that obvious. When a horse pulls a cart, for example, it moves, and the harder the horse pulls (i.e., the greater the *force*) the greater is the *speed* with which the cart moves.

The force on a falling body was seen to be connected to its weight, so that it would appear that a heavier body would fall faster than a lighter one. "The downward movement of a mass of gold or lead, or any other body endowed with weight, is quicker in proportion to its size." Taken literally, this would mean that a ten kilogram weight would fall ten times as fast as a one kilogram weight. It is quite possible that Aristotle would not have agreed to that, but some of his successors did, especially those who came some 1500 to 1800 years later.

The nature of weight and the other forces that are exerted on moving bodies was the subject of a debate that waxed and waned over the centuries. Why, for example, do falling objects increase their speed as they approach the ground? Is it because the amount of air lying below them decreases, and so offers less resistance? Is it because of their great inherent desire to reach their appointed level? Is it because the closer they get to the place where they belong, the stronger that place attracts them? Or is it that motion produces heat, and a falling body heats the air which then becomes rarefied, thereby offering less resistance? These and other explanations were offered and debated very seriously because of the misconception that a changing speed would require a changing force to produce it.

Because of his remarkably accurate and extensive researches in other branches of science and philosophy, Aristotle, along with his school, acquired a reputation which over the centuries became impeccable. These pioneering but often inaccurate descriptions of motion acquired the overwhelming authority of the rest of Aristotle's works and, at different times, were eventually accepted by Muslims, Jews, and Christians alike. The basic error lay in the attitude towards physics as opposed to the biological sciences. In the latter, classification of plants and animals is an important *first* step, and here Aristotle excelled. In physics, which deals primarily with inanimate objects, classification is not a useful exercise. Following that course led to the belief that each object obeyed its own special law, because it was its "nature" to do so. We now know that just the reverse is the case—the same laws apply to all. Nevertheless, it often seems that objects *do* obey different laws according to their nature, and at first it was reasonable to assume that such was the case.

There were of course times when, and places where, Aristotle was out of favor, and other times when to deny him was a heresy. Lucretius, again, did not believe Aristotle's description of falling motion:

> All objects, whose weights may be unequal,
> Move equally through the unstirring void.
> By no means then, would heavier bodies drop
> On lighter. . . .
>
> (L, II 238–241)

although he appreciated the effect of moving through a resistive medium:

> The things that fall
> Through porous air or water pace their flight
> According to their weight because the substance
> Of water or the nature of thin air
> Cannot by any means delay equally
> Each falling object.
>
> (L, IV 230–235)

A most remarkable criticism of Aristotle's ideas on falling bodies was offered by John Philoponos (470–540 A.D.), a Greek philosopher at Alexandria. The following passage, translated by Lane Cooper, sounds almost like Galileo:

> Here is something absolutely false, and something we can better test by observed fact than by any demonstration through logic. If you take two masses greatly differing in weight, and release them from the same elevation, you will see that the ratio of times in their movements does not follow the ratio of the weights, but the difference in time is extremely small, so that if the weights do not greatly differ, but one, say, is double the other, the difference in the times will be either none at all or imperceptible.

It would seem from this quote that Philoponos may have actually conducted an experiment, dropping two unequal weights, more than a thousand years before Galileo was born. This is not the only example of Philoponos's insight. He argued that motion could take place in a vacuum, he recognized that interference by a medium would require a moving object to take an additional time for a given distance, and he criticized the role that air was supposed to play in providing a forward force (one idea he wisely rejected was that, when an arrow flies through the air, the air displaced by the arrow gets behind it and pushes). He also recognized that the air has little effect on such motion and that the bow and archer impart a "motive force" to the arrow, an idea he attributed to Hipparchos. Although some Muslim philosophers later became aware of this work, it was not followed up in any detail for a thousand years.

Recent studies at Arizona State University and other institutions have shown that many entering freshmen really have an Aristotelian view of motion, despite what they have learned to verbalize. Confronted with examples of motion and asked how that motion would continue, many students chose what Aristotle would have predicted over what the laws of Newton tell us. Poor high-school education? Not really. After taking and *passing* a year of university physics, the students were tested again, and the results still favored Aristotle! In this case it was not the authority figure that they were obeying—many of them had never heard of him. It was just his common sense.

A particular case of the theory of motion occurs when there is no motion at all, i.e., when the forces balance and everything is in equilibrium—the study of statics. Best known to us of the ancient Greeks for contributions to modern science is Archimedes of Syracuse, Sicily (287?–212 B.C.), the founder of both statics and hydrostatics, the Archimedes of the law of the lever and of Archimedes' Principle, that the apparent loss of weight of an object when it is immersed in a fluid is equal to the weight of the fluid it displaces.

Archimedes developed the concept of specific gravity, measured the apparent diameter of the sun, and is reported to have determined the distances to the planets. He noted the magnification of an object as it sinks under water, and with his research on numbers and geometrical figures revealed that he was indeed a great mathematician. He invented and described the construction of an orrery to demonstrate the relative motions of the components of the solar system. Compound pulleys and a hydrostatic screw for lifting water were Archimedes's inventions, along with burning mirrors and military machines used in the defense of Syracuse. Four hundred years later, Athenaios of Naucratis, Egypt, described some of Archimedes's contributions to the fleet of Hieron, king of Syracuse (270–216 B.C.).

> A wall with battlements and decks athwart the ship was built on supports; on this stood a stone-hurler, which could shoot by its own power a stone weighing one hundred and eighty pounds or a javelin eighteen feet long. This engine was constructed by Archimedes. Either one of these missiles could be hurled six hundred feet. . . . The ship was ordered to be launched in the sea that it might receive the finishing touches there. But after considerable discussion in regard to the method of pulling it into the water, Archimedes, the mechanician, alone was able to launch it with the help of a few persons. For by the construction of a windlass, he was able to launch a ship of so great proportions into the water. Archimedes was the first to invent the construction of the windlass.

These inventions were to no avail, however, since the Roman troops took over and one of them killed him.

According to the famous English historian Lord Macaulay (1800–1859), Archimedes "was half ashamed of these inventions which were the wonder of hostile nations, and always spoke of them slightingly as mere amusements, as trifles in which a mathematician might be suffered to relax his mind after intense application to the higher parts of his science." Macaulay was undoubtedly right, since the attitude of Greek scholars, from Plato on, was to disdain practical matters, not even thinking of doing an experiment to check their ideas. Aristotle had written, "Since men philosophize to escape ignorance, it is clear that they seek scientific knowledge in order to know, and not for the sake of any practical end." Macaulay also had the words of the early historian Plutarch (46?–120 A.D.) in his *Parallel Lives,* a number of character studies of famous Greeks and Romans: Archimedes regarded "as ignoble and sordid the business of mechanics and every sort of art which is directed to use and profit; he placed his whole ambition in those speculations the beauty and subtlety of which are untainted by any admixture of the common needs of life."

But there were others who did not disdain practical applications. Ctesibios of Alexandria may have been a contemporary of Archimedes or perhaps lived in the next century. He invented the cylinder, plunger, and valve of the pump, used it for his hydraulic organ, and greatly improved the water clock by recognizing that the water level should be kept constant and that the hole through which the water drips should be made of stone, or agate, or some other material that is not easily corroded. Ctesibios saw that using clean water helps too.

He was followed by Philon who, like him, worked in Alexandria and also in Rhodos. While he devoted much of his attention to military engineering, he probably deserves credit as the first experimental physicist, apart from his early predecessors of the school of Pythagoras, with their measurements of the ratios of the lengths of strings on a harmonious monochord. Philon noted that water cannot be poured out of a nearly closed container unless another hole is provided for air to get in and replace the water as it pours. This was surely proof positive of the seemingly sensible idea of Aristotle that a vacuum could not exist. More than that, Philon held over water an inverted vessel with a light inside it, the rim of the vessel being submerged, and recorded that the water was drawn up inside! One man's ideas and observations, left to be rediscovered and expanded by Torricelli and Pascal in the seventeenth century, and by Priestley and Lavoisier in the eighteenth.

The tradition of invention started by Archimedes was carried on at Alexandria into the first century A.D. by Heron, whom we call "Hero," a Greek mathematician and inventor of devices operated by steam, water, or compressed air. "Hero's Fountain" is now a toy in which water is poured down a vertical tube into a container which is closed except for a tube that allows air to escape upwards into another chamber that contains water, forcing that water upwards through another opening to form a fountain. These men had discovered some of the remarkable, useful, or playful things that can be done with water and air under pressure. Hero is even reported as having invented the first coin-operated machines, one of which, on insertion of a coin, delivered holy water into a font. But this is not like a novel, where we are sometimes expected to "suspend our disbelief." We mustn't believe everything that is reported, nor can we always be certain about not believing either. Hero was a very inventive and whimsical fellow. If he had been more practical, he could possibly have changed history by inventing the steam engine.

6

Animals, Vegetables, and Minerals

> There was Aristoteles, a very distinguished writer . . . whom it took centuries to learn, centuries to unlearn and is now going to take a generation or more to learn over again.
>
> —Oliver Wendell Holmes,
> *The Autocrat of the Breakfast Table*

Biology got off to a good start. Animals in the wild can distinguish one plant from another, enough to know which ones are good to eat and which are poisonous. Those who didn't know this are no longer with us. Early man—actually it was mostly women—refined the human diet and acquired knowledge of a long list of plants that could be used to ease wounds and illnesses. The anatomy of animals, and for that matter of humans, was revealed in the process of cutting them up for food, but it would be a long time before the functions of the individual parts of animals and plants were understood. Among the earliest written records of food—in this case fish—there is a cuneiform tablet from Larsa, a town on the Euphrates, across the river from Ur, and twenty-five miles upstream. The 4,000-year-old cuneiform document lists about thirty kinds of fish that were on sale there at the time, along with their prices, some being as much as ten times as expensive as others. In other tablets, the names of hundreds of animals and plants are given in two languages, Sumerian of the local area and the Semitic language, Akkadian, of their immediate northern neighbors. The Akkadian people liked the Sumerian life-style, which they adopted into their territory, while some of them moved south to enjoy it there, the whole region forming the kingdom of Sumer and Akkad. That life-style remained, but the Sumerian language died, to be used later only in religious ceremonies, the way Latin is used today. However, the names we give to some plants can be traced to their Sumerian origin—chicory, myrrh, and crocus, for examples—but we don't know for sure that they are really the plants that the Sumerians called "kukru," "murru," and "kurkānu."

These people, so long ago, knew about the sexuality of date palms, if not explicitly, then at least practically, removing flowers from the male plants and attaching them to the female flowers. Professor George Sarton, whose books on the history of science are pinnacles of scholarly achievement, suggested that they discovered the sexuality of plants the hard way. Noticing that some of his palms were not bearing fruit, perhaps someone cut his lot down, only to discover that on the ones that he had left standing, he didn't get any dates at all the following season.

North of Akkad lived the hostile Assyrians, who founded the empire of Nineveh at a later time period. Nineveh is familiar to us from the Old Testament, as also is Sennacherib, its king at the turn of the eighth into the seventh century B.C. Lines from Byron's "The Destruction of Sennacherib" have given us a vivid image:

> The Assyrian came down like the wolf on the fold
> And his cohorts were gleaming in purple and gold
> And the sheen of their spears was like stars on the sea,
> When the blue wave rolls nightly on deep Galilee."

They were indeed a warlike and cruel people, but that is one reason—an ugly one—why they are mentioned here. The Assyrians kept what may have been the first zoo (at least since Noah!), but it was not from any motive of scientific curiosity, or from a desire to protect the animals. It was for "sport." Lions were kept in the king's private park, so that he could hunt them to demonstrate his prowess, might, and virility. Bas-reliefs showing details of lions bleeding, vomiting, and dying are now in museums, taken from the ruins of Nineveh. Nevertheless, before it was destroyed in 612 B.C., Nineveh was a haven for some scholars. Its king, Ashurbanipal† (668-625 B.C.) learned about the Sumerian cuneiform tablets that preceded him by as much as two thousand years, and he decided to have experts collect, translate, and study them. Texts in medicine, astrology, and astronomy have been found at Nineveh, ignored by the Chaldeans who followed, and covered by the sands for centuries. In the summer of 1989, Iraqi archaeologists discovered a hoard of Assyrian treasures at Nimrud, 50 kilometers down the River Tigris from Nineveh. Beautiful gold jewelry, studded with turquoise, lapis lazuli, malachite, and other precious stones reveal some of the lost riches of the Assyrian empire. Nimrud and

Nineveh were overrun and destroyed by the Medes, Chaldeans, and others in 612 B.C. Nebuchadnezzar, coming to the throne of Babylon seven years later, apparently never discovered that his father Nabopolassar, who led this sack of Nimrud, had passed up untold treasures in his haste to move up the river and do the same thing to Nineveh.

Three hundred years later, Aristotle was preparing a list of over 500 species of animals, describing and classifying them, being very careful in his observations, wary of being fooled by superficial similarities. He studied their anatomy and their physiology, their organs and what the organs did. His philosophy was that there is a great universal purpose behind each part of the body and each animal. They were there to perform a particular function—for Aristotle the physiology determined the anatomy, the performance determined the form. The amount of information on animals and plants that he and his assistants collected was enormous and was put together in a sound scholarly manner that required modification only when new information became available. He was helped by his former pupil Alexander the Great† who ordered some of his men to send back to Athens samples of fauna and flora that they encountered on their expeditions. Aristotle performed a simple and neat experiment by incubating a number of fertilized chicken eggs and breaking open one each day to learn how the embryo grows. His theories are not as good though. He insisted that the heart, not the brain, was the central organ for the senses of animals, although some of his predecessors and his immediate successor thought otherwise. He also argued that baldness was caused by lack of a hot greasy fluid, noting that some greasy plants don't lose their leaves. He then went on to attribute the deficiency of the "hot greasy fluid" in bald men to their loss of it through intercourse, which also explains why women *don't* go bald! (Actually, Aristotle wasn't so very far off track. Baldness is caused by androgen, the male hormone, of which men naturally have a larger supply than women.)

Aristotle's research illustrates the difficulty of understanding nature, even when conditions are very favorable, in this case a first-class mind with a lot of support, many disciples and assistants, but a wrong idea. Would we have argued the same way, more than 2300 years ago? We probably wouldn't have even thought of the question. Yet shining through this error is a great idea, that man is not completely different from animals and plants, and that his physiology is determined by similar laws. Aristotle even recognized that man is an animal: "Nature has made the buttocks for repose, since quadrupeds can stand without fatigue, but man needs a seat." There is no record that he dissected a human corpse, but he knew a lot about the anatomy of animals. He classified them into the *enaima*—blooded—and the *anaima*, not just anaemic, but bloodless, and described every animal he knew about, from the elephant to the bee. It is true that he thought that elephants can be afflicted by only flatulence and catarrh, but his description of the life of the bee shows very careful observation, except for the sex of the chief bee of the hive—he thought it was a king.

"Next after lifeless things in the upward scale comes the genus of plants," Aristotle said, "relatively lifeless as compared to animals, but alive as compared with corporeal objects. There is in plants a continuous scale of ascent towards the animal. There are certain objects in the sea concerning which one would be at a loss to determine whether they be animal or vegetable." (GS 1, 534) It would be an exaggeration, even a distortion, to regard this as a precursor to the theory of evolution or that his statement about objects that "would go on moving forever," noted in chapter 5, was a preview of the laws of Newton. Aristotle was offering a static picture, not an evolutionary one, with each species fulfilling its intended role in nature. "It is evident that plants are created for the sake of animals and animals for the sake of man." This view of a fixed world, with the plants and animals put there for the use of man, ultimately became a central theme of Christian theology, solidly supported by the statement of God on the sixth day of creation (Genesis 1:26): "Behold, I have given you every herb bearing seed, which is upon the face of all the earth, and every tree, in which is the fruit of a tree yielding seed; to you it shall be for meat. And to every beast of the earth, and to every fowl of the air, and to everything that creepeth upon the earth, wherein there is life, I have given every green herb for meat." We occupy such a small fraction of this wonderful world; it is comforting to believe that so much of it was made just for us.

The ancient folklore about the properties of plants as food, medicine, or poison became recorded in writing. The hit or miss "experiments" with them—a better word is "experiences"—ultimately led to a body of knowledge which became the property of the rhizotomists, i.e., the root-removers. Superstition and magic surrounded them, as they invoked spirits and practiced ancient rites while they were gathering the plants for use. Rhizotomists regarded changes in the appearance of vegetation as portents of supernatural events, although the man who could be called the father of botany, Theophrastos (372–288 B.C.), argued that these were natural events, with natural causes that he did not understand. To have faith that there is a natural cause, even though you don't know what it is, is the mark of a scientist, but to many people it was not as satisfactory, nor is it now, as ascribing the effect to the will of a god or the God.

Theophrastos was a student of Aristotle and ultimately became his successor at the Lyceum in Athens. The great master, with his mind filled with practically everything else in the way of "impractical" knowledge, left botany and mineralogy to Theophrastos, and he described and studied approximately five hundred cultivated botanic specimens. Alexander the Great died of a fever in Babylon the year before the death of Aristotle, and his generals began to fight among themselves. The supply of plants from India was brought to a halt as the control of northern India and Afghanistan was seized by Chandragupta, but Theophrastos still had plenty of material to work with.

He was very clear about the sexuality of plants, in particular the fig and the date, but, for reasons that perhaps Sigmund Freud could have explained, this discovery was repressed for two thousand years. Theophrastos described the various parts of plants and flowers, the planting and propagation of trees, timbers, herbs, cereals, juices from plants, horticulture, plant diseases and damage from insects and worms, even how plants smell. He noticed that mistletoe would sprout only on the bark of a live oak. Along with all this were discussions of a question that Aristotle would, and probably did, ask: "What were the intentions of nature in producing this plant?" In the early nineteenth century, Heinrich Heine showed his appreciation of the way that Theophrastos had classified flowers:

> It vexes me every time when I remember that even the dear flowers which God hath made have been, like us, divided into castes, and like us are distinguished by those external names which indicate descent as in a family tree. If there must be such divisions, it were better to adopt those suggested by Theophrastos, who wished that flowers might be divided according to souls,—that is, their perfumes. As for myself, I have my own system of natural sciences, according to which all things are divided into those which may or may not be eaten!
>
> (*The Brockenhaus,* UA 22, 149)

Theophrastos wasn't just a scientist—he wrote over two hundred treatises on subjects ranging from religion to education, astronomy to natural history, and a series of sketches on the foibles of humankind. One of them, on superstition, quoted in full by Professor Sarton, begins:

> Superstitiousness, I need hardly say, would seem to be a sort of cowardice with respect to the divine, and your superstitious man such as will not sally forth for the day till he have washed his hands and sprinkled himself at the Nine Springs and put a bit of bayleaf from a temple in his mouth. And if a cat cross his path he will not proceed on his way until someone else be gone by, or he have cast three stones across the street.
>
> (GS 1, 549)

Two centuries after Theophrastos, five illustrated books about plants and roots were prepared by Cratevas, physician to Mithridates Eupator (c. 132-63 B.C.), a strong military opponent of Rome, and King of Parthia, the region southeast of the Caspian Sea. The illustrious king also participated in the studies, more with an interest in practical applications—he specialized in the poisonous varieties of plants. The Romans, always the pragmatists, were also interested in herbal toxicants, but extended their knowledge of plants with the more constructive aims of medicine (to which we return later), and of improving agriculture and husbandry. Cato the Elder (234–149 B.C.) had written a short volume packed with information on how to make money by farming, from the farm's original purchase, through ploughing, manuring, planting and transplanting, caring for olives, vines and vegetables, to marketing. This was followed in the last century before Christ by the poem "Georgics" by Virgil—same subject, the cultivation of the earth, but in a style so very different from that of Cato. In eloquent verse Virgil described the sowing of seed, its cultivation, weeding, and harvesting; treatments for animal diseases; the raising of cattle, horses, and sheep; and, of course, the keeping of bees. Above all, Virgil emphasized, the hard work of farming was good for the soul: "Labor omnia vincit"—"work conquers all," the essence of the work ethic of some churches, especially Protestant churches in recent times.

With it all, the Romans' interests in other animals were limited to the chariot races at the Circus Maximus and the spectacles at the Colosseum. The ancient Circus Maximus was the center of a fire, possibly set at the command of Nero, that ravaged Rome on July 18, 64 A.D. As in the Great Fire of London, sixteen centuries later, infectious diseases were abated, and reconstruction led to a healthier and cleaner city. The Colosseum was completed in the year 80 A.D. and became the scene for spectacles beyond the wildest dreams of the kings of Nineveh. Private zoological gardens housed animals from all over the known world—lions, tigers, apes, panthers, elephants, crocodiles, birds, and many more. Some were trained for tricks that we now see in circuses; mostly they were kept to fight each other in front of the crowds, or worse, to torture and kill humans, including Christians, for entertainment. Sometimes a classic touch was added; a slave, dressed as Orpheus and carrying a lyre, was seated with a background of trees and flowing water, hungry animals suddenly appearing to tear him apart. Men became lower than animals.

In chapter 7 there is a short description of the wondrous medical properties of the hyena, as seen by the Roman naturalist Pliny the Elder. Despite his duties as head of the western Roman fleet, and as a soldier and lawyer, he found time to write an enormous amount about the knowledge of the time, meaning of course his perception of it. As his nephew,

Pliny the Younger, reported, "he counted all time lost that was not given to study." Unfortunately, Pliny the Elder did not exercise his critical faculties very seriously; his writings would have been more helpful with less quantity and more quality. A snake can be killed by spitting in its mouth, he wrote; the mares of Lusitania are impregnated by the west wind; if a menstruating woman looks at a swarm of bees, they will die immediately; and "cranes place sentinels on guard, each of which holds a little stone in its claw; if the bird should happen to fall asleep, the claw becomes relaxed and the stone falls to the ground, and so convicts it of neglect." (UA 6, 285) These are some of the wonders of nature that Pliny the Elder tells us about, along with the *chenalopex* (a goose), the *tetrao,* and other birds and animals, half myth, half real. He continues with more incidental information, sometimes accurate and always picturesque, on the habits of birds in the wild and on their tastes when served for dinner. He assures us that "young storks will support their parents when old," that "swans will eat the flesh of one another," that the *tetraos* are almost as big as ostriches and, when captured, "by retaining their breath will die of mere vexation," and that "the birds which in Spain they call the *tarda* and in Greece the *otis* [the great bustard] are looked upon as very inferior food; the marrow, when disengaged from the bones, immediately emits a most noisome smell." It is easy to see how his reports were believed, making it difficult for those who followed to separate fact from fiction. For example, Claudius Aelinanus, coming a little later—how much later is not certain—wrote, "Only one thing will cure a sick lion; eating a monkey is a cure for his disease"; "The poets . . . say that the swan is the minister of Apollo. It is believed by the elders that, having sung its swan-song (as it is called), it then dies." And, back to the hyena again, "The hyena, Aristotle says, has a soporific power in its right paw and creates a stupor by its touch alone." (UA 6, 403).

But Pliny was not always wrong. He was certainly correct when he wrote that British marl (a mixture of clay and lime) was a very effective fertilizer, that the wines of Bordeaux and Burgundy were superb, and that the planets moved without being pushed by gods. Despite his errors, Pliny deserves an *in memoriam* too. Located in the Bay of Naples with his fleet, his attention was drawn to a large cloud shooting up into the sky above Mount Vesuvius. Curious, he ordered a small boat to take him across the bay to get a closer look, but word came that people were in danger. He then ordered the galleys to be put to sea, to rescue as many as they could, but he was killed, probably asphyxiated. His works were read by medieval scholars, referred to in the Elizabethan *Holinshed's Chronicles, Of Savage Beasts and Vermin* and, on several occasions, the better ones were cited by Charles Darwin himself.

Volcanoes and earthquakes, originally attributed to evil gods, had caused some Greek philosophers to speculate on their nature and origin. The followers of Pythagoras supposed that there was a huge fire at the center of the earth, later identified with the underworld and the place of eternal punishment for the wicked. The science of geology hardly got started at that time, but we note in the chapter on alchemy that for thousands of years men had mined various metals and learned to work them, had collected precious stones and learned to decorate statues and people with them. Indeed, metals and gems had been there "from the beginning."

"The gold of that land is good; there is also beryllium and the onyx stone" (Genesis 2:12). We have noted that Theophrastos, the father of botany, was also the father of mineralogy. He directed and expanded the Lyceum after Aristotle's death, and, in addition to his many other interests, found time to write the first book on the scientific properties of gems and other minerals. He studied and described their colors, weights, hardness, transparency, and he listed some of their prices and where they could be found in the parts of Africa, Asia, and Europe that he knew about. Of the elements earth, air, fire, water, he picked the first as the origin of stones, and the last as the origin of metals. His theorizing was limited by the knowledge of the time, but his observational techniques were intact. From the writings of Theophrastos, it is possible to identify many of the minerals he had examined: types of quartz (rock crystal, amethyst, jasper, prase, and chalcedonies—from Chalcedon in Asia Minor—like agate, carnelian, and onyx); the iron oxides magnetite (Fe_3O_4), or loadstone, and the very different hematite (Fe_2O_3); red ocher for painting, along with malachite, lapis lazuli, alabaster, emerald, and more. The list may be compared with that given in Exodus (28:17-20) "a sardius, a topaz [possibly a peridot], and an emerald, . . . a carbuncle, a sapphire and a jasper . . . a jacinth [zircon], carnelian and an amethyst . . . a beryl, an onyx and a jasper" to adorn holy vestments for Aaron.

Later in this book (chapter 16) we discuss the nineteenth-century revolution in our understanding of paleontology and geology, but here we should mention the first attempts to understand the phenomena that these disciplines examine. Living early in the development of ancient Greek culture, Xenophanes of Colophon, a contemporary of Pythagoras, was effectively crowned the first paleontologist and geologist by Hippolytos, a later writer:

> Xenophanes is of the opinion that there had been a mixture of the earth with the sea, and that in the process of time it was disengaged from the moisture, alleging that he could produce such proofs as the following: that in the midst of earth and in mountains shells are discovered; and also in

> Syracuse he affirms was found in the quarries the print of a fish and of seals, and in Paros the print of an anchovy in the bottom of a stone, and in Malta parts of all sorts of marine animals. And he says that these were generated when all things were imbedded in mud, and that an impression of them was dried in the mud. . . .
>
> (GS 1, 180)

Our mineralogist/botanist friend Theophrastos thought two centuries later, "that these fossils were developed from fish spawn left in the earth, or that fishes had wandered from neighboring waters and had finally been turned into stone." But the real breakthrough in understanding geology came nearly three hundred years after that from the geographer Strabo, whose works, as we saw in chapter 5, were mostly ignored. He attributed the rise of mountains to an internal force and believed that the valley of Tempē, in Thessalia, was formed by an earthquake. In fact, he recognized the important role of earthquakes and volcanoes in shaping the surface of the earth and producing some islands in the Mediterranean. The fossil shells found in Egypt proved to him that the land had formerly been under the sea. Strabo wasn't just a volcanologist, however, because he recognized the importance of water in shaping the land. He died just a few years before the crucifixion of Jesus. Strabo would have been surprised and disappointed to learn that, eighteen hundred years later, world experts on the subject were still arguing about the forces of geology—the Vulcanists versus the Neptunists. Was the earth's surface fashioned by volcanoes and earthquakes, or was it due to water, in particular the deluge that Noah had survived, along with his family and seven pairs of clean animals and two pairs of the unclean?

7

Miracles, Surgery, and Medicine

> If there be an opportunity of serving one who is a stranger in financial straits, give full assistance to all such. For where there is love of man, there is also love of the art. For some patients, though conscious that their condition is perilous, recover their health simply through their contentment with the goodness of the physician.
>
> *Precepts,* late school of Hippocrates

The power of faith! Power of the mind! Call it what you will—even a miracle—throughout history it has been established beyond doubt as a strong force that can be used to fight mental and physical disease. The primitive belief in gods and demons appeared to be justified when the demons moved inside people and had to be exorcised by magic or religious rites. If that wasn't enough, you could always chase a hostile alien spirit away by feeding abominations to its host—e.g., lizard's blood, excrement,

> "Eye of newt, and toe of frog
> Wool of bat, and tongue of dog"

and much more. Techniques changed over the ages, but as Oliver Wendell Holmes wrote:

> There is nothing men will not do, nothing they have not done, to recover their health and save their lives. They have submitted to be half-drowned in water and half-choked with gases, to be buried up to their chins in earth, to be seared with hot irons like galley-slaves, to be crimped with knives like codfish, to have needles thrust into their flesh and bonfires kindled on their skin, to swallow all sorts of abominations, and to pay for all this as

> if to be singed and scalded were a costly privilege, as if blisters were a blessing and leeches a luxury.
>
> (Du 1, 81)

The reason practices like these became popular was because some people had faith that they would work, and because of that faith, they often *did* work. The same idea applies to many cases of faith healing at holy places—if you don't believe in it, you will get nothing out of it.

We often use the word "miracle" to describe those cases in which a person is healed primarily because he *believes* that he will be healed. A miracle is "an event that deviates from the known laws of nature or transcends our knowledge of these laws," according to the successors of Noah Webster. "Miracle" is therefore a relative term. When nobody knew anything at all about the laws of nature, everything was a miracle that was worked by a god. As more was learned about the planets and the weather, it became no longer necessary to regard them as deviating from these laws. Quite the contrary. We use some of the laws of nature that we understand in order to describe and predict events, with enormous success for the motions of the planets, less so for the weather because of its complexity. The most complicated items that we know about, however, are our own minds. What the human mind is capable of doing often does "transcend our knowledge." In that sense, some genuine healings are miracles *at present*. It's no wonder so many people feel that they have a God inside themselves. When this belief is used for positive purposes, it is superbly powerful. When it is used to claim an inner track to some Supreme Being, it can be evil. The more that science explains phenomena in terms of natural causes, the less room there is for old miracles, but the more aware we become of things that we don't yet understand, making room for new miracles. We never will understand them, though, if we regard them as *unfathomable* miracles.

Yet "faith without works is dead." (James 2:26) Three thousand years before James (possibly Jesus's brother) wrote these words, there lived in Egypt a remarkable astronomer, physician, and architect, by the name Imhotep. It was probably Imhotep who designed the first pyramid, and certainly it was he who was worshipped later as a god of medicine. He and his successors recognized the importance of a careful medical examination, asking the patient to move parts of his body on command, taking careful notes, learning from seeing, smelling, and touching, observing the pulse, and making a diagnosis based on facts. This was really the beginning of science, and it continued for centuries.

Somewhere around 2800 B.C. an early dentist drained an abscess under a patient's first molar by piercing an outgrowth—we know this because the patient's jaw has been preserved. Physicians began to specialize. There was a "palace eye physician," a "palace stomach and bowel physician," an "internal fluid physician," and many more. After a while, the physicians had patients resting in the temples and being healed by their faith in Isis. A papyrus from the sixteenth century B.C. is a medical treatise about diseases and their symptoms, with hundreds of what we now call "prescriptions." Earlier than that, another papyrus describes forty-eight cases of clinical surgery, with very careful attention to detail. Earlier still, another gives a description of a suppository, apparently for contraception. But to many people, diseases were due to demons, so magic charms and incantations were preferred to treatment by the physicians. A lot of their prescriptions were more magic than medicine anyhow, designed to appease the gods. When these didn't work and the patient died, magic spells helped him on his voyage to the other world.

Egypt was prosperous. Her war chariots invaded as far as the Euphrates river in Syria. (Later, according to Exodus, some of them were caught at high tide in the northern tip of the Red Sea.) There followed a war with the Hittites, who were centered in Turkey, and an invasion of the whole area with "sea-people" whose origin is obscure. Egypt eventually lost her empire, but her use of magic survived.

Her medical science did not—or did some of its practitioners move to Greece? If not the actual physicians, certainly some of their ideas did. Perhaps the great physician Asclepios of ancient Greece heard about them or perhaps he independently dreamed the same dreams. In any case, he was so successful that, like Imhotep two thousand or so years earlier, he became a god in the minds of those who came later. Asclepios's followers also came to the temples to bathe in holy water and to sleep in comfort, and they told their dreams to one another and to the priests. Some Egyptian drugs were used, as Homer reported in book 4 of the *Odyssey*. "Into the bowl in which their wine was mixed, she slipped a drug that had the power of robbing grief and anger of their sting and banishing all painful memories. This powerful anodyne was one of many useful drugs which had been given to the daughter of Zeus by an Egyptian lady, Polydamna, the wife of Thor." Hundreds of roots and plants had been collected over the millennia, their physiological effects noted and their magical effects, first postulated, then accepted.

Philosophers more famous for their work in other areas—Pythagoras, Philolaos, Anaxagoras, Empedocles, Democritos, Aristotle—added their theories of medicine. Some members of the school of Pythagoras recognized the importance of the brain for the senses; but, as noted in chapter 6, the great thinker Aristotle, coming later, did not believe that his thinking or anyone else's was centered in the brain. The atomist Democritos was interested in problems that touch both psychology and medicine—the therapy of music, why some people were cured by visiting the temples, and the effects of drugs, including that of hellebore on the heart. But the main center of medicine was on the island of Cos, dominated during the fifth century B.C. by Hippocrates and his family (grandfather, father, two sons, and a son-in-law).

We cannot be completely sure which member of the family made which contribution, but certainly Hippocrates II was the star (his grandfather had the same name). In the spirit of the early Egyptian followers of Imhotep, medical science was restored and developed. Superstition was rejected, while the importance of resting in bed when necessary and exercising when possible was recognized. A very sympathetic bedside manner was required. Puzzled by the seizures of epilepsy (others had named it the "sacred disease"), Hippocrates made it clear what he thought of them:

> I am about to discuss the disease called "sacred." It is not, in my opinion, any more divine or more sacred than other diseases, but it has a natural cause, and its supposed divine origin is due to men's inexperience and to their wonder at its peculiar character. . . . Those who first attributed a sacred character to this malady were like the magicians, purifiers, charlatans and quacks of our own day, men who claim great piety and superior knowledge. Being at a loss, and having no treatment that would help, they concealed and sheltered themselves behind superstition and called this illness sacred in order that their utter ignorance might not be manifest.
>
> (GS 1, 355)

Other Hippocratic writings include many case histories, meticulously recorded. After performing an abortion, one physician wrote in his notes, "Did the woman speak the truth, I wonder?" Surgery on men, at least, began much earlier, since for centuries warring men had performed amateur primitive "surgery" on each other, necessitating repairs if it wasn't too late. We learn from Hippocrates and his family how to bandage wounds, set fractures, massage joints, exercise the hands for greater dexterity as a surgeon, and set up an operating table. We learn also about healthy diets, hygiene, food for the sick, the medical effects of the climate, infant teething, and, perhaps above all, that the physician should make rational judgments based on the evidence at hand. Then of course there was the Hippocratic Oath taken by beginners in the field of medicine, embodying a code of ethics and a commitment to teach the master's children without a fee, and to teach his own and no others except a few who had also taken the oath. Now students in medical school learn from so many experts that the obligation to teach their masters' children is impractical—or their own for that matter. Like Aristotle, many physicians' children excel in other areas, but with Hippocrates medicine was very much a family affair.

In the latter part of the following century the tradition of surgery at Cos was carried on by Praxagoras. He examined the connection between the brain and the spinal cord, noted the distinction between veins and arteries, studied the arterial pulse, but somehow believed that the arteries carried air. Like Hippocrates, Praxagoras became an expert on the "humors" of the body, later made famous by Galen.

Ptolemaios Soter† (the preserver) (367–283), a general in the victorious army of Alexander the Great, was given Egypt and Libya as a reward. He spent much of his time defending his territory from the other generals, but he is remembered mainly because he established a library and museum at Alexandria and invited scholars to come and study there. Two of these were Herophilos (a student of Praxagoras), who founded the science of systematic anatomy, and Erasistratos, who did the same thing for physiology. Two years before he died, the king began a dynasty by handing over the reins—and the reign—to his son Ptolemaios Philadelphos†, who thereby became Ptolemy II, a very wealthy patron of the arts and sciences. In addition to the distinguished physicians, Alexandria was the home of the best philosophers, mathematicians (including Euclid), artists, architects, and poets of the time. Philadelphos got so involved with Alexandria that he left much of the running of the empire to his second wife, Arsinoe II, who also happened to be his sister.

The dissections at Alexandria provided for the first time a detailed view of human anatomy—the eye, the brain, the difference in function between tendons and nerves, between arteries and veins, the vital organs, and the glands. Physicians practiced vivisection on animals, and were accused of practising it on condemned criminals. They may have done so. Aulus Cornelius Celsus, a contemporary of Jesus, included this passage in his *Encyclopedia of Medicine*: "Nor is it, as most people say, cruel that in the execution of criminals, and but a few of them, we shall seek remedies for innocent people of all future ages." The Christian Fathers rightly looked back on this as pagan behavior, just what one could expect from a bunch of heathens, forgetting about John (8:7) and Luke (6:41–2) which refer respectively to casting the first stone and seeing the beam in one's own eye. We are fortunate to live in an age, despite all its cruelties, that no longer accepts either human vivisection or burning at the stake as legitimate punishments.

The medical school at Alexandria flourished for centuries—it had a first-class reputation as late as the fourth century A.D., but by then it had begun to deteriorate. About a hundred years after it had begun to operate, the Roman Cato the Elder—the expert on farming of the last chapter—wrote (in Latin, of course) that Greek physicians "seduce our wives, grow rich by feeding us poisons, learn by our suffering, and experiment by putting us to death." He was opposed to anything Greek and, for that matter, anything Carthaginian as well, his oratory being largely responsible for the Third Punic War between Rome and Carthage. Despite Cato's unfortunate chauvinism, much sensible medicine, practised and taught largely by Greek physicians, began to flourish in Rome. Asclepiades of Bithynia (c. 130–40 B.C.) studied philosophy and medicine in Greece, practiced at Parion (presently Parium, Asia Minor), and came to Rome in 91 B.C. He tried to apply the atomic theory to medicine, attributing disease to interference with the motions of the bodily atoms. On the more positive side, he was a pioneer in demanding humane treatments for the mentally ill and the physically old; he may have invented the first shower bath, and he was the first person to distinguish between acute and chronic disorders. Later in the last century B.C., Antonius Musa was a slave of Emperor Augustus, whose gout he cured with cold baths. He was also the physician of the poets Virgil and Horace, and perhaps the first to prescribe hydrotherapy.

In the next century (the first A.D.), the Romans as well as the Greeks and Egyptians in the Empire had access to surgical methods such as the removal of tonsils or bladder stones; to dentistry, including gold teeth and wiring and some bridgework; and to an enormous variety of poisonous and nonpoisonous medicaments. In the first century A.D., a Greek physician, Pedanius Dioscorides was a military surgeon in the army of Nero. He gave a systematic list of hundreds of plants, studied them as sources of drugs, and came up with almost a thousand drugs for medical purposes. He appears to have been the first physician to use the mandrake for anesthesia. His *De Materia Medica* was a standard textbook in the Muslim and Christian worlds for an incredible sixteen hundred years.

The teaching of medicine was finally made official under Nero's successor, Vespasian (9–79 A.D.). He knew the value of a field hospital and having a cadre of surgeons attached to the army. Doctors, including civilian doctors, were appointed and paid by the state so they could give free treatment to the poor. This was nothing new—two thousand years earlier the laws given to Hammurabi by the sun-god Shamash included a sliding scale for surgeons' fees, based on the patient's ability to pay. Similar practices became common later than Roman times in Muslim and Christian hospitals. Socialized medicine, that bugaboo of the United States, is very much older than socialism and almost as old as medicine itself.

A contemporary of Jesus, Aulus Cornelius Celsus (not to be confused with the Platonic philosopher Celsus of the second century A.D.), introduced the Hippocratic methods to Rome. Celsus wrote about law, military science, philosophy, and medicine; only his eight books on medicine have survived. These were based partly on the work of Asclepiades, particularly in his treatment of mental illnesses. For those who wouldn't eat, he recommended chains and flogging until they agreed to accept food, but for others he recommended sports, reading aloud, music, rocking in a hammock, and listening to the sound of a waterfall.

Celsus studied diseases of the digestive and urinary systems, dermatology, hygiene, tonsillotomies, and heart problems, but his main contribution was perhaps his attempt to see how neuropsychiatric disorders were related to other ailments, and to treat their victims as sick patients. Themison of Laodicea (31 B.C.–14 A.D.) had also been influenced by Asclepiades, and he recommended baths, fomentations, and a liberal diet for the mentally ill. Themison also studied chronic illnesses in a systematic way; he introduced new drugs and applied leeches to his patients. His writings were completed by Thessalos of Tralles who tried to simplify medical theory, but there wasn't enough experimental information available to allow good theories to develop.

Another Greek physician, Aretaeos of Cappadocia (now in Turkey) (81?–138 A.D.) was also a follower of Hippocrates who came to live in Rome. Aretaeos wrote eight volumes on the causes, symptoms, and treatments of acute and chronic illnesses, and he described diseases such as diphtheria, pneumonia, asthma, and tetanus. He also observed and reported on some cases of insanity. He was followed by another Greek, Soranos of Ephesus (2nd century) who wrote a score of books on medical history, gynecology, hygiene for midwives, and infant care. Soranos could be regarded as the founder of obstetrics and gynecology, at least in the West, where he was widely read.

As it was with astronomy and geography, the last of the important Greek writings on medicine, with few exceptions, were published during the second century A.D. They were then ignored for a very long time and were ultimately reinstated as being almost as true as the Bible. With the physical sciences, it was Ptolemy (the scientist, not the kings of Egypt of the last three centuries B.C.) and with medicine, it was Claudius Galenus from Pergamum (now Bergama, west Turkey), who was briefly called Galen (130–200?). Galen studied medicine in all of the important centers, including the school for gladiators at Pergamum, and practised medicine in Rome. His great success—he even prescribed by mail—led to jealousies, particularly when he exposed the ignorance of Roman physicians, and so he had to leave Rome. Nobody knows where Galen went, but some 20,000 pages of his writings have survived. As might be expected, they provide an enormous mixture of sensible, and (to us) nonsensical medical practices. Galen believed very strongly in experimental research, "to trust no statements until, as far as possible, I have tested them for myself." He observed a heart continuing to beat outside the body and recognized that the arteries contain blood, not air. He undoubtedly had ample opportunity to observe this as a physician to the gladiators. He described many serious diseases, including cancer and tuberculosis, recognizing that the latter is infectious; Galen studied cranial nerves and the loss of function that comes when they are damaged, and he recognized the crossover from damage on one side of the brain to loss of function on the other side of the body—and much more.

It is fortunate that Galen was such a great experimental physician, because his theories were wild, greatly influenced by the beliefs of the time. He believed in the four humors of the body—blood, phlegm, black bile, and yellow bile—as advocated by Hippocrates six hundred years earlier. By combining them with the four elements of Empedocles, he tried to make a theory of mental and physical diseases. Excess black bile made you melancholy; excess yellow bile made you choleric, or hot-tempered. If blood was the humor that dominated, you were sanguine, optimistic; if phlegm was in excess, you were obviously phlegmatic—cold, sluggish, apathetic. If everything was in balance, not too much of one or the other, then you were in a "good humor," a notion we still use today. These four basic "temperaments" were then correlated with the four elements and with the four properties of moist, dry, hot, and cold. When Galen's writings, along with those of other ancient scholars, were translated from Arabic into Latin a thousand years later, he became another Authority, and thus another wrong idea was projected into the future. One circumstance had curtailed his experimental studies—in his time the Roman Empire would not allow the dissection of corpses, so Galen had to be content with dissecting dogs, cows, apes, pigs and any other animals that he could get his hands on. However, his reports on animal anatomy appeared to relate to human anatomy, because he was not careful to distinguish between them. In the sixteenth century, when Vesalius got into trouble for dissecting the human body, it was noted that men's legs did not have the configuration that Galen had reported from dissecting pigs. The conclusion was reached that over the years men's legs had become thinner because they had been confined to tight breeches!

For the few hundred years before and after the birth of Jesus, the Roman Empire, which stretched from England to Arabia, was riddled with belief in magic, witchcraft, demons, astrology and other flights of fancy, and these spilled over into medicine as well as into alchemy and the theory of human behavior. There is much reference to the casting out of demons in the Gospel of the physician St. Luke, particularly in chapter 11. It was natural to assume that a cure by faith is accompanied by the ejection of a devil. Flights of fancy were the specialty of the compulsive Roman writer, Pliny the Elder, who left for posterity an immense "Natural History," some excerpts from which appeared in the previous chapter. Pliny was particularly fond of the magical medical effects of the hyena. The best time to capture one is when the moon is in Gemini. Every part of the hyena's body, it would seem, "worked like magic" as a treatment for some malady or other. Its teeth would cure toothache or stomach ache, its liver would cure fever, and "the ashes of the vertebrae, applied in the skin of the hyena with the tongue and right foot of a sea-calf and a bull's gall, the whole boiled up together, is soothing for gout." (UA 6, 289) Pliny continues, on and on, sometimes to his credit suggesting that he doesn't agree that all of these wonderful benefits could accrue from various parts of a hyena's body. Whether he did or not, many people went around practicing a "medicine" that was based on these and similar beliefs. The revolting prescriptions, designed to disgust the resident demons and cause them to leave, were not forgotten. And on the positive side, the accumulation of thousands of years of experience with naturally occurring drugs completed the pharmaceutical offerings. These botanical cures would ultimately be expanded through alchemy to include inorganic drugs, but it would be another millennium and a half before Paracelsus pushed this idea.

Meanwhile, Chaldean astrology, supported philosophically by the teachings of Plato and the Stoics, invaded medicine, as indeed it had invaded and distorted so many of men's struggles to understand nature. The mystical significance of numbers, derived from Pythagoras and, again, Plato, also led people astray, the number seven being especially important. There is a treatise of uncertain origin and date that emphasizes the role of this number in medicine—some diseases mature in seven days, the embryo supposedly assumes human form on the seventh day, and this all had to do with the seven "planets." When this numerology was applied to yield a specious new understanding of the Pentateuch, as it was in the last centuries B.C., the "science" of gematria was born. In the next chapter we examine these intellectual road blocks.

There was some hope for a while that this progress in medicine would be used in Christendom, and that perhaps it might provide a basis for further study. The enlightened Christian philosopher, Nemesius, Bishop of Emesa in Syria during the latter part of the fourth century, wrote a treatise on *The Nature of Man*. Nemesius accepted the truths that these physicians had uncovered, even though they were pagans, and recommended that they be studied. His contemporary, Oribasios of Pergamum (325?–395?) made a long summary of Galen's writings, at the request of the Emperor Julian. He accompanied Julian on his expeditions, but was banished for a time when Julian was killed in battle.

In the following century, Caelius Aurelianus, a physician in Sicca, Numidia (present El Kef, near Tunis, Algeria), described many diseases such as encephalitis, gout, and phthisis—a wasting disease such as tuberculosis. He also distinguished between epileptic seizures and hysterical attacks, and studied speech defects, trying to cure them. Above all, perhaps, he recommended humane treatment for the insane.

This need for mild treatment for these sad cases was echoed in the seventh century by the physician and surgeon Paul of Aegina (on the east coast of Greece) who worked in Alexandria. Basing his medicine on that of Galen and Oribasios, he wrote at a time when the Muslim explosion was about to occur, when Muhammad was inspiring his followers to spread Islam throughout the world. Paul of Aegina was a great surgeon in his own right, and he described operations such as tonsillectomy, herniotomy, breast amputation, and ophthalmic surgery. A lot of the tradition of Greek medicine was transferred to the Muslims by his writings, helping to launch the Muslim renaissance in knowledge, to which we return in chapter 10. He also encouraged the Muslims to be merciful in their treatment of insane people, a procedure that became a Muslim tradition, in contrast for more than a millennium to the tortures prescribed by Christians to drive devils out of the afflicted.

8

Astrology, Alchemy, and Numerology

Much learning does not teach understanding.

—Heracleitos of Ephesos

Ever striving to learn, and never able to come to the knowledge of the truth.

—2 Timothy (3:7)

Astronomy, chemistry, and mathematics have their distorted counterparts in astrology, alchemy, and numerology. One might have expected that by the end of the twentieth century the word "have" in that sentence should have been replaced by "had." This is essentially true for alchemy, partially true for numerology, but not at all true for astrology. This pseudo-science is concerned with the influence of the stars on human affairs and purports to foretell events on earth by planetary configurations.

The contrast between the extreme regularity of the heavens and the extreme irregularity of the terrestrial environment is abundantly evident. Those gods up there must be telling us something. Night after night, unpolluted by smog or street lights, the visibility of the sky used to be magnificent everywhere. The challenge was to find out what the heavens were saying. Over the centuries, Egyptian and Sumerian priests had studied these puzzling motions and had used them for setting up calendars and holy days. They also began to observe the stars very carefully, convinced that they offered keys to the human predicament. This challenge to understand the stars was met in some detail by Chaldean wise men of the second Babylonian Empire (626–539 B.C.). The king of the Chaldees, Nebuchadnezzar II, was famous for his rebuilding of Babylon, and infamous for his destruction of Jerusalem in 587 B.C. Nebuchadnezzar had a puzzling dream which his wise men could not interpret, so he called on the prophet Daniel. We read in Daniel (4: 6,7) that the king announced to him: "I made a decree to bring in all the wise men of Babylon before me, that they might make known to me the interpretation of the dream. Then came before me the magicians, the astrologers, the soothsayers and the Chaldeans."

Evidently astrology and magic were already flourishing 2600 years ago, and these wise men made astrology as accurate as possible. One way of improving the accuracy was to have an assistant wait by an expectant mother and at the moment of birth clash cymbals together so that an astrologer up on a roof could record the positions of the stars and the planets. Later, St. Augustine tells of people who carefully noted the positions of the constellations at the births of their domestic animals. It is not recorded in either case how this was handled for babies and animals born during daylight hours. Two astrologers in the third century B.C., Antipater and Achinapolos, believed that the horoscope should be based on the moment of conception rather than the time of birth; but, even if prospective parents did not object to the necessary invasion of privacy, it would have been difficult to determine the moment unambiguously. Apparently there was some debate about it though, because there is a horoscope in the British Museum based on the date of a birth, December 15, 258 B.C., and the date of conception, March 17 of the same year, derived by counting backwards from the birth. One is reminded of modern legal, ethical, and religious debates about the ambiguous question "When does life begin?"

Astrology became a religion. Those planets and stars were driven by real gods, and contemplation of them was a religious experience. In the early Greek civilization, Pythagoras, and later Plato, had embraced mystical feelings about the stars. With Pythagoras, study of the stars, accompanied by mathematics, was the basis of a religion, as we have seen. With Plato planets and stars were divine embodiments of his Ideas. Astronomy and astrology became a theology of nature, at least of the "divine" aspects of nature, up there in the aether. Published after Plato's death by his former secretary Philip, the *Epinomis* recommended a state religion based on the stars, a religion of order and intelligence, reflecting the order of the stellar motions and the intelligence of the stellar gods. It emphasized the importance of astronomy for attaining true wisdom. We have seen that Plato's method of pursuing astronomy was to think about ideas and not to make measurements. He wasn't even a good astrologer; they at least saw the need for accurate observations. Modern astrologers don't work at observatories.

Partly because of Berossus, a Babylonian priest of the early third century B.C., the Chaldean astrology, now considerably expanded, made its way to Greece and, later, to Rome. Berossus or his followers set up a school of astrology on the island of Cos—certainly the physicians there incorporated astrology in their treatments. By the end of the second century B.C., astrology had conquered the Roman world. Astrology, magic, and sorcery from Babylon combined with the worship of Mithras, Cybele, and other gods to provide a universal atmosphere of superstition and divination. It was difficult for rational thinking to flourish in that environment, and even the great astronomer Hipparchos succumbed to the pressure of the times and gave his scientific blessing to astrology. Epicuros and his followers had been outspoken against it, but they and their atomic ideas had become discredited, being replaced by the philosophy of the Stoics, who supported astrology. Poseidonios, the Stoic astronomer (and a good one), who had the wrong idea about the size of the earth, visited Rome in 87 B.C. to preach the doctrine of astrology.

Despite strong opposition later from Cicero and Lucretius, the conviction that the heavenly bodies controlled the lives of mortals became widely imbedded in people's minds. Eventually Ptolemy followed Hipparchos in this as in other ideas and built extensively on what others had done in optics, trigonometry, and geography, as well as astronomy. His model of the solar system had great predictive power, but he regarded it as a mathematical model, an imperial clockwork, circles moving on circles according to the mysterious rules of heaven. All told, it was learned from Ptolemy's writings that the earth is at rest, that India is only a few thousand miles west of Gibraltar, and that astrology is a valid discipline. Three strike-outs!

Ptolemy deserved better than that because he was a careful scientist. He recognized two entirely different aspects of astrology, and it is important to distinguish between them. On the one hand, there was astrology as a religion—the study of the wonders of the heavens and the deep religious feelings that accompany it. For many astronomers today that type of "astrology" becomes part of their motivation. On the other hand, there are the "priests" of astrology—those who cast horoscopes, make well-publicized but less well-verified predictions, and in these modern times write vacuous columns in the newspapers.

Astrology began to take over all areas of knowledge. Used at first to predict the fate of a country, it narrowed, first to predicting the fate of the king, and then to the fate of other individuals. The days of the week were named after the planets, sun, and moon, especially noticeable in the languages of the Latin countries, and lucky and unlucky days were defined. The first horoscope of record was cast in 410 B.C. in Babylon, but it was a few hundred years before individual horoscopes became popular. Let us recognize that astrology gave, and still gives to people a sense of order, a sense of being different, a comforting feeling of being able to understand and predict something about that most complicated of all phenomena: human behavior.

While Roman citizens generally believed in magic, omens, astrology, and other superstitions, there were times when belief in astrology could get you into trouble. The Roman historian Tacitus (55–120?) reports that a young man, by the name of Libo Drusus, was charged with "attempts against the state" and, in order to get evidence against him, his deadly enemy Firmius Cato pretended to befriend him and induced him "to try the predictions of the Chaldeans, the mysteries of magicians, and even the interpreters of dreams." He then arranged for witnesses to hear Libo state his new beliefs, and when the case came before Emperor Tiberius (42 B.C.–37 A.D.), Libo escaped judgment only by suicide. He was popular, and his surname indicated his relationship to the Caesars. Tiberius therefore got rid of him, as he had also dispatched his nephew and his own son just a few years before Jesus was crucified.

Astrological beliefs didn't come only from the Chaldean kingdom of Babylon. The astronomy of the Egyptian civilization also produced its astrology, some of which entered the early Christian world in the form of spurious "Hermetic" books, which take us all the way back to the Egyptian god Thoth. When Osiris was weighing the hearts of supplicants against a feather, it was Thoth who kept the records, these weight measurements being made later by St. Michael the archangel. Thoth was the god who had invented writing, and it was he who became the god of wisdom and magic. The Greeks identified their god Hermes with the Roman Mercury, and later Hermes Trismegistus (thrice greatest) was identified with Thoth, and sacred writings on magical, astrological, alchemical, and cabalistic subjects were attributed to him. Presumably to boost sales, the authors of some later books on these matters also ascribed the authorship to Hermes. These *Hermetica* included extensive writings on neo-Platonism, offered as the wisdom of centuries but actually composed mainly during the first few centuries A.D. They were not recognized as spurious until the beginning of the sixteenth century, and by that time they had influenced many scholars because they were thought to be very old, and if they were old, it was believed they must be very wise.

Astrology extended its influence by the association of the constellations with parts of the body—all the way from Aries (the head) to Pisces (the feet). In between: Taurus (the neck), Gemini (the shoulders and arms), Cancer (the breast), Leo (the flanks), Virgo (the bladder), Libra (the buttocks), Scorpio (the genitals), Sagittarius (the thighs), Capricorn (the knees), and Aquarius (the limbs). This idea caught on. In the fourteenth century the great surgeon Guy de Chauliac wrote a manual for physicians that was used as a standard textbook for several centuries. In it de Chauliac noted that "If anyone is wounded in the neck when the moon is in Taurus, the affliction will be dangerous."

In the earlier chapters of this book there have been a few references, like this one, to medieval and more recent writings, but essentially all of the ideas, problems, and struggles we have considered so far were encountered by the year 200 A.D. In the remainder of this chapter we pursue superstitions and pseudo-sciences into the Christian era, and in later chapters examine early Muslim science, the long conflict of Christian theologies with each other and with the sciences, and the persistence of the pseudo-sciences.

The early Christian Fathers were opposed to astrology because it subtracted from the freedom of will, and the Church continued its opposition for centuries. Some laymen recognized the quackery that went with it, and we have this from Agathias, a lawyer born in Alexandria and living in Constantinople in the latter part of the sixth century A.D.:

> Calligenes the farmer, when he had cast his seed in the land, came to the house of Aristophanes the astrologer and asked him to tell whether he would have a prosperous summer and abundant plenty of corn. And he, taking the counters and ranging them closely on the board, and crooking his fingers, uttered his reply to Calligenes: "If the cornfield gets sufficient rain, and does not breed a crop of flowering weeds, and frost does not crack the furrows, nor hail flay the heads of the springing blades, and the pricket (a buck in his second year) does not devour the crop, and it sees no other injury of weather or soil, I prophecy you a capital summer, and you will cut the ears successfully, only fear the locusts."
>
> (UA 8, 21)

Perhaps a better-known legacy from Agathias is his epigram "A Kiss Within the Cup" that a millennium later inspired Ben Jonson, who probably also got from him the idea of lampooning astrology. Later in this chapter we look at a sample of Ben Jonson's writings on astrology and alchemy.

Despite protestations from the church, belief in magic and supernatural beings was common to the point where in the ninth century, Archbishop Agobard of Lyons said that "things of such absurdity are believed by Christians as no one ever aforetime could induce the heathen to believe." The evil spirits were put out of business by repeated incantations and magic charms, which were then transformed into holy incantations and Christian rites and symbols.

In the Muslim world, astronomy flourished for centuries, as we discuss in chapter 10, but there were also some astrologers, and considerably more alchemists. One astrologer who had a major impact on medieval scientific thought was Ibn al-Farrakhan, who flourished during the first part of the ninth century. His works were translated into Latin some seven centuries later, but some of them were available to Frederick II much earlier than that. As part of the wisdom of the Muslim world, they stimulated a wide interest in astrology among the Latin-speaking scholars. Later in the same century the astrologer Albumasar (c. 790–886), advanced the view that the world was created when the seven planets were in conjunction in the first degree of Aries and that the world would come to an end at a similar conjunction in the last degree of Pisces. Better known, in some part because of Robert Browning's poem that he inspired, was the Jewish scholar Ibn Ezra (c. 1092–1164), a great traveler, an early translator into Hebrew of Arabic texts, and author of books on astrology, numerology, and neo-Platonism. Ibn Ezra also advanced the rational scientific viewpoint, wrote on mathematics, grammar, the calendar, and the astronomical tables prepared earlier by al-Khwarizmi. Browning has him say (or is Browning speaking for himself?):

> Now who shall arbitrate?
> Ten men love what I hate,
> Shun what I follow, slight what I receive;
> Ten who in ears and eyes
> Match me. We all surmise,
> They this thing, and I that; whom
> shall my soul believe?
>
> (Rabbi Ben Ezra, v. 22)

With the translation of Arabic texts into Hebrew and Latin, astrology and many other methods of trying to understand the many-varied parts of nature spread throughout the West. In his *History of Florence,* Niccolò Machiavelli mentions Guido Bonatti, who "flourished about 1230 and 1260. Though a learned astronomer, he was seduced by astrology, through which he was greatly in favor with many princes." Guido da Montefeltro, lord of the city-state of Forli (on the coast southeast of Bologna), would not go into battle except at a time that was recommended by Bonatti, who had become his personal astrologer. Dante placed Bonatti in Hell along with the other astrologers, with their heads reversed so that they had to walk backwards. However, astrology was recognized by many as a very honorable profession and in 1322, a year after Dante had died, Cecco d'Ascoli* (1269–1327) was appointed professor of *astrology* and mathematics at the University of Bologna. D'Ascoli ran a horoscope of Jesus, denied freedom of the will, believed that there could be people living on the other side of the earth, and criticized Dante's *Divine Comedy.* For these heresies he was condemned and burned at the stake. During the fourteenth and fifteenth centuries, astrologers became very important people in the courts of Europe, provided that they did not get too far out of line.

In 1391 Geoffrey Chaucer wrote "For Lytel Lowys my Sone," a treatise on the astrolabe, an instrument later replaced by the sextant for observing the positions of the stars and planets. He described in detail how it works and what you can measure with it, concluding his introduction with the statement: "The fifthe partie shal be an introductorie, after the statues of oure doctours, in which thou maist lerne a gret part of the generall rewles of theorik in astrologie." Chaucer's "Canterbury Tales" tells us how the seven bodies are identified with the seven metals:

> The bodies sevene eek, lo! hem heer anoon;
> Sol gold is, and Luna silver we threpe,
> Mars yren, Mercurie quik silver we clepe,
> Saturnus leed and Jupiter is tin
> An Venus coper, by my fader kin!
>
> (The Canon's Yeoman's Tale, lines 106–110)

The "seven" bodies were also identified with the "seven" colors and the "seven" parts of the body—Mars was not only iron, but red and blood; Saturn not only lead, but grey and brain. When you imagine the planets, with these qualities attached to them, moving according to Ptolemy's rules through the zodiacal signs (which as we have seen were identified with parts of the body), you have a complete system of astral influences, or so it seemed. But there was also the viewpoint that we are responsible for our own destinies, as we learn from Shakespeare's *Julius Caesar:*

> Men at some time are masters of their fates;
> The fault, dear Brutus, is not in our stars,
> But in ourselves, that we are underlings.
>
> (*Julius Caesar,* I, iii, 134)

Pelagius, whose ideas we consider in the next chapter, would have liked that.

In France, Michel de Notredame, or Nostradamus (1503–1566) was a physician, but he is more famous to this day for his astrological predictions. Nostradamus drew horoscopes for the sons of Catherine de Medici, became physician to the second son when he became Charles IX, and achieved extra fame by accurately predicting that Catherine's husband, Henry II, would die in a tournament. The knowledge that it was prophesied by such an expert could well have unsteadied Henry's arm, and in 1559 the prophecy became a fact. That prophecy was remembered, but the one he made in 1564, that Charles IX would live to the age of ninety tended to be forgotten.

Charles IX died ten years later at the age of twenty-four.

During the next century, in England, William Lilly (1602–1681) peppered the realm with annual pamphlets of prophecy and he tied astrology and religion together in his book *Christian Astrology*. This was published five years after the death of Galileo and the birth of Newton, so that Kepler's ellipses and Galileo's revelation of the satellites of Jupiter were already known to a few scholars, but detailed understanding of the motions of the planets was yet to come. Nevertheless, Lilly's contemporary, Samuel Butler (1612–168),0 recognized that astrology was nonsense and he lampooned Lilly and his associates in his poem *Hudibras:*

> They'll search a planet's house, to know
> Who broke and robbed a house below,
> Examine Venus and the Moon,
> Who stole a thimble or a spoon; . . . .
> They'll feel the pulses of the stars
> To find out agues, coughs, catarrhs;
> And tell what crisis does divine
> The rot in sheep or mange in swine,
> What gains, or loses, hangs, or saves
> What makes men great, what fools, or knaves.
>
> (UA 14, 348)

At least, during the time Lilly lived, there was *some* excuse for these astrological fantasies. Now there is not. The stars are so far away that they cannot possibly have a measurable effect on our psyches, unless of course we *believe* that they do. This denial of our ability to reason has a great appeal to many people, and it will continue whether or not it is supported by a President of the United States. However, you cannot help worrying about what such a powerful man could believe next.

With the identification of the seven heavenly bodies with the seven known metals, a connection was established between astrology and alchemy. Before there were any theories about the nature of matter, people experimented with whatever chemicals they could find. The Altamira artists had used crayons made of colored iron and manganese ores, sometimes mixing them with grease to form paints. As noted in chapter 1, red primitive pigments were also used by Stone Age people to rub on the bones of their dead to simulate blood and therefore life. This custom persisted for thousands of years. In 1975, an archaelogical find in Chinancheng in Central China revealed the body of a man buried in 167 B.C., well preserved and immersed in a red embalming fluid. More well known to us is the use of bitumen, spices, and sodium carbonate for the preservation of Egyptian mummies.

The big discovery was that the brilliant red mercuric sulphide that we call vermilion, or more precisely cinnabar, decomposed upon heating into mercury and sulphur. What a transformation! The blood-red separated into a dirty yellow mass and a gleaming silvery drop. With the Chinese scholars, shining mercury became the male principle, the Yang, and the dowdy sulphur became the female principle, the Yin. Cinnabar was thought to prolong life, and up to the last century it was listed as an official remedy in both Britain and the United States. It is not known how many people throughout the world and throughout the centuries died of mercury poisoning by consuming this so-called elixir of life.

In Mesopotamia, chemical apparatus for distillation existed more than five thousand years ago, and later cuneiform tablets refer to the production of mercury and describe other chemical processes. Here is part of a cuneiform recipe for making glazes, from the seventeenth century B.C., as quoted by Dr. Sarton:

> To a mina of zuku-glass (thou shalt add) 10 shekels of lead, 15 shekels of copper, half (a shekel) of saltpeter, half (a shekel) of lime: thou shalt put (it) down into the kiln (and) shalt take out copper of lead. . . . Property of Liballit(?)-Marduk, son of Ussur-an-Marduk, priest of Marduk, a man of Babylon.
>
> (GS 1, 81)

It had been discovered much earlier that when copper and tin are melted together, the resulting metal, bronze, was stronger, more durable, and with its lower melting point, more malleable. Primitive metallurgy expanded to include the Egyptian electrum (an alloy of gold and silver) and, a few hundred years after the above recipe was marked in the clay, the use of iron began to transform the effectiveness of the ploughshare and the sword. As populations grew and became established in towns, agriculture developed and war took on a new dimension, one example being the siege of Troy. In the beginning of the *Odyssey,* the goddess Athena visits Odysseus's home in disguise and tells his son, Telemachus, that she is Montes, the captain of a ship loaded with iron that is to be traded for copper. She carried a bronze spear and wore magic golden sandals. Thus were metals referred to by the earliest Greek writers whose works have survived.

Artists also continued to experiment with different materials. Paintings with blue, green, and brown colorings are preserved from 2600 B.C. (but they still are only one third as old as the Altamira paintings.) Frescoes suspended in lime and applied to wet limestone plaster, and painted limestone statues (of Prince Rahotep and his wife Nofrit, for example) are from this time or earlier. Small pieces of marble, colored glasses, lapis lazuli, and other materials were used for decoration of jewelry, statues, and coffins.

There was, then, a long tradition of working with many different types of material, an activity very widely separated from the theories of the composition of matter devised by Greek scholars. A kind of empirical chemistry developed in Egypt and Mesopotamia. Metal workers, jewelers, and goldsmiths learned from necessity how to measure precious materials more accurately. They worked with gold, silver, copper, tin, lead, mercury, iron, sulphur, carbon, and compounds of zinc, antimony, and arsenic. Their work was practical, and without mystical qualities because they were respected technicians, not priests. However, alchemy was flourishing in Alexandria by the fourth century A.D.

This heritage was picked up by the Saracens who added a flavor of experimental science. They improved the precision, conducted experiments, and anticipated the modern laboratory notebook by keeping careful records. They studied a number of chemical compounds, including *al-kohl,* a very fine powder, lead sulphide, for painting eyebrows. (Outside of Arabia the word has assumed a different meaning, the distillation of alcohol being discovered in Italy in the middle of the twelfth century.)

A theory developed and persisted in one form or other until the seventeenth century that all metals were so very closely related that with one right extra ingredient—a "philosopher's stone"—they could be transformed into one another. In particular it was believed that all of the cheaper metals could be turned into gold. Some alchemists took the view that all metals were mixtures of mercury and sulphur, and that transforming one metal to another could be achieved by finding some aethereal substance, the essence, or soul, of mercury.

Alchemy was more than a preoccupation to manufacture gold and silver; indeed early alchemy of the third century A.D. includes recipes for making *imitation* gold. The most famous Arabian alchemist, Jabir ibn Hayyam (702–725), later called Gebir in Europe, achieved much of his fame from the many treatises on the subject that were attributed to him during the following centuries. Alchemy became popular among European philosophers as Arabic texts were translated into Latin. In the late thirteenth and early fourteenth centuries Cappochio, who had probably been a fellow student of Dante Alighieri, developed into a Sienese alchemist. Dante didn't think much of him or, for that matter, of anyone from Siena:

> Was ever race
> Light as Siena's? Sure not France herself
> Can show a tribe so frivolous and vain.
>
> (Inferno, Canto XXIX; HCl 20, 122)

In his *Divine Comedy,* Dante encounters the tortured Cappochio in Hell and hears him speak:

> So shalt thou see I am Capocchio's ghost
> Who forged transmuted metals by the power
> Of alchemy and if I scan the right
> Thou needs must well remember how I aped
> Creative nature by my subtle art.
>
> (Inferno, Canto XXIX; HCl 20, 122)

The Dominican scholar, Vincent of Beauvais (1190?–1264) wrote a comprehensive encyclopedia on the scientific knowledge of his time, derived from the recent translations of Arabic works into Latin. He covered medicine, the physical and life sciences, mathematics, and alchemy, testimony to the contemporary interest in this subject. He added his own ideas, of course, being naturally concerned about the interaction between alchemy and his Christian beliefs. He decided that Noah's ability to beget children when he was five hundred years old was a sure sign that he had some secrets of alchemy, presumably lost since his time. We return to Vincent's other beliefs in chapter 11.

Arnold of Villanova (1235?–1312) was an alchemist, astrologer, and physician who taught in Barcelona and Paris. His experiments led to genuine discoveries—e.g., the poisonous natures of carbon monoxide, and of decayed meat. He also appears to have added his Christian faith to alchemy, because in one treatise that has been attributed to him he urges the student to recite the sixty-eighth Psalm while he is mixing his chemicals! He also advised that the last words of Jesus on the cross should be inscribed on some of the vessels used in alchemical research. Under different circumstances, Arnold might have contributed a lot to the basic understanding of nature. Soon after his death came the last of the important Arabian alchemists, al-Jildaki, whose works included titles like *The Brilliant Moon on the Secrets of Elixir, The Scattered Pearls, Removing of the Veils,* and *Introduction to the Secrets of Alchemy,* according to the *World Who's Who in Science.*

It was impossible to distinguish the endings of alchemy from the beginnings of chemistry, so that, understandably, the Church disapproved of both, officially condemning alchemy in 1307. In anticipation of this, some Christian alchemists wrote anonymously, some perhaps also attributing their work to Gebir. Despite this opposition, however, alchemy and astrology continued to be essential components of the thinking of philosophers into the sixteenth century, at which time alchemy began to move in a new direction, pioneered by Paracelsus (1493?–1541), a Swiss alchemist and physician. The preparation of medicines became an important part of the alchemist's goals, and pharmacology was born, though of course natural herbs had been used as medicine for centuries.

The ancient idea discussed in chapter 1 that there were gods in everything still flourished. Not pagan gods, of course, but angels and devils. Theological arguments were therefore seen to apply to material processes—transmutation of metals was related to transubstantiation of the sacramental bread and wine into the body and blood of Christ, and to the belief in resurrection of Christ's body. Ever since the second century A.D., because at that time Saint Clement of Alexandria had said so, it was commonly believed that noxious gases in caves, mines and wells were due to devils. These evil spirits of metals supposedly took possession of some mines and caused them to be abandoned.

According to some sources, the real cause was discovered during the time of Columbus by the alchemist Basil Valentine, who was more chemist than mystic, but fear of persecution from the Church caused him to work in secret. His treatise on alchemy was found some years after his death, and people learned from it to ventilate mines and burn off some of the gases by lighting fires. Or is that really the way it was? Basil is supposed to have been born in 1394, and his manuscripts were not "discovered" until 1600, at a time when it was even more dangerous to express any opinion that could lead to the serious charge of heresy. Some unknown author of the time may have saved his own life by inventing "Basilius Valentinus"—almost certainly it was his "editor," the German chemist Johannes Thölde.

Thölde's books reveal an extensive knowledge of the chemistry of gold, of the precipitation of one metal by another, and of compounds of arsenic and antimony, with application of the latter to medicine. He added a third material, salt, to the theoretically basic mercury and sulphur, which is not surprising, since Johannes was in charge of the boiling vats at the salt works in Frankenhausen. And the ventilation of mines? By that time Paracelsus had discovered the real cause and had spread the word in his inimitable way—he was very outspoken. It would not be until 1818, however, that Sir Humphrey Davy would include among his many discoveries the invention of the miner's safety lamp. Worse than that, as far back as the first century B.C., the Roman architect Vitruvius had advised, "Let down a lighted lamp and if it keeps on burning, a man may make descent without danger."

Reginald Wolfe, the Royal Printer of Queen Elizabeth, had the idea of publishing a "universal Cosmography of the whole world" and placed Raphael Holinshed (d. 1580?) in charge of the chapters on the history of the British Isles. These *Holinshed's Chronicles* were the only parts published (1577 and 1587); they served Shakespeare as source material for his English historical plays, and for parts of *Macbeth, King Lear,* and *Cymbeline.* Holinshed assigned the description of Elizabethan England to William Harrison, and from his very complete account we quote here only his comments on alchemy:

> All metals receive their beginning of quicksilver and sulphur, which are as mother and father to them. And such is the purpose of nature in their generations that she tendeth always to the procreation of gold; nevertheless she seldom reacheth unto that her end, because of the unequal mixture and proportion of these two in the substance engendered, whereby impediment and corruption is induced which as it is more or less doth shew itself in the metal that is produced.
>
> (Book III, HCl 35, 320)

The alchemical theory of the mercury-sulphur composition of metals was very persistent.

Early in the seventeenth century, a magnificent satirical play, *The Alchemist,* appeared by Ben Jonson, who was friendly to neither astrologers nor alchemists. We quote part of his introductory acrostic:

> A cheater and his punk, who now brought low,
> Leaving their narrow practice, were become
> Coz'ners at large; and only wanting some
> House to set up, with him they here contract,
> Each for a share, and all begin to act.
> Much company they draw, and much abuse,
> In casting figures, telling fortunes, news.
> Selling of flies, flat bawdry, with the stone,
> Till it, and they, and all in fume, are gone.
>
> (Argument; HCl 47, 541)

For "punk" read mistress, for "coz'ners" swindlers, for "figures" horoscopes, for "flies" familiar spirits, for "fume" smoke, and of course the "stone" is the philosopher's stone.

The Flemish physician and chemist Jan Baptista van Helmont (1577?–1644) isolated a number of gases—and indeed adapted the word "gas" from "chaos"—but he thought that the gases he discovered might be living spirits. Johannes Kepler was a dedicated astrologer. Isaac Newton set up a laboratory of alchemy at Cambridge and left a collection of unpublished manuscripts on the subject. Robert Boyle's *The Sceptical Chymist,* published in 1661, was the turning point from alchemy to genuine chemistry, although thirty years later he left to his executors a quantity of red earth with instructions for turning it into gold. It makes us wonder what false scientific idols we may be cherishing today.

Astrology and alchemy were based on the sun, moon, and five known planets—seven bodies in all. As late as 1801 the widely respected philosopher Georg Wilhelm Friedrich Hegel "proved" that there could not be more than seven of them. The sacredness of the number seven showed up over and over again: the seven wonders of the ancient world; the seven days of the week; the seven angels, trumpets, seals, and vials of Revelation; the seven gates of hell; the seven garments of Ishtar; the seven champions of Christendom (Saints George, Andrew, Denis, Patrick, James, Anthony, and David); the seven wise men of Greece; the Seven Seas; the seven deadly sins; the seven ages of man; the seven liberal arts; "seven for the seven stars in the sky" (Pleiades, Dipper, or the sun, moon and five planets); and many other heptads.

Beginning with Pythagoras and his worship of the beauty of numbers, some of the integers acquired special meanings, a belief strengthened by Plato's mystical mathematics and by the use of letters to represent numerals in the clumsy Roman system. During the last centuries before Christ, there arose a tradition among rabbinical scholars to define the rules by which the Mosaic Law should be interpreted. According to some there was a hidden language behind every word, every letter, every number, a language of gematria—Biblical numerology—which led to the most ridiculous conclusions. They "proved" by this method that every part of each Law has three score and ten meanings, making it necessary for God to spend three hours a day studying them all.

The Christian saints developed a different version of biblical numerology. St. Clement of Alexandria (150?–210) subscribed to the belief that the Creation occurred in 6000 B.C. because it took six days to make Adam, and a day in the sight of the Almighty is as a thousand years, thus setting the date for the coming of the second Adam, Jesus. St. Clement had read Psalm 90:2, "For a thousand years in thy sight are but as yesterday when it is passed." St. Augustine (354–430) believed that "the world was created in six days because six is a perfect number"—the sum of its factors or "aliquot parts" being equal to itself: 1+2+3 = 6. The next perfect number is 28, which must have something to do with the moon.

In John (21:11) it is recorded that after getting instructions from the resurrected Jesus as to where to cast their nets, Peter and the other apostles caught one hundred and fifty-three fish. Why 153? St. Augustine noted that this number is the sum of all integers from one to seventeen—the sum from one to any number is one half of that number multiplied by the next integer above the number. But why 17? Because it is ten plus seven, ten for the Ten Commandments plus the magic seven! But we shouldn't call it "magic" seven. That is pagan. Seven stood for the four elements plus heart, mind, and soul, or maybe Father, Son, and Holy Ghost. Later, Pope Gregory the Great (540?–604) was deeply impressed by the fact that Job had seven sons, which therefore typify the twelve apostles. Why? Because $7 = 3+4$ while $12 = 3 \times 4$. With Gregory the three was for the Trinity and the four was for the four corners of the earth.

As St. Augustine put it, "ignorance of numbers prevents us from understanding such things in Scripture." He argued that you should fast for forty days because 40 is four times ten. Four represents time, the four seasons. Ten of course is 3 plus 7, seven we know about, and three is for the Trinity. John (6:19) tells us that the disciples "rowed about 25 or 30 furlongs" but for St. Augustine that meant "twenty-five typifies the law, because it is five times five, but the law was imperfect before the gospel came; now perfection is comprised in six, since God in six days perfected the world, hence five is multiplied by six that the law may be perfected by the gospel, and six times five is thirty." Q.E.D. Under different circumstances a great mind like that of Saint Augustine could have added to the legacy left by the Greek philosophers. He too was struggling to understand. He worked hard, he studied with meticulous detail, he was imaginative, and he genuinely believed that he was contributing to the world's precarious store of wisdom. That can happen to anyone who starts from a wrong hypothesis.

Years later, the same sort of thing happened to a really good German mathematician, Michael Stifel (1487–1567). He gave us the signs used in arithmetic today, studied the concepts that led later to logarithms and exponentials, and became a leading expert on number theory. He was an Augustinian monk who converted to Lutherism and, unfortunately, to astrology and gematria as well. From these last two he (incorrectly) predicted that the world would end on October 3, 1533. Some of the earlier cabalistic writings had just been translated into Latin, and, as a mathematician and mystic, Stifel was intrigued. We read in Revelation (13:18) of "the code number of the name of a man," the Beast of the Apocalypse. This number, 666, represents the Aramaic letters that spell Nero Caesar. By a ridiculous miracle of gematria, Michael "proved" that it was really Pope Leo X†, who had issued the bull excommunicating Luther. In more recent times, made more believable by the immense evil that he generated and by the number of letters in each word of his name (using the English spelling of his first name), Führer Adolph Hitler was identified with the Beast of Revelation.

In the century after Stifel's fallacies, this mystical interpretation—the "cabala"—had spread among Jewish rabbis and Christian clergy. Uriel Acosta* (c. 1591–1647) fought a losing battle against the rabbis of Amsterdam and their belief in the value of gematria. Any word in the Old or New Testament was regarded as a cryptograph, the letters revealing the numerical value of its hidden meaning. For example, the word for "in the beginning" in Hebrew acquired the value 913 which was also the value of the Hebrew phrase that means "in the Law it was made." Therefore, the Law existed from the beginning! As the asterisk attached to his name denotes, a tribute to Uriel Acosta appears in the section "In Memoriam."

From the Last Supper comes fear of the number thirteen—a superstition so powerful it causes the builders of some high-rise hotels to avoid that number when designating the floors. There are numerologists on the lecture circuit today contributing to the thoughts of the New Age movement. Tuesday is a lucky day because on the third day of Creation, "God saw that it was good" twice (Gen. 1:10 and 12). The number three is very special as well—the Holy Trinity; the Hindu Trimurti (Brahma, Vishnu, and Shiva); the three virtues (faith, hope, and charity); the Three Sisters (Clotho, Lachesis, and Atropos, fates of human destiny, spinning, measuring, and cutting the thread); the three gilded balls of the pawnbroker (derived from the coat of arms of Lombardy); the three R's (reading, 'riting, and 'rithmetic); the three dimensions of space; and many more.

This superstition also pervades the chapter titles of this book—which naturally has seventeen chapters.

9

Emperors, Saints, and Heretics

Better heresy of doctrine than heresy of heart.

—John Greenleaf Whittier

Before Julius Caesar was assassinated in 44 B.C., he had written a will, appointing as his successor his grand-nephew Octavian (63 B.C.–14 A.D.), but Marc Antony (83–30 B.C.) had other ideas. There followed an uneasy cooperation between the two as they began a reign of terror in Rome, forcibly taking money from the people in order to finance their war against Caesar's killers, Brutus and Cassius. The line of the Ptolemies of Egypt, begun almost three hundred years earlier, had led to Queen Cleopatra (69–30) mother by now of Caesarion (47–30), son of Julius Caesar. Marc Antony ordered her to meet him in Tarsus to answer charges that she had aided the armies of Cassius. (Tarsus is a town near the south coast of present-day Turkey where, years earlier, Alexander the Great had nearly drowned in its river Cydnus. A few decades after Marc Antony's command, a Jew named Saul, later called Paul, would be born there.) As Antony awaited her, Cleopatra began her spectacular and seductive passage up the river, sending messengers to invite him to her luxurious barge. Her crew was formed of beautiful female slaves, appropriately undressed, but no match for their queen, adorned with gold and partly covered with transparent Seric silk. It is not clear that the subject of Cassius ever came up.

A war with Octavian was inevitable, and ultimately the fleet of Antony and Cleopatra was defeated by the Octavian fleet under Agrippa (63–12) at the Battle of Actium, off the west coast of Greece. Antony and Cleopatra ended their lives, and Caesarion was killed by Octavian, who was now in charge and soon was to be called Augustus—"holy." Until his declining years he was a great emperor. There was peace and prosperity, except on the empire's expanding borders; Augustus freed many slaves and he felt that Rome should return to ancient standards of morality and religious faith. He passed a law against adultery that would come to haunt him, since his only daughter, Julia, married Agrippa in 21 B.C., and led a very scandal-laden life while giving birth to five children. When asked why they all looked like Agrippa, she replied, "I never take on a passenger until the vessel is already full." After Agrippa died she really got out of hand and was then married to Tiberius who was to become the next emperor of Rome. Augustus had had enough, and she was banished, probably because of her relationship with the poet Ovid, who was also banished—to another place—at the same time.

Augustus wanted to change the moral behavior of the Romans by making laws, not only against adultery but for the encouragement of families. The father of the largest family was to be preferred in making an appointment to an office, and after three children the mother was to be emancipated from the power of her husband. Many other rules were violated too, laws which also didn't go down very well with the people, To Augustus it was then religion, not legislation, that was the answer to societal ills. He had asked Virgil and Horace to encourage religion in their poems —Virgil in particular told of the Golden Age, of the piety of Aeneas, fabled ancestor of the Romans, of his love of family, his courage and gratitude, his support of justice, and his other fine qualities. John Dryden's translation of the *Aeneid* (Book 2) tells of Aeneas's words as he leaves the burning Troy:

> Haste my dear father, ('tis no time to wait)
> And load my shoulders with a willing freight.
> Whate'er befalls, your life shall be my care;
> One death, or one deliv'rance, we will share.
> My hand shall lead our little son; and you,
> My faithful consort, shall our steps pursue.
>
> (HCl 13, 124)

The heroic Aeneas, son of Venus and Anchises (blinded for having bragged about their affair) was believed to be a god from whom the Caesars were directly descended. Augustus encouraged worship of these "Roman" gods, as opposed to imports like Cybele and Mithras, while privately he was both skeptical and superstitious.

As we saw in chapter 6, Virgil's *Georgics* encouraged use of the land. Augustus was delighted with this poem because he wanted to demobilize many of his troops and have them turn to much-needed agriculture. This takes us back to the Greek goddess Demeter and other gods and goddesses of the earth and harvest. The Romans had their own god for that—Saturn, related to the Latin word "sator," for one who sows or plants. Saturn had other qualities, more like Cronos, father of the chief gods, but he was also Father Earth. In the sixth book of the *Aeneid* (1077–1081), Virgil extols the virtues of Augustus, putting him in the same league as Saturn:

> But next behold the youth of form divine,
> Caesar himself, exalted in his line;
> Augustus, promis'd oft, and long foretold,
> Sent to the realm that Saturn rul'd of old;
> Born to restore a better age of gold.

While Augustus hoped for a transformation of morality through religion, there was no way for him to know that he was born B.C., and would die in an A.D., and that we would read in Luke (2:1-7) (Lamsa translation), "And it happened in those days that there went out a decree from Caesar Augustus to take a census of all the people in his empire. . . . And every man went to be registered in his own city . . . Joseph also went up from Nazareth . . . to Bethlehem, with his purchased bride Mary while she was with child—and she gave birth to her first-born son."

Here is not the place to outline the life or the message of Jesus, described in so many sacred and secular writings and addressed each year in a million sermons. Nor shall we follow the theology of St. Paul or St. John, except to note a contribution from someone else who influenced them and their successors—the Jewish philosopher Philo of Alexandria. Paul argued with "philosophers who were of the teaching of Epicuros, and others who were called Stoics" (Acts 17:18) and Philo had also looked back to the Greek philosophers, in particular to Plato. To Philo, Plato's ideas were a manifestation of the mind of God, what the Stoics and Gnostics called the "Logos," the rational principle of the universe. It is sometimes translated as "Word," and we read about it in the first and fourteenth verses of the first chapter of St. John's gospel. As with the theophany of the Greek god Apollo, the earliest concepts of Christian theology were based on the belief that "the Logos became flesh and dwelt among us," ultimately becoming the second member of the Holy Trinity. There would be many arguments about that. Philo worked hard to show that the Jewish Scriptures could be reconciled with the teachings of the Greek philosophers, at a time (38 A.D.) when the Greeks in Alexandria had invaded the synagogues and conducted an organized massacre of any Jews found outside the ghetto, destroying thousands of Jewish homes and businesses. Philo headed the Jewish delegation to the insane Emperor Caligula to present the Jewish cause; after Caligula was murdered, his successor Claudius I ordered Jews and Greeks to live together in peace.

In order to dovetail the Jewish Pentateuch with Plato, something had to give, and with Philo it was the literal meanings of even the simplest parts of the Jewish Scriptures. Like the exponents of gematria mentioned in the last chapter, he saw allegories in everything—e.g., the golden candlesticks in the tabernacle really represented the planets, the robe of the high priest represented the universe. That type of thinking started a trend in which the Old Testament and, later, the New, were seen to offer a theological description of nature. There had been other Jewish scholars such as the conservative Shammai, and the more liberal Hillel with a much deeper understanding of the Mosaic Law, but, while inheriting that Law from the Jewish people, early Christian scholars looked more to Philo for its interpretation.

In earlier chapters we have referred to Pliny and Seneca, Roman writers of the first part of the first century A.D., but not to Epictetus who was born in the year 50. He was a slave who naturally was opposed to slavery, but who taught that a slave can be spiritually free, while an emperor can be spiritually enslaved. Epictetus reminds us of what Xenophon said about Socrates, as quoted in chapter 4—he was more interested in "what is godly, what is ungodly" than in fine points of philosophy or science. He had a Stoic wonder at the order in the universe and a belief in a divine Creator which, pagan or no pagan, led to the acceptance of many of his teachings by the Christian leaders who were to emerge later. Epictetus taught that good should be returned for evil, that criminals should be treated as mentally sick, and that we should wonder at the way that wild animals live in harmony with nature. Had he heard of Jesus and his teachings?: "The foxes have holes and the birds of the air have nests" (Matthew 8:20). "Consider the lilies of the field, how they grow; they toil not neither do they spin" (Matthew 6:28).

The first major disagreement about Christian theology is recorded in 2 Timothy (2:16-18). "Shun empty and worthless words, for they only increase the ungodliness of those who argue over them. And their word will be like a canker eating in many; such are Hymenaeus and Philetus, who have strayed from the truth, saying that the resurrection of the dead is already passed, thus destroying the faith of some." (Lamsa translation) The idea that the resurrection was a spiritual reality that had already taken place with Christians, that to be born again was to be resurrected, was common among Gnostic Christians. Some of them found the ways of the world too much of an impediment to their spiritual needs, and withdrew to a simpler life-style, precursors of the monastic movements of St. Benedict and others. In the middle of the second century Valentinus taught

Gnosticism to the illustrious students Clement and Origen, but, like Hymenaeus and Philetus, he was regarded as too unorthodox and was excommunicated. *The Treatise on Resurrection*, one of the tracts found in 1945 as part of the Nag Hammadi Library, is full of the mysticism and symbols of Valentinus, but it states very clearly the gnostic viewpoint of resurrection that was so opposed by Paul in his letter to Timothy.

At times the early Christians were persecuted without mercy, through painful death and malicious slander. St. Ignatius Theophorus was martyred in the year 108, the strength of his religion and his conviction being evident in his words: "Suffer me to be eaten by the beasts, through whom I can attain to God." This was during the reign of the emperor Trajan, the supporter of Mithraism, from which, ironically, Christianity was to derive so many of its ceremonies and doctrines.

The philosophy of Plato was studied and taught at that time by St. Justin (100?–166) who opened the first Christian school in Rome. He tried to persuade the emperor and his associates that there was no need to persecute the Christians—they would "render unto Caesar the things that are Caesar's, and unto God the things that are God's" (Matthew 22:21; Mark (2:17). Justin was outspoken, he made enemies, and in the year 166 he too was put to death, along with some of his students. The Roman emperor at the time of this martyrdom was the Stoic philosopher, Marcus Aurelius, very learned and, for an emperor, very gentle in many ways, but he was strongly opposed to Christianity. However, one of his legions was composed of Christians, and in a battle in 174, in what is now Hungary, a violent thunderstorm worked against the enemy, the Quadi. The Christians had prayed to their Jehovah and Christ and gave them the credit, while their Roman leaders ascribed it to Jupiter and his thunderbolts, while some others thanked Mithras.

Outrageous charges had been made against the Christians—atheism, for worshiping a different god; incest, because of their love of one another; and cannibalism, for their drinking and eating of the body and blood of Christ. The Christian apologist Athenagoras wrote to Marcus Aurelius defending the Christians against these charges, but St. Irenaeus, the Bishop of Lyon, decided that it was time to unite against the common enemy. He stressed, however, that Christianity could break up not only under these external pressures but from its own internal disagreements of doctrine. In 186 he exhorted all Christians to follow the doctrines that the Church councils approved, on pain of being found guilty of heresy. He, too, appears to have been martyred.

The pragmatic Romans had never been too fond of philosophy, although they had captured Greece in 146 B.C., and were thus greatly influenced by many aspects of Greek culture. In the early centuries of the Christian era, however, Roman emperors began to rebuild Greece (along with setting up a magnificent building program in Rome) and established schools of philosophy at Athens, in the same spirit of Plato, Aristotle, and the Stoics. The high level of morality and belief in God preached by the Stoics, together with the pious convictions of Plato, fitted well with the revival of pagan religions and later with Christianity. The science of Aristotle did not have that much appeal at first, nor did his philosophy. That would have to come later.

Despite the warnings of St. Irenaeus, differences of opinion on the theology of Christianity continued. It was a long and bitter struggle for followers of a particular ecclesiastic to have his ideas of Christianity sanctioned by the Church leaders, since those ideas that were not sanctioned rapidly became heresies. There were so many opinions that all we can do here is to describe some of them briefly, remembering that if their proponents had won the power struggles, some of our Christian churches would now be preaching *their* opinions as dogmas instead, while others would at least be paying them lip service. Winners and losers alike were for the most part seriously struggling to understand what religion means, what is the relation of God to man and man to God, but the losers and their followers in these debates were hounded and persecuted by the winners, their fellow Christians. They would deserve an *In Memoriam* section too, except that, if they had won, they would almost certainly have subjected the losers to the same outrages.

A Christian prelate, by name Sabellius, in the late second and early third century maintained that Christ was indistinguishable from God the Father, being a form that God took. Thus God felt the pain of the cross. Sabellius gathered a number of followers, but he was excommunicated around the year 220. Another form of this monarchianism, as it came to be called, was based on the belief that Christ was the *adopted* son of God, a mere man, supernaturally chosen and inspired. A competing concept from the second century on was due to the Christian Gnostic Marcion and his followers, who established churches throughout North Africa, Gaul, and Asia Minor, based on the belief that the father of Jesus was not the stern Jehovah of the Old Testament, but a God filled with the love that Jesus had revealed. They were excommunicated, but their message lives on.

The philosopher Titus Flavius Clements (150?–210) was converted to Christianity and laid much of the foundation of metaphysical Christian doctrine. He was known as Clement of Alexandria, later Saint Clement until the middle of the eighteenth century, when Pope Benedict XIV removed his name from the list of Catholic saints. He and his student Origen (185?–254?) continued the attempt to incorporate Greek philosophy and astronomy into Christianity. St. Clement's most lasting legacy to science was his argument that the Jewish tabernacle was a symbol of the earth, placed at the center of the universe. This dovetailed nicely with Ptolemy's model of the solar system, which became the first important "pagan" idea to become entrenched in the required beliefs of the Christian.

Both Clement and Ptolemy spent many years in Alexandria during the second century, although they probably never met since Ptolemy made most of his observations at Alexandria before Clement was born. Clement also believed that the apocryphal Second Book of Esdras, which helped Columbus on his voyage, was divinely inspired. Whether it was or not, Clement himself was inspired as a writer. He was a very learned and articulate man, and words such as these could not fail to make an impression:

> Some there are, who, like worms wallowing in marshes and mud, in the streams of pleasure feed on foolish and useless delights—swinish men. For swine, it is said, like mud better than pure water, and according to Democritos, 'dote upon dirt.' . . . None of these (sculptors) ever made a breathing image, or out of earth molded soft flesh. Who liquefied the marrow? or who solidified the bones? Who stretched the nerves? who distended the veins? Who poured the blood into them? or who spread the skin? Who ever could have made eyes capable of seeing? Who breathed spirit into the lifeless form? Who bestowed righteousness? Who promised immortality? The Maker of the universe alone; the Great Artist and Father has formed us, such a living image as man is.
>
> (UA 7, 136)

Unfortunately, like others who preceded and followed him, Clement saw allegories in the Bible where there were none, and he failed to see allegories, particularly in Genesis, where they existed. As we saw in the previous chapter, he combined a metaphor with numerology and manufactured his own allegories. That kind of thinking led to indescribable fantasies among his successors, good men, most of them, struggling to deduce a deep understanding from their false premises. Around that same time, Gaius Julius Solinus gave a description of the world as known to the Greek philosophers, with comments on natural history, religion, and social topics. It was Solinus who first used the term "Mediterranean"—the center of the earth. He commented on the absence of snakes in Ireland but was generally unreliable, partly because he had used Pliny as his source. He represented the last gasp of the short-lived Roman scholarship, but John Milton, in his *On Education*, put him on his required reading list.

In addition to persecution and slander against Christians, sober philosophical arguments were raised, in particular by a second century Roman by name Celsus, who felt that Christ's miracles, the resurrection of the body ("the hope of worms"!), the ultimate burning of all non-Christians, and the loving nature of Jehovah were just too much to be believed. First to respond to Celsus was St. Clement's student Origen, admitting that these are things hard to believe, but arguing that paganism is also absurd. Origen's father had been arrested and beheaded for his Christian faith, and Origen devoted the rest of his life to advancing Christianity by an enormous—even compulsive—outpouring of tracts on, and translations of, the Scriptures. Sex apparently reared its distracting head while Origen worked, and to avoid interference with his studies and writing he castrated himself, taking very literally the passage in Matthew's Gospel: "There are eunuchs who make themselves eunuchs for the sake of the kingdom of heaven" (Matthew 19:12). Origen's struggle to understand the Bible led to a treatise on prayer and an exhortation to martyrdom, and he was himself tortured during the terrible reign (249–251) of the Roman emperor Decius, who persecuted all of the Christians he could find.

Like his teacher, Origen had some very unfortunate ideas that became part of the Christian theological world system. Comets portend catastrophes, the stars are living beings and have souls (deduced from Job 25:5, "the stars are not pure in his sight," with shades of Plato), demons hover in clouds and are responsible for famines and plagues, God gave Adam the Hebrew language, there are four gospels because there are four elements, and much more. Origen thought that Creation had been instantaneous, but it is hard to argue that he anticipated the "Big Bang" theory, since his estimate of the time since the beginning was too short by a factor of three million. There were other defenders of Christianity in the third century, notably Minucius Felix, who wrote the earliest known work of Latin Christian literature—an argument between a Christian and a pagan that refuted all of the charges that had been made about this new religion.

In these early years of Christianity the sacred books of the Jews were beginning to assume enormous significance among the Christians, and it wasn't very much longer before the story of Noah began to cause a problem. More and more animals had become known, and there didn't seem to be enough room for them in an ark 300 cubits long, 50 cubits wide, and 30 cubits high. A cubit, from the Latin word for elbow, *cubitum*, and a similar Greek word, is the distance from the elbow to the tip of the index finger, about eighteen inches, so that the Ark would have been sizeable, 150 yards long, but it still would have been a *very* crowded zoo. Origen, in charge of Christian instruction at Alexandria and later in Caesarea, decided that the cubit was six times as long as people had thought, possibly because (Genesis 6:4) "there were giants on the earth in those days," making the Ark about half a mile long.

An unknown author of about that time—probably another Christian in Alexandria—wrote a book that described the peculiar nature of many things, especially animals. The book is referred to as *Physiologus*, and by some reports that is the name of its author. The whole idea of the book was to use allegories to clarify and support the Scriptures. Instead of Aristotle's magnificent work on animals, Christian scholars read about many exciting and nonexistent animals, some mentioned in the Bible—e.g., the unicorn, the dragon, the basilisk with its fatal breath and look—and were reminded that "the weaned child shall put his hand on the cockatrice's den." (Isaiah 11:8) The word "cockatrice" is translated as "asp" by Lamsa and as "adder" in the new Oxford Bible, but however it is regarded now, it was thought then to be hatched by a serpent from the egg of a cock, and to rival the basilisk with its deadly glance. Perhaps it *was* a basilisk.

The belief in the cockatrice and its peculiar breeding habits arose from an error in translation, as did the belief in the ant-lion. In Job (4:11) we read in the King James version that the *old* lion perishes for lack of prey. In the new Oxford Bible it is the *strong* lion, but in much earlier editions of the Bible it was the "ant-lion." The first translators of the Bible from the Aramaic to Greek had a problem with the original word and rendered it in transliterated form as *myrmekoleon*. But *myrmeko* means "ant," and the story developed of this creature, the product of the unlikely conjugation of a lion with an ant—presumably a large one, probably a gi-ant. With the head of a lion and the rear-end of an ant, it perished because its ant nature prevented it from eating proteins, and its lion nature stopped it from consuming carbohydrates. And if you doubt the story, you wouldn't expect to find one anyway, because, as Job has told you, it "perished." These writings had an enormous impact throughout Christendom for a millennium and a half, influencing very strongly later books—e.g., *The Bestiaries*—on the same general subject. People believed what they read, although (perhaps because) it was offered with the fidelity to fact characteristic of the worst of yellow journalism.

A younger contemporary of Origen, but coming from a different place and a different religious background, developed an alternative interpretation of the basic theology of Christianity. Mani (215–273), a Persian sage, founded a religious movement based on the Zoroastrian eternal conflict between good and evil, coupled with the Christian belief in salvation. His followers, the Manichaeans, spread as far as India and China, and for ten years included Augustine (later Saint) as one of their members. Mani was crucified by the Persian Magi, fanning the belief of many that he was indeed a Messiah sent by God as he had claimed. Mani's message spread in time as well as space. Manichaeans propagated for a millennium, despite the fact that they believed that in the eternal fight between good and evil, women were the work of Satan.

While Mani was preaching at Ctesiphon, in the province of Baghdad, and in other parts of Persia, Plotinus (205–262) was developing the philosophy/theology of Neo-Platonism, first at Alexandria, then during extensive travel, and finally in Rome. At Alexandria he was a fellow-student of Origen under Ammonius Seccas, trying to develop a philosophy that would unite the teachings of Plato with those of Christian thinkers. But Plotinus decided to take off to Persia to learn what he could about Persian and Indian beliefs. He may have even met Mani; certainly he visited Mesopotamia about the year (242) when Mani began his preaching there. Plato's contemplation of Ideas became for him a spiritual quest to return to God by eliminating all matter that stood in the way. Man had fallen from God and only by abjuring the pleasures of the senses could he partake of divine ecstasy. The Hindu influence is evident in Plotinus's belief in reincarnation, providing other chances for the soul to progress upwards to God. God is good, and God is unity, and one should progress away from emphasizing material beauty towards identifying with the divine unity of the spirit.

Plotinus's pagan ideas were accepted by the Christian fathers and have become part of Christian teaching. A passage from his writings, as given in Stephen Mackenna's *The Essence of Plotinus* and quoted by Dr. Durant, reads as follows: "Withdraw into yourself and look. And if you do not find yourself beautiful, yet act as does the creator of a statue—he cuts away here, he smoothes there, he makes this line lighter, the other purer, until a lovely face has grown upon his work. So do you also . . . never cease chiseling your statue until . . . you see the perfect goodness established in the stainless shrine." One is reminded of Norman Vincent Peale's power of positive thinking, or of the works of Ernest Holmes, founder of the Church of Religious Science—superficially only, however; by no means did they regard matter as the obstacle to the spirit that Plotinus did, but they do share a common belief in the ability to improve oneself.

Could a person who had committed a mortal sin since being baptized be readmitted to the Church? Arguments about this caused several schisms of the Church during the third century. Hippolytus (*not* the saint of the same name and time) argued that such a sinner could not be readmitted. He was murdered in 235; but two others, Novatian in Rome and Novatus in Carthage agreed with him. When Cornelius was appointed twenty-first pope in the year 251, Novatian objected very strongly, arguing that he was too lenient. It even got to the point where Novatian was set up by his followers as another pope in Rome—an anti-pope—and the sect so formed continued for some centuries.

Early in the next century there arose two more major conflicts within the Church. The first of these was initiated by Donatus, the Bishop of Carthage. In 311 the Donatists protested against a *traditor* bishop, by name Felix, and the Primate of Carthage, Caecilian, whom he had consecrated. They had good reason to complain, for to them *traditor* meant "traitor"—one who had cooperated with the Romans during their periodic purges of Christians. They denied the validity of sacraments that were administered by any priests who themselves were unrepentant of their sins, and they demanded that those who joined the sect would have to be baptized again. Partly for this and partly because they wanted Carthage to get out from under the power of Rome, the Donatists were driven from the churches and indeed were forced to hand over their own churches to the Catholic faith. The Church Fathers did not want to alienate Rome; they had other plans for it.

The other split was started by Arius, a Greek ecclesiastic who died in 336 A.D. He taught that God is alone and unknowable, that Christ was born at a particular point in time, and therefore could not be co-eternal with God, and should be worshipped as a secondary deity. The Trinity was a small hierarchy of Gods, not "Three in One and One in Three." The furor against Arius was led by St. Athanasius (293?–373) both at the Council of Nicaea (325), at which Arius was pronounced an anathema, and during the rest of his life since Arianism refused to go away. The emperor Constantine (280?–337) had thought that he had prevented a real split in the Church by convening the Council of Nicaea and by banishing Arius and burning his writings, but it didn't work out that way.

When Constantine died in 337, his three sons succeeded him, but the middle one, Constantius II (307–361) lived longer and was more aggressive than the others. He was a supporter of the doctrines of Arius and decided that, despite the decision of the Council, the people should be Arians too. As far apart as Alexandria and Constantinople, thousands of Christians were killed by other Christians, especially during the years 342–3, fighting for their particular choice of theologies. Athanasius never stopped preaching against Arius, despite persecution, and in his turn Augustine added his weight to the dispute. The doctrine of the Trinity, derived from Egyptian sources, eventually won and became enshrined as dogma. The hom*o*ousian doctrine—that the Son is of the *same* substance as the Father—won out over the hom*oi*ousian doctrine of Arius, that the Son was *like* the Father, but not of the same essence. The magic number three overcame the single letter i, with probably more Christians killed in the conflict than had been killed in the persecutions by all of the pagan Roman emperors.

A few years before he died, Constantine had established a physical and political stronghold for Christianity when Byzantium was declared the capital of the Eastern Roman Empire on May 11, 330. With much of the new city built at his command, it was naturally "christened" the city of Constantine—Constantinople. Next year his nephew Julian (331–363) was born, later to rule the empire. Basically more a philosopher than a ruler, Julian was very much impressed by the knowledge and religion of the Greek civilization. Indeed, those ancient rites imported to Greece by Epimenides in 596 B.C., as mentioned in chapter 4, were very moving, and Julian as Emperor tried to bring this religion to his empire. As he saw it, it was a religion with all the care and love of the best of Christianity, without its self-contradictory gospels and its uncertain theology. He instructed his priests to follow the example of the Christian clergy, by sharing money, food, and clothes with the poor, and inspiring the people by living good clean lives. He regarded the story of creation in Genesis as blasphemy unless it was interpreted allegorically: God is "represented as ignorant that she who was created to be a helpmate to Adam would be the cause of man's fall. Secondly, to refuse to man a knowledge of good and evil (which knowledge alone gives coherence to the human mind) and to be jealous lest man should become immortal by partaking of the tree of life—this is to be an exceedingly grudging and envious god." (quoted by Durant)

Like Ikhnaton who 1700 years earlier had tried to impress his religion on Egypt, Julian died very early in his thirties, as indeed had Jesus. A javelin pierced Julian's liver while he was leading a fight against the Persians, hoping to conquer and convert them. No Persian claimed the reward that had been offered for killing him, however, and it was suspected that a Christian was responsible. By then Julian's religion was not gaining ground anyhow, despite his dissemination of the myths associated with it—he had recognized that no religion can succeed without the presentation of its myths as historical facts.

Meanwhile, Christian hospitals were opened in Caesarea (369 A.D.), Edessa (375), and Rome (400). Some wealthy Roman ladies, converted to Christianity, founded more hospitals and convents. Orders of monks and nuns dedicated themselves to care for the sick; the Order of St. Lazarus ministered to the lepers. Anesthesia was induced by inhaling a mixture of opium, hemlock, mulberry juice, and the magical mandrake root. But more and more, the saving of souls gained precedence over the saving of bodies.

During this period, the library and pagan temple at Serapis were destroyed (389), and much of the magnificent library at Alexandria met the same fate at the command of the Christian Patriarch Theophilus (392). Two years later, Emperor Theodosius I put an end to the more than a millennium of Olympic Games. The writings of Epicuros were banned as the works of an atheist. Preceded by Democritos, and followed by Lucretius, he was regarded as an evil purveyor of lies. And that went for his atomic theories too, if anybody cared. Few people did, however, for scholars and emperors had other things on their minds. For scholars, now rapidly becoming scholastics, the problem was to formulate a standard Christian theology and to convert succeeding emperors to the faith. For emperors it was to stay in power with the help of the Church, and to defend the ten-thousand-mile border from invasion.

Saint Augustine, his father, and his mother, Saint Monica (332?–387), lived in Tagaste, presently Sauk-Ahras in the area of Constantine, north Algeria, some hundred miles west of what was then Carthage. Monica was a devoted Christian and she really worried about her son, as well she might —Augustine reveals in his *Confessions* the excitement and torment of a young man:

> What was it that delighted me but to love and be loved? But in this love . . . black vapors were exhaled from the muddy concupiscence of the flesh. . . . How great a distance was I banished from the delight of Thy house in that sixteenth year of the age of my flesh when the fury of lust . . . had received the scepter in me, and I wholly yielded myself up to it. . . . Among my equals I was ashamed of being less filthy than others; and when I heard them bragging of their flagitious actions . . . I had a mind to do the like, not only for the pleasure of it, but that I might be praised for it.
>
> (UA 7, 341)

It was a strain on the family budget for Augustine to go to Carthage to continue his studies, but he did some tutoring to make ends meet. He kept out of trouble by living with a mistress, the same one, for fifteen years, even having a son with her and taking her with him to live in Rome and then to Milan. Momma caught up with him, however, persuaded him to listen to the serm ns of Saint Ambrose and, at the age of thirty-two, to become engaged to a ten-year-old girl. Augustus put off the marriage (indefinitely as it turned out) to wait for the child to mature, sent his mistress back to Africa, and tried to live a celebate life. It was too much for him. "Give me chastity," he prayed, "but not yet."

Augustine took another mistress but soon became converted to Christianity, was baptized, and with some friends founded the Augustinian Order. As a sudden convert, he identified with St. Paul, whose words had led to his conversion. St. Ambrose had played a part too, and a few years later that good man effected a triumph for the church over the state by forcing Emperor Theodosius I to remove his robes of office and come publicly before him to repent for having commanded a shocking massacre of the people of Thessalonica. The only weapon Bishop Ambrose had was a threat not to administer mass to the emperor, who was thus persuaded by the fear of eternal damnation. It showed the enormous power that the Church had come to wield over a Christian emperor and his subjects, a power used often for good, as in this instance, but later for shocking evils as well.

St. Augustine was persuaded to become Bishop of Hippo (present Bône, Algeria), and from that See he issued an enormous volume of theological works that dominated Christian thinking for close to a thousand years, and indeed have a large impact even today. Some of the eighty-three scientific and theological questions that concerned him: How could people who have been eaten be resurrected in the body? How could all of the animals, birds, and insects known at that time have fitted into the Ark, and how could Adam have named them all, as related in Genesis (2:19,20)? Why was the world created in six days? Why does that contradict Genesis (2:4): ". . . they were created in *the day* that the Lord God made the heavens and the earth." The last of these questions was easily answered—the matter was created in one day but it took six days all told to put it into shape. Why six days? We saw the answer to that one in the last chapter.

Augustine believed and very persuasively preached that man's sin had cast a curse not only on the animal world but on the vegetable world as well. In view of the approaching end of the world, it seemed to him and others impious and a waste of time to study nature, but he couldn't help worrying about the Creator's purpose in coming up with superfluous creatures like flies, mice, frogs, and worms. The dangerous animals and stinging insects had a purpose, though, to prevent us from cherishing and loving this life. What a shocking distortion of theology, to teach that we should not enjoy life!

He saw a solution to some of Adam's problems by adopting the old idea of spontaneous generation, allowing many insects and small animals, like frogs, to be potentially created in the beginning, but to emerge much later from carrion, water, and filth. Maggots on some rotting meat provided the evidence, not to be challenged, at least in Christian countries, until the Italian physician and poet Francesco Redi reported his experiments on the subject in 1668. St. Augustine was also puzzled by the appearance of animals in places very distant from the landing-point of the Ark. The islands were a special problem—did the animals swim there? Not too likely. Probably God commanded the angels to carry them there. Much later this was still a problem, especially when kangaroos were discovered in Australia. But Australia, and the antipodes generally, could not exist. Augustine strongly opposed the idea that the earth could have a population on the other side, asserting that those who believe that could not be saved. Referring to the Nineteenth Psalm, "Their good news has gone out through all the earth and their words to the end of the world," he argued that such people made King David into a liar and also St. Paul, who had referred to the same psalm.

Augustine thought that pagan gods still existed as demons, and that the satyrs and fauns of ancient mythology were real. According to him, the serpent in Eden was forced to eat dust to make him "penetrate the obscure and shadowy," for to do so reveals the sin of curiosity. Noah's Ark was "pitched within and without with pitch" (Genesis 6:14) to show how the Church is protected from the heresies that are trying to leak in. Much more serious was his interpretation of the words from one of Christ's parables (Matthew 22:9) as a warrant issued by God for religious persecution: "Go out into the highways and byways and compel them to come in." The text was later translated as: "Go therefore to the main roads and whomever you may find, *invite* them to the marriage feast." *That* was what Jesus said. Of course, Augustine was working from a Latin translation, with no apparent interest in finding out if it was accurate. What a tragedy that such a smart, earnest, and dedicated scholar, struggling hard to understand the words of Scripture, should be led to regard them as a command to persecute those who did not adopt the official version of Christianity.

Augustine's main interest and impact, however, lay in the area of psychology/theology, and he was just as wrong here too. He must have missed his mistresses a lot, probably without realizing the reasons why he wrote so eloquently and persuasively on the evils of sex. He wasn't the first, of course; as we have seen, sex, woman, and sin had been linked together for centuries. In Job 25:4, we read (Lamsa translation), "How then can man be justified with God? Or how can he be declared blameless, he who is born of a woman?" We know that sex can be mistreated and can bring some serious evil, but to say that we are all born in sin because of the dirty thing our parents did to form us is an outrage that has been propagated through the centuries. It takes us back to the chapter on Creation and the Fall, since Augustine was inspired by the Old Testament as well as the New, and tarnishes the most wonderful thing that we know about in the entire universe—the creation of another human being.

A contemporary of St. Augustine from the school at Antioch, St. John Chrysostom (345?–407), attempted to apply a little more reason to the interpretation of the Bible, but the ideas that he and his associates were beginning to work out were overwhelmed by the myths and wonders that the other theologians were turning into dogmas. The thoughts on the naming of the animals by Adam that were advanced at the same time by St. Gregory of Nyssa (present Nish, Yugoslavia) met the same fate. Like Lucretius, he believed that man had invented speech himself. According to the others, God spoke Hebrew and taught it to Adam, and other languages were dispersed from the Tower of Babel, where a mean God had said (Genesis 11:7), "Come, let *us* go down and there confuse their language, that they may not understand one another's speech." It is not recorded who it was upstairs that their God was talking to, but the story of the tower and the name Babel probably reflect the Babylonian tradition of building tall ziggurats, temple towers used also for astrological observations. If one collapsed from neglect, lightning, or wind, it was easy to believe that an angry God had resented the potential intrusion into his territory.

The British monk Pelagius (360?–420?) had an entirely different view of theology, and traveled to Africa where he argued with Augustine, then on to Palestine and Rome. In 415 he was accused of heresy, acquitted in Jerusalem, but Emperor Theodosius I and Pope Zosimus objected strongly to his teaching and banished him from Rome. Pelagius's were real heresies—that there is no original sin, that Adam's sin has not left an inherited taint, that unbaptized infants are not damned, that man is responsible for his own salvation, and that he has complete freedom of will to make decisions for himself. How sobering to think of early heresies like these that were not just relatively minor infractions within the same belief system, but radically different, somewhat in the spirit of Unitarianism but especially overlapping the attitudes of the more recent churches of Christian Science and Religious Science. For many centuries, the teachings of Pelagius had no more chance of convincing anybody than had those of Aristarchos of Samos. Yet Pelagius was a very pious man, working hard for the cause of Christianity as he saw it; from *Holinshed's Chronicles* (1577) we have it that "There have been heretofore, and at sundry times, divers famous universities in this island, and those even in my days not altogether forgotten, as one at Bangor erected by (Pope) Lucius, and afterwards converted into a monastery . . . by Pelagius the monk." (Book II, ch. 6)

The year 415 in which Pelagius was first accused of heresy was also a year in which cruelty by early Christians reached a new peak in the disgraceful persecution and murder of Hypatia* by the hatchet-men of Archbishop Cyril of Alexandria (376–444). The good archbishop also persecuted the Novatians, expelled the Jews from Alexandria, and presided at the Council of Ephesus (431) in which Nestorius was condemned as a heretic. For these works, Cyril was promoted to Saint.

Meanwhile, the Roman Empire was fighting for its life. Rome was sacked by the Gothic chieftain Alaric and his troops in the year 410, and the Vandals plundered Spain in 420, crossing to Africa nine years later and laying siege to Hippo, where the seventy-five-year-old Augustine was still the bishop. He aroused the people to resist the siege and they did so for over a year. During that period he died. Further invasion of the Empire by Atilla the Hun and maybe as many as half a million troops was followed by the sacking of Rome by the Vandals in 455 and by the Barbarians in 476. With all the death and pain these wars brought with them came a theological question: "How could a Divine Providence allow these evils to happen to Christians?" St. Augustine had given an answer for this one as well, as soon as these invasions began. It was because many people of the Roman Empire continued to worship other deities and were promiscuous and otherwise sinful. In this Augustine spoke like some of the Hebrew prophets. But he wanted to have it both ways: Alaric had spared the Christian churches when he invaded Rome because the Christian God was the true God.

Another schism of the Church began to threaten from Constantinople a few years after the death of St. Augustine. Nestorius (died 451) preached that in Jesus Christ the divine and the human were not joined in the unity of a single individual, and that Mary should be called the Mother of Christ, not the Mother of God. Nestorius was deposed for heresy at the Council of Ephesus (431) and banished. His followers migrated to places such as Mesopotamia and even as far away as China, where they were out of reach of the Catholic Church, taking with them their version of Christianity and passing on information about the Greek culture to scholars in other lands.

A contemporary and friend of Augustine, St. Philastrius, made matters worse by his catalogue of heresies that guided so many priests and bishops a millennium later in their scramble for an entrance to heaven by torturing and killing the unorthodox. It was a heresy to deny that the stars are brought out by God from his treasure house and hung in the sky every evening, and any other view would be "false to the Catholic faith." It was a heresy to disagree with the orthodox view of the time of creation, or to doubt that an earthquake was the voice of an angry God. Some of the primitive beliefs of chapter 1 became part of Christian dogma. Yet at the time of St. Augustine's death, a ten-year-old boy in Carthage was beginning studies that would qualify him as a lawyer and in 470 as an encyclopedist. Felix Capella probably read what was available of the writings of Heraclides of Pontos; at any rate he supported the idea that Mercury and Venus circle the sun which, in its turn, along with the other planets, was thought to circle the earth. The old ideas hadn't died after all, though not much attention was paid to him.

At the end of the century the writings of Capella were completely overshadowed by those of an unknown author referred to as Dionysius, who supported St. Clement's biblical view of astronomy. One reason for his impact was that many people believed that he was *the* Dionysius—not the god of wine Dionysos, but the judge who was converted by St. Paul during his visit to Athens (Acts 17:34). He took the visions of Micai'ah (I Kings 22:19-23), of Isaiah (6:2), and of Ezekiel (chs. 1 and 10), added some Christian virtues, and established a hierarchy of thought that hovered over Christian beliefs until Dante Alighieri (1265–1321) crystallized it in his *Paradise*. Micai'ah (Micah) had reported that he had seen "the LORD sitting on his throne, and all the host of heaven standing beside him on his right hand and his left," and Isaiah had seen the seraphim, purifying ministers of Yahweh, each with three pairs of wings. Dionysius ascribed Christian love to them, and they became the highest angels of all. Ezekiel had seen the cherubim, and in Psalm 18:10, repeated in 2 Samuel (22:1–3), King David tells how the Almighty "rode on a cherub." They became the angels of light and of the knowledge of God, and ranked number two. So it went, down to the ninth rank of supernatural beings, the regular angels, the drones of the system, working hard and humming incessantly. This picture may be compared with that of John Milton:

> Where the bright Seraphim in burning row
> Their loud uplifted angel trumpets blow,
> And the Cherubic host in thousand quires
> Touch their immortal harps of golden wires.
>
> ("At a Solemn Music," lines 10–13)

The sixth-century traveler Cosmas Indicopleustes decided that angels were pushing the sun and planets around and opening and shutting the windows of heaven. His *Topographica Christiana* (547) supported the biblical account of the world's beginning and structure: The earth is a flat rectangle, four hundred days' journey from east to west, and two hundred from north to south. It is surrounded by four seas beyond which are huge walls holding up the sky ("Were you with him when he spread out the great sky, helping him hold it up?" Job 37:18)—a huge box modeled after the Jewish tabernacle. Lines from Genesis and the Psalms suggested that there was an upper story holding an enormous tank full of water. Genesis 1:7: "God made the firmament and divided the waters that were under the firmament from the waters that were above the firmament"; and Psalms 148:4: "Praise him, heaven of heavens and waters that are above the heavens." It occurred to Cosmas that the angels living up there not only pushed the stars and planets around but caused rain by opening taps in the tank every now and then. "He waters the hills from his chambers" (Psalm 104:13). Philo would have been very proud of this sophistry.

While Cosmas, working out of Alexandria, was traveling as far as India and Sri Lanka, then returning to write up these fantasies, we pause for a moment to recall some of the writings of his contemporary, John Philoponos (470–540), also of Alexandria, as reported in chapter 5. It seems a miracle, as the word was defined in chapter 7, that this lone figure of the first Christian millennium—and right in the middle of it—could criticize the physics of Aristotle and anticipate the physics of Galileo. The thinking of the time ignored not only his ideas but the very problems he was trying to solve. That someone thought of them means that others could have followed up and infused some light into the Dark Ages, but Christian monkery (as in translation Mohammed was to call it a few decades later) stood in the way.

Two fourth century bishops of Caesarea in particular, Eusebius and then Basil, made their opinions of scientific enquiry very clear with statements such as "we think very little of these matters, turning our souls to better things" and "it is a matter of no concern to me." John Philoponos was such an anomaly that doubts can be raised as to whether he existed at all. Was he the "Piltdown Man" of science history, fabricated by someone much later as a joke upon historians? Not likely at all. Was he invented by someone in the sixteenth or seventeenth century who had discovered how wrong Aristotle was about motion but knew that his life would be in danger if he published such heretical ideas under his own name? As discussed earlier, an early fifteenth century alchemist by name Basilius Valentinus was indeed fabricated as the author of some seventeenth-century scientific studies in order to protect a real author's life. However, it is more likely that Philoponos had read Lucretius. He was a real miracle too.

In the year 527 Justinian became emperor of the Eastern Roman Empire, and his laws of the land recognized the ecclesiastic leadership of the Roman Church. Roman Catholicism became the state religion and all dissenters were to have their rights and property, and sometimes even their lives, forfeited. He also closed the Academy in Athens that Plato had founded centuries before. Its philosophers, led by Damascios* and Simplicios*, fled to Persia, taking with them some knowledge from their Greek predecessors. Many years later, Dante reported in his *Divine Comedy* that, during his visit to Paradise, Justinian had confessed to him of having lapsed, not from faith but from the correct dogma, and having been set straight by Pope, and Saint, Agapetus I:

Caesar I was;
And am Justinian. . . . I did hold
In Christ one nature only; with such faith
Contented. But the blessed Agapete,
Who was chief shepherd, he with warning voice
To the true faith recall'd me. I believed
His words; and what he taught now plainly see. . . ."

(*Paradise*, Canto VI)

The direction of nearly all attempts throughout Christendom to understand anything at all about nature was thereafter determined by the Church for centuries. Doubt became a sin, Greek knowledge was replaced in most centers by Genesis, inquiries about this world were replaced by thought about the next. However, the spirit of medical research at Alexandria, if not its quality, had persisted. In the fourth century Flavius Vegetius had described pioneering studies in veterinary science, and medical books gave surgical procedures and other treatments of eye, ear, nose, and tooth problems. A publication in 395 had recommended that to avoid unwanted pregnancies women should carry a rabbit's foot! Alexander of Tralles (525–605) came from his home town in Lydia to study medicine at Alexandria. Some years later, he produced twelve books entitled *The Art of Medicine*, describing the state of knowledge of the time, emphasizing his own specialties of internal medicine—e.g., diseases of the lungs and the nature and eradication of intestinal worms. He was probably the first parasitologist.

A little later St. Isidore of Seville (560?–636) alleged that, before the Fall, the sun and moon were much brighter, but at the Second Coming the great lights mentioned by Isaiah will shine again in their former glory. The learned St. Isidore was the author of a large medieval encyclopedia. Dr. A. D. White quotes this excerpt from it: "bees are generated from decomposed veal, beetles from horseflesh, grasshoppers from mules, scorpions from crabs." We can hardly call it evolution, and it is certainly dead wrong, but at least, like earlier writings of St. Augustine and others, it got away from the idea that all living things actually appeared almost simultaneously at the time of creation. Over in England, Bede (673–735)—later "The Venerable," and even later "The Saint"—had this to say: "Fierce and poisonous animals were created for terrifying man, in order that he might be made aware of the final punishment of hell."

This belief and many others equally erroneous formed the intellectual legacy that Charlemagne would inherit in the next century. In Mecca and Medina, however, Muhammad (570–632), a contemporary of St. Isidore, had sparked a religious and political revolution that in its turn would produce a renaissance of learning. We therefore follow the Nestorians, neo-Platonists, and other refugees from Christendom to the Arab worlds of the sixth and following centuries.

1. JULIUS CAESAR 2. AUGUSTUS 3. COLUMBUS 4. SENECA 5. THEODOSIUS I 6. ATTILA

Attempts to understand the universe often involved a struggle between authoritarian despots and enlightened pathfinders. Sometimes the despots were a mix of repression and enlightenment. The Caesars, for example, were usually more interested in maintaining the status quo of tradition than in new ideas. However, Lucretius lived under Julius Caesar's reign and Augustus supported the arts (1, 2).

In the Renaissance, Ferdinand and Isabella expelled the Jews and Muslims from Spain in a frenzy of religious intolerance, yet they supported Columbus's (3) voyage to the New World. Columbus's discovery was also a mixed blessing. He advanced our thinking about the shape of the earth, but at the same time he opened up the New World to colonialism and the plundering of native peoples.

Among the truly unenlightened despots was Nero, who forced the Stoic philosopher Seneca (4) to commit suicide. Another was Theodosius I (5) who banned the books of some of the great thinkers of the classical age because they deviated from Christian tradition. And, of course, Attila (6) surpassed them all as the very epitome of destructive barbarism.

10

Caliphs, Scholars, and Linguists

One day when the Sultan was in his palace at Damascus a beautiful youth who was his favorite rushed into his presence, crying out in great agitation that he must fly at once to Baghdad, and imploring leave to borrow his Majesty's swiftest horse.

The Sultan asked why he was in such haste to go to Baghdad. "Because," the youth answered, "as I passed through the garden of the Palace just now, Death was standing there, and when he saw me he stretched out his arms as if to threaten me, and I must lose no time in escaping from him."

The young man was given leave to take the Sultan's horse and fly; and when he was gone the Sultan went down indignantly into the garden, and found Death still there. "How dare you make threatening gestures at my favorite," he cried; but Death, astonished, answered, "I assure your Majesty I did not threaten him. I only threw up my arms in surprise at seeing him here, because I have a tryst with him tonight in Baghdad.

—Edith Wharton, *A Backward Glance*

In our largely Christian society we learn so much more about the early history of the Jews than we do about that of the Greeks, Romans, or Egyptians from whom we have inherited so much; and it is as if the ancient people of Persia, the Far East, and the rest of the world never existed. Nor do we hear much about the Jews after the birth of Jesus, at least until recent times. From a public elementary school and a Methodist Sunday School in then-isolated Melbourne, Australia, I came away with the impression that there were Jews in the early days but that after Jesus came they rapidly became extinct and were replaced by Christians. There was also a mysterious group called "The Heathens," not to be confused with a place up in the sky called "Heaven," because they certainly weren't going there.

Yet some of these heathens contributed beautiful poetry that ranks with the best of the Psalms, while others had deep religious experiences that matched those of the Hebrew prophets or the Christian saints. Some worshiped a God as powerful and just as Yahweh, with more tenderness and mercy, although not as merciful as Jesus described Him. Parts of the Koran of Muhammad (570?–632) read like the Old Testament:

> In the Name of God, the Compassionate, the Merciful:
>
> All that is in the heavens and the earth magnifieth God, and He is the Mighty, the Wise.
>
> His is the kingdom of the heavens and the earth. He giveth life and giveth death. He is powerful over all things.
>
> He is the first and the last, the seen and the unseen, and all things doth He know.
>
> It is He who created the heavens and the earth in six days, then ascended the Throne; He knoweth what goeth into the earth and what cometh out of it, and what cometh down from the sky and what riseth up into it, and He is with you, wherever ye be; and God seeth what ye do.
>
> His is the kingdom of the heavens and the earth, and to God shall all things return.
>
> He maketh the night to follow the day, and He maketh the day to follow the night, and He knoweth the secrets of the breast.
>
> (Koran, Sûrah LVII, 1–6)

Perhaps this is even more familiar:

> And We said: O Adam! Dwell thou and thy wife in the Garden, and eat ye freely (of the fruits) thereof where ye will, but come not nigh this tree lest ye become wrongdoers. But Satan caused them to deflect therefrom and expelled them from the (happy) state in which they were.
>
> (Koran, Sûrah II 35, 36)

These verses could have been in the New Testament:

> Feed the hungry, visit the sick, and free the captive if he is unjustly bound.
>
> A man's giving in alms one piece of silver in his lifetime is better for him than giving one hundred when about to die.
>
> ("Of Charity")

> The most excellent of all actions is to befriend any one on God's account.
>
> (from the *Table Talk* of Muhammad)

Why do we read the Bible over and over again and yet never, never read these words? Let us recognize the common origin of the Muslim, Christian, and Jewish religions. It may not stop us from tearing each other apart, but it may help. Muhammad continues:

> We sent Noah and Abraham, and we gave their seed prophecy in the Scripture: and some of them are guided, but many are disobedient. Then we sent our apostles in their footsteps, and we sent Jesus the Son of Mary, and gave him the Gospel, and put in the hearts of those that follow him kindness and pitifulness; but the monkery, they invented it themselves! We prescribed it not to them—save only to seek the approval of God, but they did not observe this with due observance.
>
> (Koran, Sûrah LVII, 26–27)

Muhammed knew about the early Christian clergy and he didn't like what he saw.

Muhammad's father-in-law Abu-Bakr (572–634) succeeded him for a couple of years immediately after his death, and then came Omar (581?–644) who in ten years as the second orthodox caliph extended the teachings of Muhammad through conquest of Persia, Syria, Palestine, and Egypt.

Omar and his troops flourished on a vegetarian diet, mostly barley or barley-bread, salt, and dried fruits, with only water as a drink. No meats, no wine or liquor. Omar was the undisputed ruler, his symbol of office a simple stick that he carried, a stick that was feared more than a sword.

Jerusalem came under Muslim domination in 637, and within the next four years the ancient Persian empire was forced to yield. Inspired by the new religion, this was the first "holy war" of Islam. This conquest of Persia was sharply criticized later by the scientist-historian al-Biruni in the book on the history of the region that he published in the year 1000. Omar allowed many Jews, Christians, and even Persian Zoroastrians to continue to practise their various faiths, however, provided that they paid a tax for the privilege. We will not follow the details of the next twenty years of caliph assassinations and civil war that led to the split between the Shiites and the Sunnis, and to the establishment in 661 of the capital of Islam in Damascus under the control of a family by the name of Umayyad.

Before the end of the century Armenia, the island of Rhodes, and all of North Africa had also become Muslim territory, and Constantinople and Asia Minor were threatened. By the year 711, the Saracen expansion had begun to engulf nearly all of Spain, and by 732 the invaders had spilled over the Pyrenees into France. Just one hundred years after the death of the Prophet Muhammad, his teachings had been spread over an enormous area, for he had inspired what many Jews had hoped that Jesus would inspire—military might.

The Muslims' ambition to conquer the Franks was a turning point, however, because Charles Martel (689?–741) and his army thoroughly defeated them when they met in the region between the cities of Tours and Poitiers. Some battles have an enormous effect on future generations —this was one of them. Had Abd-er-Rahman's troops succeeded, perhaps Europe, England, and eventually America would have become Muslim. It makes us recognize how the choice of our particular form of religion is primarily determined by historical circumstances and the decisions of military leaders.

This was the beginning of a long struggle between the French and the Arabs, through the Crusades and on into the middle of the nineteenth century, when Marshal Bugeaud's campaign in Africa against the troops of Abd-el-Kader inspired the London *Punch* to write in 1846:

> There can't be the slightest question
> That the French are a very great nation,
> But the special line
> In which they shine
> Is the spread of civilization.
>
> . . . .
>
> Some millions of men and money
> An insignificant price is
> For instructing shoals
> Of Mahometan souls,
> In the practice of Christian vices.
>
> . . . .
>
> The world says Bugeaud's troopers
> Give no quarter at all in battle,
> But 'tis false, I declare,
> They *often* spare
> The women, and *always* the cattle.
> With these philanthropic feelings,
> Can anything be harder
> Than that glorious France
> In each advance
> Should be thwarted by Abd-el-Kader?"

(*Punch* 10: 97)

—and much more. It hadn't been very long since the battle of Waterloo, and making fun of the French military exploits was regarded as a legitimate sport.

We recognize that there was, and still is, an overlap between Muhammadanism and Christianity—a common belief in the prophets of the Old Testament and in the One God who spoke to them, and a belief in the last day of judgment, with only good Muslims or Christians, as the case may be, making it to heaven, but with all Christians and bad Muslims (or is it all Muslims and bad Christians?) bound for hell. The continued fighting of Muslims with each other, and of Christians with each other, along with Muslim-Christian battles, all of them lasting to this day, have unfortunately given to so many young men the premature opportunity to discover which eternal fate was to be theirs.

Muhammad transformed the moral behavior of all parts of the world that came under his influence. He abolished drunkenness, idolatry, gambling, and, for a while at least, astrologers and soothsayers. He established a belief in angels and devils more widespread than among Christians, and he went down in history as a worker of miracles without which it is impossible to be a god or a major prophet. As a child the radiance from his face would blind anyone who did not divert his or her eyes; he could see in the dark because wherever he looked was illuminated; he never suffered from the heat of the desert, because there was always a small cloud that placed itself to protect him from the sun. There are many other stories of his miraculous powers, believed in by Muslims as fervently as Christians believe in the miracles of Jesus.

The ambiguities of any religion always lead to dissent and to the formation of a number of fervent sects, and the theological verbal battles in Christendom had their echo in Islam. Debates had arisen about the Koran, the holy book of Muhammad. Was it eternal? Had it always existed in the mind of Allah? A group called the Mutazilites boldly answered these questions in the negative. Where the Koran contradicted reason, they asserted, it must be treated as an allegory rather than as a literal truth. They thought that it was ridiculous to ascribe human physical attributes and emotional attitudes to Allah. The difference from the Christian dissenters was that the Mutazilites received support from the Caliph of Baghdad.

Back in the seventh century, al-Abbas (567?–653), a paternal uncle of Muhammad had made a lot of money. By purchase or inheritance he obtained the right to distribute to the pilgrims the holy water from the well of Zamzam in the sacred region of Mecca. By tradition, it was water from the spring that was found by Hagar when, pregnant with Ishmael by her master Abraham, she was fleeing from Abraham's wife Sarah (Gen. 16:7). The descendants of Abbas preserved and multiplied the wealth and remembered their genealogical relationship to the Prophet. Just a hundred years after the Umayyads had established the caliphate in Damascus, these Abbasids overthrew them and established their own dynasty of caliphs in Baghdad. If we remember Edith Wharton's story, Death made his many appointments with all of the Umayyad princes in both Damascus and Baghdad, except for the nineteen-year-old prince Abd-er-Rahman, who escaped to Spain. (This was not the same man as the Abd-er-Rahman whose troops had been defeated by Charles Martel.) Under these new rulers there would be an early renaissance in Baghdad—several hundred years of Abbasid prosperity, conflicts, and then dissipation, years of superb architecture and first-class scholarship, with an independent Umayyad caliphate descended from Abd-er-Rahman rising in southern Spain and challenging Baghdad for intellectual, literary, and architectural supremacy.

It was the second Abbasid caliph al-Mansur (712?–775) who transferred the caliphate to the new city of Baghdad. He was a patron of learning, encouraging and supporting Greek and Latin classics into Arabic. He commanded Muhammad ben Ibrahim al Fazari (d. 800) to translate from Sanskrit to Arabic some then-recent works by the astronomers of India. He started the tradition of official support for knowledge and research.

Harun al-Rashid† (764?–809), the fifth Abbasid Caliph of Baghdad, the caliph of the "Arabian Nights," is included in our list of "Helpful Despots" for his support of medicine and the arts, and for his own scholarship, but he was nonetheless a very tough and cruel tyrant. He wisely left much of the administration of the realm to his tutor Yahya, so that he could devote some of his time to teaching the Christian Byzantine Emperor Nicephorus I a lesson (he had refused to pay tribute). He also ordered and witnessed the execution of his own sister, and the beheading of her husband, Jafar, Yahya's son, with whom Rashid appears to have had an intimate relationship. In these days of openness about sexual relations, and of originality of fashion, it is surprising that the Harun-Jafar cloak has not been re-invented. It was a wide cloak with the usual number of sleeves but with two collars, through which two heads could protrude. The two men sometimes went about like this, with an arm of each to address the world, and presumably an arm of each around the other, under the cloak. Such an outfit could of course be used by a man and a woman, or two women, as well as two men, reminding us of the story of creation as related by Aristophanes in chapter 3.

With it all, poetry, storytelling, and sexual activities intrigued Harun—he sometimes offered huge rewards, like money and slave girls, to a man who wrote a poem that he liked, and a woman who told a good story would be offered a bed for the night. He set the stage for his son and successor, al-Ma'mun† (786–833) who used the prosperity that he inherited to vastly expand the caliphate's support for medicine and science, and to encourage learning and freedom of enquiry. Freedom up to a point, of course. In his support of the Mutazilites al-Ma'mun issued an edict that stipulated Muslims must believe that the Koran was not eternal, with severe punishment to those who thought otherwise. After al-Ma'mun died, however, his later successor (847), al-Mutawakkil, re-established the orthodox interpretation, relying on his Turkish guards to convert or kill. Edicts against Jews and Christians were reinstated, and the relation between Christians and Muslims became even more polarized. The influence of the two great caliphs remained, however, in their emphasis on learning and the arts.

By the year 830 al-Ma'mun had set up in Baghdad an academy, an observatory, and a library full of Greek and Hindu manuscripts. With some donors, the gifts stop with tangible objects like these, and they are very helpful to society, but al-Ma'mun went further. He recognized that these buildings and resources would be useless if they were not occupied by scholars, so he issued invitations to physicians and philosophers from as far away as India to come and participate, presumably helping them financially. Science, medicine, and mathematics began to really flourish in Baghdad and other cities of the realm. Elementary education was available to all at a minimal cost, with no thought of sparing the rod or spoiling the child. From the age of six, simple Muslim theology was taught, along with ethics and history; for high schools, grammar, mathematics, astronomy and other subjects were added. At a higher level, the student was expected to travel to Mecca, Baghdad, Cairo, and Damascus to learn at first hand from the scholars in those cities.

Two books by al-Asmai (739–831), of Baghdad and Basra—*Savage Animals* and *The Making of Man*—are evidence of early studies in veterinary science and human anatomy. Later in the century al-Jahiz (d. 869) wrote the *Book of Animals,* which contains very preliminary ideas on evolution, animal psychology, and adaptation. He also discovered how to extract ammonia from animal offal, and he examined problems in anthropology and the natural sciences. His contemporary, al-Hasib (770–864), was more interested in astronomy, developing astronomical tables and making geodetic surveys at the caliph's "request." (Then, as now, sponsors of research often required a practical result for their philanthropy.) This work was followed up and greatly extended by a mathematician from Khwarazm (now Khiva, in the southwest corner of Soviet Russia). Muhammed ibn Musa (780–850) was named al-Khwarizmi, after his birthplace, and in later Latin translation it became *algorismus,* from which our word "algorithm" is derived. His astronomical tables, with revisions, became the standard for centuries throughout the Muslim world. He used Hindu numerals, derived from his familiarity with the works of the Hindu mathematician Brahmagupta (598–660), and wrote extensively on geometric and algebraic solutions of quadratic equations. It was the consolidation—al-jabr—of this work that led him unwittingly to feed another word into our language—"algebra."

To illustrate the spirit of the times, we take note of three brothers, by name al-Hasan, who also lived in Baghdad during the first part of the ninth century. They had an observatory in their home, studied and supported the sciences, and jointly wrote books on the balance, and on the geometry of the sphere. The old atomic theory had been examined for the first time since Lucretius, with an essential modification—the Almighty continually intervened to change the atoms and their motions. The writings of Ibrahim al-Nazzan, who died in 845, show that he rejected this arbitrary addition to the theory, but it would be a long time before the atomic theory would surface again.

The philosopher/scientist al-Kindi (803?–873) studied mathematics, optics, astronomy, meteorology, and anything else that was available, including the speed of a falling object. His biggest impact at the time, however, was in the areas of philosophy and religion, where he joined with the Mutazilites in their heretical opinions that nearly cost him his life. Like the earlier, but mostly later Christian scholastics, he tried to reconcile Greek philosophy with faith, in his case, of course, the faith of Islam.

At that time, Abu'l Farghani was preparing a text on astronomy that would be used for about seven hundred years, and translations from the Greek were stimulating Muslim interest in medicine and science. Ibn Masawaih (777–857) translated Greek medical works into Syriac and at the same time he wrote an early treatise on ophthalmology. He found favor with the Caliph of Baghdad (in this case al-Mutasim, immediate successor to and older brother of al-Ma'mun) who provided apes for him to dissect. His student Hunain Ibn Ishaq al-Ibadi (809–873) was a Nestorian Christian who became the head of Mamun's Academy. He was also a physician who, with later help from his son Ishaq, or Isaac, translated into Syriac and Arabic some two hundred works of the ancient Greeks—Aristotle, Plato, Hippocrates, Galen, Dioscorides, and others. These were mostly on medical subjects, but also included were Plato's *Republic* and Aristotle's *Physics,* along with the Septuagint form of the Old Testament. Life under these caliphs could be very rewarding, provided that you didn't offend them. Hunain had received from Mamun an amount of gold equal in weight to that of the books he had translated, and from Mutawakkil a year's jail for refusing to prepare a poison for one of the caliph's enemies.

All of this newly discovered knowledge, coupled with the enthusiastic support of scholarship by some of those who were in control, made ninth century Baghdad a very exciting place. Some knowledge of Greek discoveries had been brought earlier, primarily by the Nestorians and other refugees from the Christian suppression of unorthodoxy, but many more details were needed to start an intellectual revolution. Translations continued, notably by Thabit ibn Qurrah (836–911) and his students, who brought Euclid, Archimedes, and other mathematicians and physicists to the Arab world. His contemporary, al-Balabakki (d. 912), translated the works of Heron, Autolycos (who had challenged the rotating-spheres model of the solar system by Aristotle and Eudoxos), and other mathematicians. He also contributed an original paper on the astrolabe, an elegant and impressive instrument for locating the stars and planets. A contemporary was Al-Battani (858?–929) who made very significant astronomical observations from an observatory at Ragga, on the Euphrates, in particular correcting Ptolemy's value for the precession of the equinoxes. He also demonstrated that the solar apogee, the point at which the earth and the sun are at their maximum separation, had shifted since Ptolemy's time. That is very impressive, indeed.

While astronomy was developing like this in Mesopotamia, the great physician al-Razi, or Rhazes (865?–924?), came from Ray to be head of a hospital in Baghdad, having been supported by the dynasty of Saman† in East Persia. His twenty-volume *Comprehensive Book of Medicine* became a textbook that was used for centuries. Rhazes made the first accurate studies of the infectious diseases smallpox and measles, and he contributed to the study of gynecology, obstetrics, and ophthalmic surgery. He may have been the first to use plaster of Paris for casts and animal gut for sutures. He also divided the world into the classification of chapter 6—animal, vegetable, and mineral.

The miracles of the medicine of India were transmitted to Islam by Indian doctors who came to al-Ma'mun's academy in Baghdad. India had a long tradition of first-class medicine, surgery, and hygiene. From the fifth century B.C. there is a book by Sushruta on surgery, obstetrics, drugs, infant feeding, and hygiene, written in Sanskrit at the University of Benares, India. His surgery, and that of his colleagues, was spectacular. He described 121 surgical instruments, from lancets to speculums, gave patients drinks to cut sensitivity to pain, performed abdominal operations, and he pioneered the surgical reconstruction of the nose and ear by grafting on some flesh from another part of the body. He recognized the importance of dissection of human cadavers, although the Brahman religion was opposed to it.

From this superb start, medicine in India had progressed through Charaka, its second-century A.D. Hippocrates, and on through the centuries, often ahead of the rest of the world—at least in some areas. An early medical text from 540 A.D. states that doctors knew how to "take the fluid of the pock on the udder of the cow upon the point of a lancet, and lance with it the arms between the shoulders and the elbows until the blood appears; then, mixing the fluid with the blood, the fever of the smallpox will be produced." (Du I, 532) How many people died of smallpox in Europe during the twelve hundred years that elapsed before vaccination was rediscovered, improved, and spread widely by Edward Jenner? But the idea of vaccination was not injected into Muslim or Christian societies, or if it was, it didn't take. Perhaps the Brahman worship of cows and bulls had made it a sacrilege.

In India, cleanliness, tied to the Brahman religion, was of paramount importance; but in China, despite some progress in medicine, the hygiene, sewage disposal, and other public health developments were primitive. The Persians held cleanliness in high esteem, and it became a part of their religion of Zoroastrianism. The ancient Egyptians had practised circumcision and in later times became so compulsive about internal cleanliness that they made themselves sick by overdoing the use of the enema. In some areas they had sewers—their copper pipes have been found. However, it was the Jews who were ahead of all others in preventative medicine through hygiene and careful diet. These essential parts of their religious rites did much to save them from pestilence and assist them through countless conditions of slavery. In Christian Europe, as we have seen, emphasis was placed more on purification of the soul than cleanliness of the body and the environment, while for the Muslims, as for the Indians, cleanliness was part of godliness.

On escaping from the Abbasids back in 750, Abd-er-Rahman (731–788) had traveled via North Africa, picking up enough support to assume the title of Emir of Cordova. His successors, including numbers II and III of the same name, consolidated their position, beautifying the city with new mosques and other structures, and overpowering efforts by neighboring cities and individual fiefdoms to assert their independence. In 929, Abd-er-Rahman III declared himself Caliph, the first Umayyad Caliph since the rout by the Abbasids in 750, and thirty-two years later his son al-Hakam II† (913?–976), inherited the small but prosperous empire of Andaluz, with its capital at Cordova.

The new caliph lost no time in extending the work of his father, enhancing the beauty of Cordova and founding an institute of learning at the University. He acquired an enormous library, gathering manuscripts from Greece, Egypt, and India, and supporting scholars in medicine, astronomy, and mathematics. The old feud between the Abbasid and Umayyad families became less like that between the Montagues and the Capulets of *Romeo and Juliet* fame, or the Hatfields and the McCoys of turn-of-the-century West Virginia and Kentucky, and more like the competition between Oxford and Cambridge, or Harvard and Princeton.

Unfortunately, al-Hakam's son and successor, Hisham II, was less sympathetic, destroying all books in the great library that appeared to conflict with the Sunni creed. Many of the scholars fled to Baghdad or Damascus. In Damascus the physicians among them met twenty-four colleagues in a hospital that had been founded by the Umayyads in 706. Those physicians who fled to Baghdad, however, had approximately a thousand licensed doctors to work and compete with, or so it is reported. There was indeed enormous activity in medicine there, more or less presided over by Ibn Thabit (860–940), a physician to three caliphs from 908 until his death. He became the administrator of the Baghdad hospitals and held the physicians to very high standards. He insisted on examinations for would-be doctors, issuing licenses only to those who passed. Five years after he died Baghdad was invaded by the followers of a Caspian chieftain Buyayh, and for the following hundred years his descendants were really the ones in charge of Baghdad. Other despots established themselves in Cairo (Egypt), in Ghazni (present Pakistan), and elsewhere. Here we remember only these two because of their support for scientific enquiry—al-Hakim† (985–1021), Caliph of Cairo, and Mahmud† of Ghazni (998–1030), both dead in their thirties.

You had to be tough, cruel, and paranoid to rise as a military leader, and in these qualities al-Hakim excelled. This "Mad Caliph," as he was called, persecuted Jews and Christians unmercifully, destroying their places of worship, culminating this hatred with the destruction, in 1009, of the Church of the Holy Sepulcher in Jerusalem. According to tradition, this was built over Christ's tomb, and the Christian world was naturally appalled. Although it would be another ninety years before the first Crusade, the memory of this contributed towards the success and horror of that expedition. Al-Hakim set up an institute in Cairo to teach the theology of the Shiites, but he also sponsored medicine and astronomy.

Four great Arab scholars were alive when the Big Clock struck the end of the first Christian millennium—Ibn Yunus (950?–1009), Ibn al-Haytham (Alhazen, 965?–1039?), al-Biruni (973–1048), and Ibn Sina (Avicenna*, 980–1037). Ibn Yunus, working in Cordova and Cairo, contributed to astronomical theory by studying the small change in the angle between the planes of the earth's equator and the earth's orbit (called "the obliquity of the ecliptic"). He also solved problems in spherical trigonometry, applying them to observations of the stars and planets, and he summarized two hundred years of Muslim astronomical observations in the Hakimite Tables, in honor, not of the generous al-Hakam II, but of his sponsor, the Caliph of Cairo.

Al-Hakim had heard of Alhazen, too, and was naturally intrigued by his scheme to regulate the flooding of the Nile. Unable to deliver on this task, Alhazen suddenly discovered that his many talents included acting, not on the stage but in real life, for his life depended on it. He managed to avoid execution by pretending for years that he was insane, at least when anyone he wasn't too sure about was looking. Apparently al-Hakim was convinced that he was just another nut, but with it all he became the dominant figure in Arabian physics, particularly with his studies of light and optics. He explained reflection and refraction, binocular vision, and focusing with lenses and with spherical and parabolic mirrors. He studied the rainbow and invented the pin-hole camera (without the film). He understood the concepts of atmospheric pressure, and atmospheric refraction of light. He noted that, from where he was, the last flicker of sunset, scattered from the atmosphere, occurred when the sun was about a third of a radian (19°) below the horizon. Knowing the distance to the horizon, he was then able to give the first estimate of the height of the atmosphere (10 miles), actually the height above which the scattering was too small for him to see. His descriptions of his many ideas and discoveries had a major impact on European scientists who came later, Roger Bacon and Johannes Kepler in particular.

Al-Biruni was a universal scholar, into just about everything known at the time except medicine, and anticipating Francis Bacon by six hundred years with his scientific philosophy. He wrote on the history of the Arabs, criticizing their overthrow of the Persian king Yezdegird III in the year 641, and with him the Zoroastrian religion, the practice of which the king had restored to its earlier purity. He wrote on astronomy and physics, using a theorem of Euclid and his knowledge of the spherical shape of the earth to determine the earth's radius. If applied to the sphere of the earth, or to a plane section of it through its center, the geometrical theorem told him that when standing on a mountain overlooking the sea and looking to the horizon, you could calculate the diameter of the earth by squaring the distance to the horizon and dividing it by your height above sea level. Thus, if you could see a distance of 45 miles from a height of a quarter of a mile, the diameter of the earth would be 8100 miles. There are problems, of course, mainly in measuring the distance to the horizon and in the bending of light by refraction in the atmosphere, since Euclid's theorem applies to straight lines. It was a good idea, though, and, knowing the diameter of the earth, you can turn the argument around to estimate the distance to the horizon if you know how high you are above the water surface (from a height of four feet the horizon is approximately two and a half miles away, according to this simplified picture).

At present, Ghazni is a city in central Pakistan, some five hundred miles south of Balkh, now in Afghanistan, where the Saman dynasty had supported scholarship during the tenth century. As their power diminished,

Mahmud of Ghazni (998–1030) expanded the empire that had been established by some of the Samanid guards, with Ghazni as center. Mahmud was particularly interested in extending the empire to Persia in the west, and invading and ravishing India to the east. He wanted al-Biruni with him, and that learned scholar traveled throughout India and then spent the last thirty years of his life in Ghazni writing a history of India, with emphasis on the religions of the country and on the work of its astronomers. He translated from Sanskrit into Arabic and vice-versa, wrote about precious stones, and measured their specific gravity, and realized from his travels that, since the earth is round, all objects must be attracted towards the earth's center. Biruni was not the only scholar to enjoy the support of the tyrant Mahmud, who built an enormous mosque and a university in Ghazni, with a library impressive both for its appearance and for its contents.

Often throughout Muslim history poetry was regarded very highly and richly rewarded. Mahmud included on his staff the great poet Firdausi (941–1020), the Persian Homer, the author of the *Book of Kings*, sixty thousand rhyming couplets that told of the legends and history of Persia up to the fatal year 641. Although he had dedicated this tremendous work to Mahmud, Firdausi did not receive anything like the pay that he thought it deserved. After making his displeasure known, and writing a satire about his despotic sponsor, he fled to Herat, several hundred miles northwest, in Afghanistan. You didn't get away from a man like Mahmud that easily, though, and Firdausi moved to Mazanderan, near present Tehran, and then to Baghdad. He wrote another poem about the trouble that Joseph got into with Potiphar's wife Zuleika, sometimes known as Phraxanor. It was a well-known story, since Muhammad had written about it in the Koran. Muslim commentators had enriched the story with the appearance of the angel Gabriel (looking like Joseph's father Jacob), just as Joseph was about to yield to her advances. The Koran says merely that "he saw the demonstration of his Lord." Firdausi didn't get much for this poem either, even though Mahmud "forgave" him, and his last years were spent in abject poverty. Seven years after Mahmud died, his empire was overrun by another Turkish group, the Seljuks, whose leaders, operating from Baghdad as their capital, called themselves the "masters," which we translate as "Sultans."

The other genius we have referred to, and who also straddled the year 1000, was Avicenna, in many ways the most illustrious and most aggressive of the Oriental Muslim scholars. At the age of seventeen his medical knowledge had enabled him to save the life of the Emir of Bokhara, but his fame soon spread to Mahmud, in Ghazni, and the command was issued for Avicenna to join the other intellectuals at the institute there. Unlike

Biruni, Avicenna refused, but he too found out that you didn't cross Mahmud, who circulated his picture—"Wanted"—throughout Persia, and offered a reward. He escaped to the desert, traveled widely, cured the illness of the Emir of Hamadan and was appointed vizier in that town. Wherever he went, there were many who simply hated him, and, disguised, he escaped from Hamadan, ultimately finding a haven several hundred miles southeast in the court of the Emir of Isfahan, also in Persia, present Iran. With all this and much more, Avicenna found time to write many treatises on philosophy and science, as well as a very comprehensive book on medicine. This *Canon of Medicine* was a systematic compilation of the entire theoretical and practical medical knowledge of his day. He studied drugs; analyzed the causes, symptoms, and cures for many diseases; recognized that the retina, not the lens, is the basic organ of vision; and developed methods of therapy for nervous disorders. He based his work on that of Hippocrates and Galen, but brought it up to date with more recent treatments and drugs:

> One ought to obtain perfection in this research, namely, how health may be preserved and sickness removed. And the causes of this kind are rules in eating and drinking, and the choice of air, and the measure of movement and rest; and doctoring with medicines; and doctoring with the hands.

Three and a half centuries later, Geoffrey Chaucer would remember Avicenna's treatise on drugs and poison as the Pardoner tells the tale of two murderers who were tricked into drinking poison:

> But certes, I suppose that Avycen
> Wroot never in no Canon, ne in no fen,
> Mo wonder signës of empoisonyng
> Than hadde these wrecches two, er hir endyng.
>
> (*Canterbury Tales,* lines 889–892)

In 1628, William Harvey's *On the Motion of the Heart and Blood in Animals* gave credit to Avicenna's remarks about tumefaction, or swelling of the veins and arteries.

Avicenna also recorded some interesting ideas about the physical sciences. He suggested that light moves with a finite velocity, although of course there was no way to test this hypothesis at the time. Like Strabo he believed that both volcanoes and flooding had determined the shape of the earth's crust, recognizing that this process must have taken a very long time, and that the earth itself must therefore be very old. Like Xenophanes, also referred to in chapter 6, he recognized fossils of marine animals as proof that the regions where they were found had once been inundated.

Avicenna is also known to us as a philosopher because of the effect his works had on the Christian scholastics who read them when they were translated into Latin during the following two centuries. Greatly influenced by Aristotle and Plato, he involved himself in the ancient chicken-and-egg problem as to which came first—the individual thing or the Platonic Idea of the thing—man first, or the idea of man first? He argued for the idea existing before the thing, since it existed in the mind of God, while for mortals, we have first to see the thing and then get the more general idea. Avicenna also developed the argument that we hear so very often today: that intelligent life could not be in existence without an intelligent Creator, in accordance with the time-honored invention of a god whenever there is something that is not understood. The Muslim belief in the resurrection of the body was to him a pragmatic concept that would attract more followers than the more nebulous belief in the resurrection of the spirit. He may have heard that Christians were having a lot of success with the idea also.

There were so many more scientists and physicians in those intellectually heady Muslim days of the tenth and eleventh centuries. Over in Cordova, Abdul Qasim al-Zahravi (936-1013), physician to Abd-er-Rahman II, and a great surgeon, was the author of a number of books which, translated into Latin, became the standard texts on surgery for hundreds of years. He was followed by al-Iraqi (961–1013), physician to al-Hakam II and the author of a medical encyclopedia that described bone and eye surgery, ligatures of the arteries, cauterization, and much more. In Egypt, the physician al-Misiri (c. 998–1061) wrote a commentary on the works of Galen, and a book on the art of medicine.

In physics and astronomy, Abul Wafa (940–998), working at Baghdad, had introduced to trigonometry the ratios which we call "tangent" and "cotangent," and generalized the analysis to spherical trigonometry, useful for his astronomical studies. He made extremely accurate measurements of the moon's motion, studying small irregularities which we now know are due to the gravitational forces of the sun and the planets. In the next century, Abu Barakat (1077–1164), a Jew converted to Muhammadanism, primarily a philosopher, was the first person to see that acceleration can arise from a constant force, and to study the motion of projectiles with this in mind.

One could go on naming al-this and ibn-that and abu-the-other—names unfamiliar to our English-thinking minds—and we will continue to do so. At this point, however, we may pause to ask ourselves why there were so many first-class Muslim scientists, physicians, and poets during the eighth to the thirteenth century of the Christian era, with relatively few before or since. Muslims weren't any smarter during those five hundred years than at other times, and no more or less intelligent than their contemporary Christians. It all has to do with who is in charge, and what the priorities of a particular civilization are. Free inquiry was encouraged by some of the Muslim caliphs, who saw no conflict between science and their religion. The Christians, on the other hand, got so entangled in their theology and their struggle for existence that their leaders became intellectual dictators, as some of them still are today. It was, and is, exceedingly difficult for free inquiry to take place in that sort of environment.

In addition to the caliphs who supported research, the translators of Greek philosophy, medicine, science, and mathematics played an essential role in the Muslim renaissance. Unfortunately, the early Christians missed the opportunity to find out what the "pagans" had learned about these subjects, showing very little interest in launching a campaign to translate these works into the Latin that they could understand. There were exceptions, of course. We saw in the last chapter how Ptolemy's earth-centered model and parts of Plato's philosophy fitted the straitjacket of Christian orthodoxy and became dogmas. As a result, after a thousand years of Christianity, and five hundred years of Muhammadanism, the great civilization of Islam had reason to look down on the Christians, with their filthy cities, lack of medical and other scientific knowledge, and their beliefs in magic, spells, witches, and the power of the evil eye. We recall the comment made by the great St. Agobard and quoted on page 104.

In Muslim Spain, Cordova and Seville were flourishing but conditions were changing rapidly. During the tenth century, the enlightened reign of the caliph al-Hakam II had been followed by that of his young son Hisham II, a puppet of the military under the Muslim vizier Ibn-Abi-Amir. The success of his campaigns against the Christians in northern Spain led Amir to call himself "al-Mansur," the victorious, but with all of his military activities he had continued to support learning. Soon after his death, Cordova was utterly ravaged by the Berbers for a dozen years, with much damage to the people and to the precious heritage in its libraries. Later in the eleventh century, however, the tradition of support to literature was carried on by the poet-caliph al-Mutamid, but by then nearly all of the physicians and philosophers had moved elsewhere. King Alfonso VI, king of Leon and Castile for forty years or so before he died in 1109, avenged the earlier raids on his territory by capturing Toledo.

Over in Ray and, later, in Baghdad the great religious teacher al-Ghazzali (1058–1111) was studying and rapidly earning his later title of "Father of the Church of Islam." He had traveled to the University at Ray from his home in Khurasan, and on completion of his studies was returning home when Bedouins raided the caravan he was traveling with and stole everything. He went to the Bedouin chieftain and asked him to please let him have the notes that he had taken during his years of study, otherwise those years would be wasted. The notes were duly thrown onto the ground in front of him with the comment, "I thought that you went to the university to learn, not to take notes."

Al-Ghazzali decided to go back to the university to study from a new perspective. Nothing wrong with taking notes, of course, provided that the information does not come in through the eyes and ears and, uncomprehended, out through the fingers. Al-Ghazzali mistrusted scholastic theology and intellectualism, and eventually gave up his teaching in Baghdad to probe more deeply the mystical powers and experiences of religion. Since the eighth century, Sufism had been a system of Muslim mysticism, opposed to the rationalism of the Mutazilites, i.e., a more personal approach to religion, independent of which religious sect a person embraced. Sufism had led to excesses that brought executions for heresy, but al-Ghazzali's thoughts and experiences united the mystical with the rational, thereby providing Sufism with a lasting theology.

It was at this time that the first Crusade against Islam was launched by Pope Urban, resulting in the capture of Jerusalem by the Christians in 1099. The successful crusaders were unbelievably brutal, slaughtering all Muslims they could get their swords into, and any Jews who had helped them, and then holding a service to thank the loving Jesus for their victory.

With relations between Muslims and Christians strained to the limit, those were dangerous times for the English traveler and philosopher Adelard of Bath (1090?–1150?) to travel across Europe, Asia Minor, and North Africa. He learned Arabic and made the first Latin translation of Euclid's geometry and Muslim trigonometry. Aethelhard, as he was called by some, studied physical phenomena and posed seventy-six questions about nature, one of them, like one of St. Augustine's, being concerned with the interpretation of Plato's universal ideas as divine ideas. After returning from his travels around 1130 he was asked by his English friends to compare Muslim and Christian thinking. His answer: "I learned from my Arabic masters under the lead of reason; you, however, captivated by authority, follow your halter. For what else should authority be called than a 'halter'?" (Du 4, 1004) He also introduced Arabic numerals to the Latin-speaking world, along with the astronomical tables of al-Khwarizmi.

The twelfth century also saw the development of astronomy in Muslim Granada. Ibn Bajja (1106–1139) criticized the basic premises of the astronomical theory of Ptolemy and initiated a systematic research program followed by his successor Ibn Tufail (d. 1185?). Ibn Bajja's criticism was known to both Averroës and Albertus Magnus—we do not know if it could have led to a reassessment of the ideas of Aristarchos, and an anticipation of those similar ideas of Copernicus. The fact is that it had little effect.

In no way can we keep track of the many major and minor wars of the time, or for that matter of any time, but we can't ignore the Almohades, a religious sect based on the belief in the unity of the divine being, sort of Muslim Unitarianism. Coming from North Africa, the scene of so many disputes about Christian theology some six to ten centuries earlier, they gathered strength under Abd-el-Mumin (1094?–1163) and invaded Muslim Spain, capturing Cordova, Seville, and other cities. Ibn Zuhr, also called Avenzoar (1091–1162), was third in the line of six generations of physicians in Seville, and he eventually became Abd-el-Mumin's vizier and physician. He gave detailed clinical descriptions of internal tumors, pericarditis, and tuberculosis, and he wrote on hygiene, diet, and the pathological effects and therapeutic treatments of their abuse. Other books that he wrote did not survive. He was very outspoken against quackery and superstition, particularly in the medical profession, and he was a great admirer of Galen. He taught Ibn Rushd (1126–1198) whom we call Averroës.*

During the twelfth century, overlapping each other's lives by more than sixty years, there were two men living near each other who in their different ways had tremendous influences on Western thought—Averroës in Cordova, Muslim Spain, and Gerard of Cremona, in Toledo, Christian Spain. Averroës was a native of Cordova, and after spending a period of time in Morocco and Seville, he returned there as a judge in 1171. He wrote treatises on astronomy, grammar, jurisprudence, and medicine, his textbook of medicine being widely used later in Christian medical schools. However, Averroës is known chiefly for his ideas on philosophy and theology, and for the impact they had in the next century when they were translated into the Latin that Christian theologians could read. He felt that the Arabian logicians—the mutakallimun—were negating science with their concept of the continuous interference by the Creator in atomic motions, and that religion was not a branch of knowledge to be reduced to propositions and systems of dogma. Science, he believed, should be concerned with examining and reflecting on the material things of the world. More than that, though; he saw that the perception of scientific truth, based on reason rather than mysticism, could lead the mind to a union with God.

Averroës asserted that the philosopher cannot acknowledge the absolute authority of the Koran or the Bible, but he recognized that the uneducated derived much benefit from accepting either of them as infallible. He criticized Plato for describing the state of souls in another life: "These fables serve only to distort the minds of the people, and especially of children, without being of any real benefit towards ameliorating them. I know perfectly moral men who reject all these fictions and do not yield in virtue to those who admit them." For these and other heresies he was banished by another al-Mansur in 1195, and a few years later he died in Marrakesh. When translated into Hebrew and Latin, his writings had a disturbing effect on Jewish and Christian thinkers. They were widely debated, strongly criticized, and partially accepted. Averroës's belief in God as the highest reality, from which intelligence arises, became part of official Christian scholastic theology. Not so widely accepted by the theologians was his division of intellect into the active part that belonged to God, and the lower intellect—sense perception, imagination, memory, etc.—that perishes with the body. Within half a century of Averroës's death, many free thinkers in the Christian world agreed with him that the world is governed by natural laws, without any divine intervention, that heaven and hell do not exist and that there is only one immortal soul, which our active intelligence joins when we die.

In or about the year 1165, the Christian linguist Gerard of Cremona (1114?–1187) had come to Christian Toledo, less than a hundred and fifty miles north of Cordova, to which Averroës would soon be returning, but apparently they never met. Raymond, the Archbishop of Toledo, had become aware of a host of Arabian manuscripts that had been left behind when the Muslim forces retreated sixty years earlier, and just about everyone had suddenly become a Christian. Raymond then established a college of translators to provide the Latin equivalents of these Arabic and Syriac manuscripts. Gerard found support, accommodation, and colleagues, worked hard on improving his Arabic and became unbelievably productive. He must have sat in his cell, many hours every day for the last twenty years of his life, translating, translating. With a couple of assistants he managed to produce an average of more than three major translations a year.

It was a devious route by which the works of Aristotle, Euclid, Archimedes, Galen, and Ptolemy thus came to us—from Greek to Arabic or Syriac, mostly in the ninth century, and now from those languages into Latin three hundred years later. It would be several more centuries before they became available in modern languages like Italian or, eventually, English. Gerard also translated into Latin the works of the Arabic scholars al-Kindi, al-Khwarizmi, al-Razi, Avicenna, and several others. He wasn't the first translator; in chapter 8 we referred to the translations and teachings of the Jewish scholar ibn Ezra who was the first person to make extensive translations of Muslim scholarly writings into Hebrew.

Gerard traveled through Italy, France, and across to London, meeting with Jewish groups and spreading the rationalistic scientific philosophy of the Greeks and Muslims. Abraham ibn Daud (David) (c. 1110–1180) was at Toledo when Gerard arrived, translating from Arabic to Hebrew and formulating a synthesis of Aristotle with Judaism, believing that no bounds should be set to scientific enquiry. John of Seville, possibly David's son, carried on the work of translations at Toledo. From the Arabic to Latin he translated some works of al-Kindi, Ibn al-Farrakhan, al-Battani, and other early scholars. There were new translations of Plato also, but his ideas had been known to the Christian scholars for centuries, as we have already seen. Way back in the early part of the fourth century, Chalcidius, a Roman philosopher, had translated some of Plato's works into Latin and stimulated the neo-Platonist input to Christian theology.

Some Jewish philosophers and physicians fared better in Muslim countries than in Christian, but the attitudes changed with the overthrow of ruling dynasties by their conquerors. The greatest Jewish scholar of the era, Moses ben Maimon, or Maimonides (1135–1204) was born and raised in Cordova and studied medicine in peace under the Arab scholars until the Almohades invaded from North Africa. He had to pretend that he was a Muslim, and keep up the pretense after he had moved to Fez. In his early thirties Maimonides moved to Cairo, where the attitude towards Jews was more tolerant. He became physician to the son of Saladin, the sultan who later drove the Christians out of Palestine but who was defeated in the third Crusade in 1192. As a physician, Maimonides wrote about asthma, poisons, diet, hygiene, drugs, hemorrhoids, and the healing power of nature. He was also the first of that era to speak out against astrology. His main impact, however, was due to his philosophical writings, especially his effort to reconcile Aristotle with the Old Testament, and his many comments like the one about the Creation quoted in chapter 3 (p. 44).

On July 4, 1189, King Richard I of England (the Lion-Hearted, age 32), and King Philip II of France (age 24) met at Vézélay, in northern France, to bury the hatchet, not in each other as they had been doing for the past two years, but to unite forces on the third Crusade against Saladin, the Sultan of Egypt and Syria. When they paused for a moment in what we would call a photo opportunity, an unknown artist left us a record that is now in the National Library, Paris. Their heads are covered by their chain mail coifs, their bodies by their hauberks of the same metallic mesh, only their faces showing. Their long chausses run from their waists over their feet, like panty hose fashioned from small interlinked metal rings. You don't see their gambesons, which they would also need in battle—leather or quilted fabric underwear to prevent bruises. Philip is proudly displaying the coat of arms of France, adopted only ten years earlier—many white fleur-de-lis sprinkled over a blue background. Richard's chest is covered with the three British lions, gardant, on a red background. One is reminded of a verse from Thomas Gray's *Elegy:*

> The boast of heraldry, the pomp of power
> And all that beauty, all that wealth e'er gave
> Awaits alike th' inevitable hour:—
> The paths of glory lead but to the grave.

After a couple of years, Philip returned home, leaving Richard to fight Saladin. They never met in person, but they tempered their ruthless cruelty with gentlemanly restraint, respecting and trusting each other. Once when Richard's horse was killed under him, Saladin sent him another one, right in the middle of the battle. But they both wanted an honorable truce, and Richard gave his sister Joan (1165–1199), widow of William II, the former king of Sicily, an opportunity that should have won for her the Nobel Peace Prize had it been offered at the time. Marry Saphadin, Saladin's brother, and reign with him in Jerusalem as King and Queen! Richard would give her the coast of Lebanon and Israel, which he held, and Saladin would hand over Palestine. This would allow Muslims and Christians full access to the holy city and bring peace. Her response, in essence: "Marry a Muslim? No way"—and hostilities resumed. The marriage probably wouldn't have worked anyhow, because the Crusaders who heard about it thought it would be a sellout.

Back in Toledo, the Scottish translator and astrologer Michael Scot (1175?–1234) had arrived to study Arabic. He received sponsorship from Frederick II of Hohenstaufen† (1194–1250), Holy Roman Emperor, to help in the translation into Latin of the works of Aristotle, and the commentaries of Averroës on the great philosopher as well. Frederick then sent him to universities in Europe to spread these translations as widely as possible among scholars and students alike. (Because of his work in astrology, alchemy, and other occult sciences, Scot acquired the reputation of being a real wizard, complete with a demon horse and a demon ship.) Frederick II also established direct relations with the Muslim scholarly world, sending to the mathematician Kamal-al-din (1156–1242) in Baghdad the question, "How can you construct a square equal in area to a given segment of a circle?" Squaring the circle was hard enough, and this one is even harder, but Frederick did not require that it be done with ruler and compass alone, that constraining rule of Greek geometry. The problem then becomes a simple exercise in trigonometry unless you turn it around and ask, "If I have a circle and a smaller square, where should I cut the circle to obtain a segment equal in area to the square?"

The study of botany expanded in Seville under the influence of a contemporary of Michael Scot, al-Nabati (c. 1170–1239) who traveled to the Orient and came back to write about the plants he had seen there. His student, Ibn al-Baytar (1197–1248) traveled with him and later went on his own through northern Africa, Constantinople, and many other places before settling into the employment of the sultan of Egypt. His book marks him as the greatest botanist and pharmacologist of those disastrous times of Islam, for he not only knew his plants, he also knew how to prescribe them for the ills of mankind.

These were really times of disaster for Islam, because in the year 1219 the Mongul conqueror Genghis Khan (1167–1227) began an invasion that would continue for forty years, destroying all buildings and people in its way. His capital at Karakorum has been in ruins for centuries, lying there north of the Gobi desert, not far from Hsi-ku-luh in central Mongolia. Having conquered Korea and northern China, the Khan appeared to be contented, offering a conditional peace to the Shah of Khwarizm. (That is the independent state known to us mainly because of its great mathematician al-Khwarizmi, from whom, as we have seen, the words "algorithm" and "algebra" came into our language.) The offer was accepted, but pride intervened. Two Mongol merchants in Khwarizm were executed as spies, and the incident escalated into the invasion of the whole Muslim empire. On February 12, 1258, Hulagu, grandson of Genghis Khan, completed the work by capturing and destroying Baghdad and Merv. Their libraries were burned, their writings destroyed, a million or so of their populations put to the sword. Persia, Syria, Mesopotamia, and the Caucasus were all devastated by the Mongol hordes, whose practically only interest was destruction. To this day, Islam has never fully recovered. Four years later, the Christian forces pushed them out of Spain. During the last months of 1990 and the first of 1991, Baghdad, under Saddam Hussein, did to Kuwait what Hulagu had done to Baghdad 733 years earlier, killing, torturing, plundering, as well as wantonly destroying centers of knowledge —in this case the Kuwait Institute of Scientific Research.

The astronomer al-Tusi (1201–1274) was imprisoned during the Mongol invasion of 1256, but he entered the service of the conquerors, assisted in the capture of Baghdad, and was rewarded with a library and an observatory at Marragheh, in northwest Iran. Al-Tusi wrote extensively on Hellenistic astronomy and geometry, criticizing the theory of Ptolemy. Safer over in Morocco, al-Hasan continued the tradition of first-class astronomy with his detailed treatise on astronomical instruments. There is a book by al-Qazwini (1203–1283) entitled *The Wonders of Animals and the Singularities of Creatures,* which is encyclopedic but not very critical, and mineralogy was represented in a treatise on precious stones by al-Qabajaqi. The thirteenth century also brought us medical studies by ibn al-Lubudi (1210–1267?), on physiology, rheumatism, and the works of Hippocrates. The art of surgery and preventative medicine was represented by ibn-al-Quff (1232–1286). Perhaps the most interesting medical research is the work of al-Quarashi, who died in 1288. He practised medicine at Damascus, wrote about eye diseases and careful diet, and discovered that the blood circulates from the right to the left ventricle through the lungs, three hundred years before Michael Servetus was burned at the stake, partly for discovering this same fact.

There is evidence, however, that the Jewish doctor Asaf ha-Jehudi had discovered that the blood circulates *through the arteries and veins* three hundred years *before* al Quarashi, although it is not clear if he recognized that it was being pumped by the heart. Whether he did or not, a long line of Jewish physicians leavened both Muslim and Christian medical practises; in fact, they played a critical role in disseminating scientific medicine in Christendom. Salerno, on the coast south of Naples, was home to the first Latin Christian medical school, and hundreds of Jews congregated there. As it declined during the thirteenth century, other medical schools were established throughout Italy, and Jewish doctors contributed materially to their development. Prescriptions ranged from the remarkable discoveries that seaweed (rich in iodine) could help goiter and that gold could help painful limbs (Forestier's 1928 gold treatment for arthritis?) to the blood of dragons and the semen of frogs for other ailments.

These Jewish doctors did not believe that the cause of sickness was "possession" by a demon, and this led their patients to have less faith in supernatural cures. The Church naturally responded by banning Jewish physicians from attending Christian homes, and ultimately (1301) Jews were banned from the very school of Montpellier (in the south of France) that earlier Jewish physicians had built. Two centuries later Johannes Geiler (1445–1510) was warning the congregation at Strasbourg Cathedral to keep away from Jewish doctors. The Jewish population has always included more than its share of scholars, and the continuous unconscionable persecution of the Jews over millennia has made it so much harder, often impossible, for many of them to make their full contribution to scientific knowledge. Despite this, they have excelled, but the story of their struggle to do so must be left to some of their number to tell.

With the loss of Baghdad and southern Spain, Muslim scholars in the fourteenth century made their homes in Cairo, Damascus, or Mecca. Ibn al-Duraihim (1312–1360) gave descriptions of domestic animals and birds, making more than two hundred miniature paintings of animals on gold backgrounds, a very accurate Muslim rendering à la John James

Audubon. Around the same time the Sultan of Yemen wrote a book on the *Useful Trees and Aromatic Plants* of his area, and al-Damiri (1344–1405), working in Cairo and Mecca, listed nine hundred and thirty-one animals that are mentioned in the Koran and in Arabian literature. Astronomy was continued in Damascus by al-Shatir (1306–1364?), evidenced by his treatises on astronomical theories and instruments, and on the obliquity of the ecliptic, together with his astronomical tables. Ibn Ahmad al-Mizzi (1291–1349) wrote treatises on astrolabes and quadrants, while carrying on the long tradition of the muezzins of the Great Mosque at Damascus. Some Arabian physicians stayed in Spain, as can be seen from the work from Granada on the contagious nature of some diseases by Ibn al-Khatib (1313–1374). His was one of the first accounts of the Black Plague, which reached Granada in 1348.

There were many more Arabian scholars during the period 800 to 1400; we have named some, judged to be among the most important, but to examine all of their contributions even briefly would more than double the length of this chapter. We have not discussed the array of Muslim geographers from those times, or paid any real attention to the theorems of the mathematicians. We know so little about the Muslim people and tend to look on them with the scorn that characterized the attitude of their ancestors to our ancestors a thousand years ago. We can compare this chapter with the last; two sets of people living next to each other during the same times, but with entirely different priorities. From those times and later, as the destinies of Christians and Muslims have unfolded, we can see the disastrous effects that happen when the followers of any particular religion have any control over the government of a country. We can also note the disastrous effects that follow when the teachings of some great religious leaders—love of nature, ethical behavior, wealth without compulsion, use of reason, concern for others, self-realization—are forgotten.

11

Faith, Reason, and Doubt

Faith is the substance of things hoped for, the evidence of things not seen.

—St. Paul, Hebrews 11:1

Since it is Reason which shapes and regulates all other things, it ought not itself be left in disorder.

—Epictetus

If a man have a strong faith he can indulge in the luxury of skepticism.

—F. W. Nietzsche

The English scholar Alcuin (735–804), also known as Ealhwine or as Flaccus (flap-eared), had met Charlemagne† (742–814), King of the Franks, and had so impressed him that he received an invitation to stay at the royal court in Aachen in order to teach Charlemagne and his family. Alcuin had studied at the cathedral school that Archbishop Egbert had founded at York, and he became master of the school and curator of its library. In 782 he accepted Charlemagne's invitation and spent the next eight years teaching the king's family grammar, Latin, Christian theology, and the Christian "astronomy" of the time. He encouraged study by quoting from the Scriptures and the church fathers: "Better is wisdom than all precious things, and more to be desired" (Proverbs 8:11). "Nearly all the bodily powers are changed in the old, and only wisdom can grow" (St. Jerome). Alcuin's influence expanded beyond the palace school to the monasteries of the Franks, for whom he wrote some instruction manuals. He fought the newly revived Spanish heresy, that Jesus was God's son by adoption, and he helped to have these adoptionists condemned at the Council of

Frankfurt in 794. Eventually, Alcuin became abbot of the Benedictine house of St. Martin at Tours. By his teaching and from his influence with Charlemagne, he made Tours a model school, the best in Christendom with the exception of his alma mater, York. He wrote to Charlemagne, "Your servant lacks the scarcer books of scholarly learning which I was wont to have in my own country," asking if he might "commission some of our youths to procure certain needful works for us and bring the flowers of England with them to France." (UA 8, 204)

This was written during the closing years of the eighth century, at a time when Harun al-Rashid was making Baghdad the center of Arabic culture when he wasn't fighting the Christian emperors in Constantinople. He had no desire to fight Charlemagne also, the grandson of Charles Martel who had repulsed the Arabs so decisively at Tours earlier in the century. The Caliph and the King-Emperor exchanged gifts instead of fighting each other. The year 800 was the year that the King became Emperor of the Holy Roman Empire (Western Division) and that Alcuin, at the direction of Charlemagne, completed and published his revised version of the four-hundred-year-old Vulgate Bible of Saint Jerome. This beautifully illuminated manuscript eventually came to the British Museum. One illustration shows a naked Adam and Eve invoking their Redeemer, Adam's left hand holding a circular leaf in the appropriate place, his right arm extending across his chest, the index finger pointing in accusation at Eve on his left. Her right hand holds a leaf; the left hand is caressed by a snake, who stands upright and balances on the end of its tail. As with modern commercial and political propaganda on television, the visual image made a much bigger impression than did the words. With illustrations like this, the myth of Adam and Eve became indelibly printed on the minds of men and women, not as a delightful story but as a historical fact of supreme significance.

The establishment of schools and the encouragement of learning continued in France under Charles the Bald† (823–877), and in England under Alfred the Great† (849–901). Compared to the research institutes and medical schools at Baghdad and Damascus, and those that would be developed in the coming years at Cordova, Cairo, and Ghazni, these were more like elementary and secondary schools, but they were a start. Many monasteries encouraged and implemented this new attitude toward learning. Unfortunately, their occupants defined scholarship in terms of their limiting conviction that the only path to truth was through the set of documents that an earlier group of clerics had enshrined as the Word of God. However, it is a miracle that anyone had the opportunity to teach or to learn anything at all. The ninth was a terrible century for Europe—the Saracens attacked Rome, the Norse attacked France and Italy, the Danes invaded England, the descendants of Charlemagne fought among themselves. Many monasteries were pillaged, their members dispersed, their manuscripts destroyed. With it all, there were two Christian men who spoke up against the Christian beliefs and practises of these troubled times—Johannes Scotus Erigena "Born in Erin" (810?–877?), and St. Agobard of Lyons (779–840), quoted in chapter 8.

Erigena saw the universe as an all-pervading unity, somewhat in the spirit of the early Stoics, but with a Christian flavor. He tried the near-impossible task of reconciling the authority of the Church with reason, and with his view that the world was developing in a rational way. With this he initiated the debate about the conflict between faith and reason, a conflict that was to rage for centuries. He was aware of Heracleides's idea that Mercury and Venus move around the sun—he suggested that Mars and Jupiter also rotate around the sun. Charles the Bald heard about him and invited him to France; by 847 he was appointed as head of the school at Charles's court. Erigena was a strong and articulate proponent of the freedom of the will, joining Hincmar (806?–882), Archbishop of Reims, in opposing the teaching of predestination by the German Benedictine monk Gottschalk (805?–868). At Charles's request, Erigena translated into Latin what everyone thought were the writings of Dionysius the Areopagite, who was converted by St. Paul. These exciting descriptions of the hierarchy of the heavenly hosts, noted in chapter 9, were shown later not to have the implied authority of St. Paul, but by that time everybody believed them.

In England, the epileptic Alfred the Great became king of the West Saxons in 871, and king of England fifteen years later when he captured London. Over the past hundred years there had been wave after wave of invaders from Denmark. Nothing and nobody were safe from them. Four years before the day that King Ethelred I, Alfred's older brother, was killed in battle against them, the Danes had invaded Northumberland, killing and destroying everyone and everything that stood in their way. The cathedral school at York where Alcuin had been the director was hit badly, its manuscripts scattered, its people slaughtered. That spark of knowledge and culture was extinguished by the ruthless pillaging by the Danes just as surely as, nearly four hundred years later, the culture of Baghdad would be destroyed by the ruthless Mongols. Alfred fought the Danes for thirty years and finally defeated their fleet two years before he died. With it all, he tried very hard to restore and encourage learning. Like Charlemagne, he studied at the palace school that he had founded, and invited scholars from abroad to come there and teach. Soon after becoming king of the West Saxons, he invited the aging Erigena to his school.

Erigena had caught a glimpse of the evolutionary process in the universe, as we have seen. He argued for the use of reason in theological debates by stating that "reason and authority come alike from the one source of Divine Wisdom." He added that, since the church fathers often contradicted one another, reason must be used to settle any arguments between them. For this, his writings were banned by two church councils, and nearly four hundred years later, in 1225, Pope Honorius III ordered them to be burned, teeming, as they were, with what he called "hereditary depravity." Finally, Gregory XIII, pope from 1572 to 1585, paid them the ultimate compliment of putting them on the *Index.* Erigena had started something—a school of scholasticism supportive of the idea that the God-given faculty of reason can help man in his struggle to understand, something that had been known to the religious Pythagoras some two thousand years earlier, but had been forgotten.

St. Agobard was more interested in correcting Christian customs than in arguing about theology. He compared the filthy Christian cities to those of the much more hygienic Muslims. He was appalled at the ordeals by fire that were supposed to lead to justice, and he spoke up against persecutions for witchcraft, little dreaming that it would be a disgraceful Christian practice seven hundred years later. He objected strongly to gematria and other forms of numerology that had become common among some Christian ecclesiastics, and he argued against the cruel treatments for mental illness, so different from the gentle approach used by Muslim doctors. He was also an iconoclast, condemning the use of paintings and sculpture in the service of religion, fearing that people would be led to worship the images rather than the true God of the Trinity, and mindful of the Second Commandment.

The question of "graven images" had escalated into a major conflict between the eastern emperors of the Roman Empire and the western popes. It had started with Leo III (680?–741), Eastern Roman Emperor, who had saved Constantinople in 717 from the attack by the Umayyad caliph Suleiman. (This Leo III died a few years before the birth of Pope Leo III, who crowned Charlemagne.) Emperor Leo was thoroughly opposed to image worship. The Greek Church was overrun with them. There were images of the saints, pictures of Mary and of Christ, holy relics, statues, all cherished and worshipped for their magical powers. Leo reminded the people of the Lord's exhortation recorded in Deuteronomy (4:15–18): "Take therefore good heed to yourselves . . . lest you corrupt yourselves, and make for yourselves images and the form of any figure, the likeness of male or female, the likeness of any beast that is on the earth, the likeness of any winged fowl that flies in the air . . . of anything that creeps on the ground . . . of any fish that is in the waters beneath the earth." In 726 Leo succeeded in getting an edict passed that banned from the churches all of these statues and paintings. The populace revolted against the destruction of what they saw as their gods, and the fury escalated. Leo's decree, coupled with some of his actions (like sinking a fleet that had come to depose him), led to a complete rupture with Popes Gregory II and III (730–732).

Leo's son and successor, Constantine V (719–775) had been named Copronymus from the Greek *copro* (excrement) and *onyma* (name). He imposed the decree against images in a ruthless, sickening fashion, with unspeakable tortures to those who opposed him. He would have none of this monasticism and celibacy either. As far away from Constantinople as Ephesus, monks and nuns were forced to marry each other or die. Leo's grandson, Leo IV, married the Athenian woman Irene (752–803), a staunch believer in the use and adoration of images. When her husband died in 780, Irene assumed power and in 787 she convened the Second Nicene Council which supported and even defined the veneration that is due to images. She had to give way to her son (771–797?) when he reached the age of nineteen and became Emperor Constantine VI, but she plotted against him. She had him arrested, imprisoned, blinded, and put to death in his twenty-sixth year. She reassumed her title of empress, only to be deposed a few years later (802) by Nicephorus who, as we saw in the last chapter, was soon forced to pay tribute to Harun al-Rashid.

The crowning of Charlemagne as Emperor in Rome (800) was precipitated partly by Irene's barbarity in Constantinople. The magnificent religious art of the Renaissance owes its existence, also in part, to Irene, possibly the most vicious woman known to history, though not without some competition. The later Protestant objections to the icons of the Catholic Church had their origin with Irene's grandfather, Emperor Leo III, and, of course, so much earlier in the Book of Deuteronomy.

St. Agobard was brave to speak up against the images that the Nicene Council, under the influence of Irene, had approved when he was a boy. In 838, two years before his death, the Council of Kierzy declared his teachings heretical. The attitude towards the matter depended, as so many attitudes do, on who was in power. The death of Agobard came two years before that of Emperor Theophilus, whose wife Theodora (d. 867?) also took over as co-regent with her brother Caesar Bardas (d. 866). She brought back image worship and expelled all of the iconoclasts, paving the way for images not just to the glory of God and Jesus but, by prestige drainage, to emperors who had themselves depicted being crowned by Jesus himself. Coins became an important element of the propaganda—a bust of Christ on a coin of Constantine X (1059–1067), a silver coin showing Christ crowning King Levon of Armenia at the end of the twelfth century, and many more.

One of the faculty members at the ancient University of Constantinople during the emperorship of Theophilus was also named Leo, an expert in mathematics, astronomy, astrology, philosophy, and medicine. As Jesus had said, "A prophet is not without honor save in his own country" (Matt. 13:57) and Leo wasn't appreciated by the Emperor Theophilus until his fame had spread to Caliph al-Mamun in Baghdad. The caliph made a very enticing offer to Leo to join the group of scholars that he had brought together. Leo responded by indirectly letting Theophilus know about it, and was rewarded with a state professorship. One way to early promotion in universities today remains the same—to get an offer from a better institution. Al-Mamun kept pressing, sweetening the pot with an offer of much gold to Theophilus himself. The emperor replied by appointing Leo as Archbishop of Salonika.

St. Ignatius Nicetas (799?–878), Patriarch of Constantinople, the highest dignitary in the Eastern Church, disapproved of the fact that Caesar Bardas had divorced his wife and was living with his son's widow, and excommunicated him. The emperor (now Michael III, known as the Drunkard) replied by banishing Ignatius and replacing him with Photius (820?–891), a scholarly layman. Photius had lectured on philology, philosophy, and science at the University of Constantinople, and had attracted many students. But the clergy was naturally suspicious of its new nonecclesiastical boss. Ignatius appealed to the Roman Pope, who ordered Michael to reinstate him.

Despite his lay background, Photius must have impressed the clergy, because he was confirmed in his office by a meeting of the Council of Churches. This was followed by another meeting at which the Council excommunicated the pope! Not only that, the Council of the Greek Church objected to the fact that Latin priests shaved their beards and were required to be celibate. No wonder, they said, that there are so many children in the West who do not know who their fathers are.

Finally, there was the matter of a single word in the church doctrine. The Latin Church had decreed that the Holy Ghost proceeded from the Father and the Son, according to the Nicene Creed of western Christendom dating from 589. The Greek Church would have the Holy Ghost proceed from the Father *through* the Son, a creed from the fourth century. However, by the time these old disputes had resurfaced, Bardas had been murdered by Michael, working with a newcomer, Basil I, who then got rid of Michael, too. The two branches of the Christian Church decided to cooperate with each other for a while. Photius continued to encourage education by writing a summary of the works of nearly three hundred classical scholars. He denounced Basil as a murderer and refused him the sacraments. For this he was replaced by Ignatius, but he was reinstated when Ignatius died in 878.

During the following century, and halfway into the eleventh, the Roman papacy was such a disgrace to the name of Christianity that it is a wonder that more people didn't expect Christ to return at the millennium and correct the mess. In 897, Pope Stephen VI had his predecessor Pope Formosus tried before an ecumenical council. Not so extraordinary, perhaps, except that Formosus had been dead since the year before; his corpse, dressed in purple robes, sat in the prisoner's seat and was found guilty, and the assembled Christian Council saw to it that vengeance was wreaked upon it!

Until Leo IX assumed the papacy in 1049, the papacy was a continual story of simony, intrigues, and murders—with one exception. Gerbert (940?–1003) became Pope Sylvester II during the last four years of his life. He worked with the young Holy Roman Emperor Otto III (980–1002) to set up a new and less corrupt Roman Empire. Otto was German, and the Italian nobility suspected his motives and supported Crescentius who rose up against him. Otto, still in his teens, had Crescentius executed and then made the big mistake of his short life—he took to bed Stephania, Crescentius's widow. By the time he was twenty-two, she poisoned him, and Gerbert died in the next year, probably in the same way. After that, it was back to business as usual for the papacy for another fifty years.

All that remained from Gerbert were his textbooks of advanced learning in mathematics and the sciences, his books on music and theology, his letters, and the clocks and astronomical instruments that he had built. He seems to have been the first Christian to give a description of Spanish-Arabian numerals (without the zero). He had learned about them, and much more, from his studies at the Muslim schools in Spain. He was an initiator of the renaissance of learning among Christian scholars, and it would have come earlier if he had lived longer. The Church missed an excellent opportunity to pioneer science rather than oppose it and persecute those who believed that scientific reasoning could help understanding. However, it was encouraging that at this time Bishop Fulbert of Chartres (960?–1028) directed a cathedral school in which theology, medicine, science, and classical literature were taught—and began a rebuilding program for Chartres Cathedral.

Otto III's father, Otto II (955–983), had fought and beat the Danes in 974, and forced their king, Harold Bluetooth (d. 985) to become Christian. Overnight the Danes were converted, at least officially, but it didn't stop their raids on England. Bluetooth was killed in a war against his own son, Sweyn Forkbeard (d. 1014) who became the father of Cnut (994?–1035). (Alhazen, Avicenna, and Cnut died within a few years of each other.) Ethelred II (968?–1016) became king of England because his mother had seen to it that King Edward the Martyr was assassinated at the age of fifteen. After much fighting, Cnut succeeded Ethelred and married his widow, Emma, cousin of William of Normandy. Cnut's illegitimate son Harold Harefoot temporarily beat out his legitimate son Hardecanute for the throne, but died in 1040. (It was in that year also that Macbeth murdered Duncan and took the throne of Scotland.) Hardecanute died soon afterwards and Edward the Confessor, son of Ethelred and Emma, became king. When he died early in 1066 the happy family went to war with itself.

Edward had promised his cousin William that he would be his successor, but later he opted for his brother-in-law Harold, who was crowned on January 6, 1066. Harold's brother Tostig was his bitter enemy and arranged with King Haardrade of Norway to invade the northern part of England. Harold force-marched his troops north, caught the invaders by surprise, and Tostig and the king were killed. By that time the wind had changed, so that William's fourteen-hundred-ship fleet had crossed the Channel, and his men were busy harassing the villages of the south coast of England. Harold moved south to meet him, establishing his position on the hill of Senlac, six miles northwest of the coastal town of Hastings.

As he moved forward, William could see a hastily built stockade and a freshly dug trench, with marshy ground in front of him to the left. To the right were assembled Harold's *huscarls,* bodyguards in full armor, each carrying in gloved hands a huge battle-axe and a large shield, either circular or shaped like a kite, wide at the top and tapering to the bottom. Each head and face was protected by a conical steel helmet with a built-in noseguard. Each man wore his *hawbeck,* a knee-length tunic of chain-mail fastened between his legs and extending up to cover ears and jaws. A broad sword hung from each belt. The *lang-fyrd,* half-armed serfs, were everywhere, unable to afford full armor, lucky to have a steel cap, a sword, and a tunic. Harold's position at the top of the hill was proclaimed by the fluttering Golden Dragon of Wessex and Harold's Royal Standard—a fighting man wrought on gold and studded with diamonds. It was October 14, 1066.

To Taillefor the minstrel was given the honor of leading the charge of William's foot-soldiers. Singing excerpts from the "Song of Roland," and juggling his sword as if it were a cheerleader's baton, he was the first to strike a blow and the first to die. The Norman knights followed but couldn't break the stockade; William's left flank, troops from Breton, advanced but were caught in the marshy ground. Someone started the rumor that William had been killed, so he removed his helmet so that all could see him. He called for another assault and then a retreat, so that Harold's bodyguards would follow for the kill. Away from their entrenched position, they were surrounded by the French troops coming up on William's right, the reorganized Bretons on the left, and the Norman knights in the center.

William gave the order to his archers to shoot into the air so that the arrows would descend on the heads of Harold and his men. One of them found the only chink in Harold's armor—one of the holes through which he could see. His two brothers who supported him, Gyrth and Leofwine, had also been killed, Gyrth was felled by a mace swung by William himself. There were no more serious competitors to the throne, and William became William I, king of England. His campaign had been blessed by Pope Alexander II; Harold, on the other hand, had visited the anti-pope

Benedict X. "Such bickerings to recount, . . . what more worth is it than to chronicle the wars of kites or crows flocking and fighting in the air?" wrote John Milton in his *The History of Britain.* Here are three points of view on the battle and its outcome:

> "Men who bravely strove to save their country from the calamity of foreign servitude." Winston Churchill (1620?–1688)
>
> "Without the Normans, what had it ever been! A gluttonous race of Jutes and Angles—lumbering about in pot-bellied equanimity." Thomas Carlyle (1795–1881)
>
> The Norman nobility was "given over to gluttony and lechery." William of Malmesbury (1090?–1143)

The town of Malmesbury is in Wiltshire, eighty miles west of London, and during these times of intrigue and battle of the eleventh century, a remarkable Benedictine monk named Oliver lived in Malmesbury Abbey. In his youth Oliver fitted wings on his hands and his feet, ascended a tower in order to get help from the wind, threw himself off, and is reported to have flown for a furlong (two hundred and twenty yards) or more. He broke both of his legs on landing and always walked with a limp, berating himself for not having added a tail. He saw the great comet of April 24, 1066, and is reported by William of Malmesbury to have said, "Thou hast come, thou hast come, bringing sorrow to many mothers. Long ago have I seen thee, but now more terrible do I behold thee, threatening the destruction of this country." The events of that year gave indisputable credence to the role of comets as divine warnings of impending calamity.

While this was happening contacts with the Muslim world were beginning. Constantine the African (1015–1087) traveled to India, Babylon, and elsewhere, studying medical techniques wherever he went. He finally settled down in the Monte Cassino monastery in Italy and wrote a report on the medicine that he had learned about during his journeys. He added a translation into Latin of the works of Hippocrates and Galen. The Christian doctors were exposed for the first time to the medical practices of the ancient Greeks and the contemporary Arabs. In the preceding chapter it was noted that, early in the next century, Adelard of Bath expanded this contact with the Arab world and also thought and wrote about the divine nature of Plato's universal ideas.

Is the whole more than the sum of its parts? An ambiguous question always leads to divergent answers, and often to a great deal of fruitless argument. Let us take an example: Are Plato's universals more than their constituents; is the concept of Man more than individual men? Plato had argued that the universal concept can outlive and outlast the individual cases; that the universal is therefore more real; but why is this so important? Let us take another example: Is the Church more than the sum of its members? The realist would say yes, the nominalist would say no, the universal is always a mere word, just the name (nomen) of the collection, an argument more in the spirit of Aristotle than of Plato.

Jean Roscelin (1050?–1120?), a canon at the Compiegne Cathedral, preached the nominalist philosophy. Universal "beauty" was just the name of the class of beautiful things, a "book" just the name of all books, and —here came trouble—a "church" is just the name of the group of people who are its members. The Church hierarchy could not allow this—the Church is more than its members, just as, centuries later, the Communist state would claim to be more than its individual people. Was, then, the Trinity more than the individual Father, Son, and Holy Ghost? By the nominalist arguments, the Trinity would be just the name for three Gods. That was too much; in 1092, Roscelin was summoned to appear before the synod at Soissons and told to retract or be excommunicated. He retracted, went to England, and returned to France to teach. His star pupil was Peter Abelard* (1079–1142), the articulate, handsome lover of Héloise, the ruthless destroyer of the arguments of all who opposed him, and the proponent of the use of reason in understanding theology and in understanding the world.

Two prominent clerics espoused the cause of faith against reason: St. Anselm (1035–1109), Abbot of Bec in Normandy and later Archbishop of Canterbury, and the much younger St. Bernard of Clairvaux (1091–1153). Anselm's idea of understanding was first to have faith which helps you to understand; you could then support this new understanding with reason. He advanced the argument that the idea of supreme perfection implies the existence of that of which it is the idea. The idea of a perfect God by definition means that the God exists, otherwise He would not be perfect. We could regard this type of reasoning as the ultimate psychological ontological argument for the existence of God, or we could use the standard trick taught to freshman physics students when confronted by a formula—apply it to some particular cases and see if it works for them. If it does, proceed cautiously, because it may still be wrong. If it doesn't work, correct it or forget it. A monk by the name of Gaunito applied it to a "perfect island," thereby proving that the island existed. Thomas Aquinas, determined to use reason in support of theology, didn't buy the argument either.

Bernard of Clairvaux was an eight-year-old boy when the first Crusade had resulted in the capture of Jerusalem by the Christian forces. When he grew up he became the leader of France in stirring up the passion that led to the second Crusade in 1147. He founded the Cistercian monastery at Clairvaux, in the French province of Aube, about a hundred miles east of Paris. Clad in his white monk's habit of the Cistercian Order, he sat in his cell writing sermons, epistles, and evangelical hymns like:

> No voice can sing, no heart can frame
> Nor can the memory find
> A sweeter sound than Jesu's Name
> The Savior of mankind.

He was very pious, and he didn't like the teachings of Peter Abelard at all, speaking forcibly against them, for he considered them to be heresies. In the ongoing argument about the relative roles and merits of reason and faith in the Christian version of religion, Peter Abelard made his position clear: "Doubt is the road to inquiry, and by inquiring we receive truth."

On April 2, 1866, Thomas Carlyle gave his inaugural address as Rector of the University of Edinburgh, speaking on the origins of universities: "I dare say you know, very many of you, that it is now some seven hundred years since Universities were first set up in this world of ours. Abelard and other thinkers had arisen with doctrines in them which people wished to hear of, and students flocked towards them from all parts of the world." (HCl 25, 362)

Abelard did indeed impress people, although many of them very negatively. Like Avicenna, he irritated a lot of his colleagues, because he was very quick to argue and to destroy the arguments of others. George Henry Lewis, a contemporary of Thomas Carlyle, was not very impressed with Abelard: "He presents the figure of a quick, vivacious, unscrupulous, intensely vain Frenchman." (UA 9, 228) English Protestants were not very fond of French Catholics, particularly during the nineteenth century.

Abelard's eloquence in philosophy brought him fame which spread all the way to Rome, the Pope sending messengers to hear him and report back. The local populace made way for him as he walked the streets; the ladies were impressed, and one of them, Héloise, was very impressed. She was in her teens, Abelard was twenty years older. She lived nearby under the care of her uncle Fulbert, the canon of Notre Dame, and Peter offered him free tuition for her in exchange for room and board. "Hereupon," he wrote, "I wondered at the man's excessive simplicity, with no less amazement than if I had beheld him entrust a lamb to the care of a famishing wolf." She was a good student, excelling him in Greek and Hebrew, and she learned quickly, as he put it, "The books were open before us, but we talked more of love than philosophy, and kisses were more frequent than sentences." With that sort of thing going on right under his nose, it wasn't too long before Fulbert found out. They were forced to separate, they met secretly, she became pregnant, Peter arranged for her to visit his sister, he offered to marry her and did so—a theme that has been re-enacted by so many couples since. Fortunately, Fulbert's next move has not been as common—he and his friends caught Abelard and castrated him.

Peter wrote many love songs to Héloise; they became popular and through them their love was common knowledge. After the cruel mutilation she wrote to him: "When you formerly sought me for worldly pleasures, you visited me with thick-coming letters and by frequent songs you put your Héloise in the mouths of all. How much more righteously should you now urge me on to God than then to pleasure." The two of them, separate, sought solace in their religion.

Later, at the Council of Sens (on the Seine, a hundred kilometers upstream from Paris), Abelard was now sixty-one years old. He challenged Bernard to repeat his charges of heresy. Bernard, no great debater, was afraid to respond, asking only that some of the passages from Abelard's writings be read aloud. To the surprise of everyone Abelard also refused to debate the issue, saying, as he left the assembly, that he would appeal to the authorities in Rome. Once so popular with the crowds, he was now in danger of being lynched. His enemies had publicized his questioning of the reality of the Holy Trinity, and perhaps he decided that he should get out of town in a hurry. The Council issued its report to Rome: "Peter Abelard makes void the whole Christian faith by attempting to comprehend the nature of God through human reason. Already has his book on the Trinity been burned by the order of one Council; it has now risen from the dead. . . . It is time therefore to silence him by apostolic authority."

St. Bernard had written about Abelard to one of the cardinals: "Though a Baptist without in his austerities, he is a Herod within." He continued his campaign against Abelard by using the technique of suggesting guilt by association. He implied that, in writing about the Trinity, Abelard was like Arius, that in the matter of divine grace he was like Pelagius, that in speaking about Christ he was like Nestorius—three condemned heretics from the past. St. Bernard was canonized just twenty years after he died, for he was a very honorable and highly influential man.

On the way to Rome, Peter Abelard was taken ill. He stopped at the Abbey of Cluny, about two hundred kilometers south of Clairvaux, and stayed there for two peaceful years, away from his enemies, until his death. All Christians had been given a lesson on the unorthodox use of either their minds or their libidos. But Abelard left us his lessons too. As the German philologist Max Müller wrote many years later: "Abelard was persecuted and imprisoned, but his spirit revived in the Reformers of the sixteenth century, and the shrine of Abelard and Héloise in the Pereha

Chaise is still decorated each year with garlands of *immortelles*."

William of Conches (1080?–1150 or later) was born about a hundred kilometers west of Paris, in almost the same year as Abelard. He taught in the school that Fulbert had founded at Chartres, also advocating reason as a check on faith—he did not believe that Eve was literally formed from Adam's rib. He had read Democritos and believed that all matter consisted of a combination of atoms. His complaint about teaching is relevant today; there were those who gave easy courses in order to increase their popularity, thereby gaining more students and fees, while lowering academic standards. He studied the atmosphere, recognizing that the density and temperature of the air decrease as one moves to higher elevations, and he compared the circulation of the air to the currents in the ocean. For these ideas we might regard him as the first Christian scientist (with a small s) or, at least, physical scientist.

On at least two occasions, William of St. Thierry, a zealous monk of Reims, had urged St. Bernard to condemn Peter Abelard, and he also urged him to do the same for William of Conches. At that stage of history William's ideas about the physical sciences didn't cause alarm; it was his espousal of reason that did. He was forced to retract his heresies, to agree that Eve really did come from Adam's rib, and to denounce the use of reason in the determination of Christian beliefs. William moved south to Anjou to get away from these outrages; there he became tutor of the young Henry Plantagenet, later to be Henry II, King of England.

A different view of scholarship was advanced by his contemporary Hugh of St. Victor (1096?–1161), a man who spent many years in the Parisian abbey that had been named for the fourteenth pope, Victor I. Hugh wrote books with titles such as *The Union of Body and Spirit* and *The Ostentation of the World,* offering arguments of scholastic research and a neat description of its approved method: "Learn first what is to be believed" and starting from that, look for passages of Scripture and other texts to confirm it. He argued that the universe was created in an instant, and also in six days, and that Jerusalem is set in the center of the world. Another contemporary, the Italian theologian Peter Lombard (1100?–1160?), maintained that no created things would be hurtful to man if he had not sinned. His collection of the opinions of the fathers of the church codified this and other beliefs and practices, and later the great St. Thomas Aquinas gave his blessing to many of them. As late as the eighteenth century, John Wesley asserted that, before Adam's sin, "the spider was as harmless as the fly." Indeed, it was widely believed into the nineteenth century that, before the Fall, serpents stood erect and walked, just as Peter Lombard had written.

Later in the twelfth century there was a "voice of one crying in the wilderness" (Mark 1:3). Alexander Neckham (1157–1217) not only wrote a popular encyclopedia of scientific knowledge, but, more important, he had a glimpse of what a scientist must do—hard, time-consuming work, and intense application of his mind to a problem. True, of course, but the same could be said of the many who struggled with the sophistries of the scholastics. Neckham, later the Abbot of Cirencester, studied and taught at the University of Paris. He was the first, outside of China, to record the use of the magnetic compass.

Eight years after Abelard's death, the abbess of the abbey at Bingen, a town on the Rhine river, published a book titled *Causes and Cures.* Perhaps without knowledge of the healings that had occurred in the temples of Imhotep and Asclepios, some monks and nuns cared for and housed the sick in their monasteries. They preserved the knowledge of medicinal herbs and of the miracles of faith healing, which the good Abbess duly recorded. Soon afterwards, the Church forbade this activity, leaving the practise of medicine entirely to the physicians. One such physician gave some advice to others that has not been completely forgotten today: "Imply that all patients who come to you are really sick, so that if they recover it will mean more glory for you, and if they don't you won't get the blame. Always prescribe something, so that it will look like you have earned your fee, and do not take the opportunity offered by the patient's bed-ridden condition to make love to his wife, daughter or maid."

Physicians are the most immediately useful among the scientists, and their struggle to understand their discipline was primarily determined by the complexity of the subject matter. However, as always, there were external factors that delayed development. Chief among these was an apparently natural reluctance on the part of various populations throughout history to allow human dissection, leading to some laws forbidding it. In some circles it was feared that it would interfere with the resurrection of the body, although cadavers seemed to perish anyhow. Then, of course, as in other disciplines, there were the superstitious and religious theories, in this case of resident demons, astrological significances, and therapeutic charms. Scientists who aim toward healing people of their ailments, however, naturally attract less negative attention than those whose work disturbs entrenched philosophical and religious beliefs. At times when science was banned in medieval Europe, an exception was often made for the field of medicine.

Peter of Spain (1210–1277) wrote a number of books on medicine and surgery, especially on diseases of the eye, and he wrote comments on Hippocrates, Hunain Ibn Ishaq, and other physicians. By no means was the Church opposed to this activity; on the contrary, a year before he died Peter became Pope John XXI. In Milan, Italy, Lanfranchi (Lanfrancus Mediolanensis, d. 1306) was practising medicine and becoming the first person to distinguish between breast cancer and hypertrophy. On moving to France in 1290, Lanfranchi wrote books on surgery, stimulating better practises in both countries. He must have known his compatriot, Aldobrandino of Siena (d. 1287), a physician who commented on the works of Galen, Avicenna, and Hippocrates and who wrote the first book in Italian on how to improve and preserve one's health. A little later (1301) Henri de Mondeville (1260–1320) was appointed as physician to the king of France, and he wrote the first book on surgery in the French language. These and other men show that medicine, and surgery in particular, also participated in what could be called the First Renaissance, the revolution in knowledge that came to the Christian world during the thirteenth century, stimulating new ideas, especially in theology and science.

In the early years of that century, Domingo de Guzman (1170–1221) had founded the Dominican Order of monks in Southern France, while, in Italy, Giovanni Francesco Bernadone (1182–1226) had initiated the Franciscan Order. The Black Friars (the Dominicans) saw part of their way to the truth through poverty, frequent fasting, and vegetarianism. Later in the century, as translations from the Arabic of early Greek works became more available, they tended to be proponents of the philosophy of Aristotle, insofar as it did not appear to be in conflict with Christianity. The Grey Friars (the Franciscans), about whose founder many miraculous stories were being told, went shoeless and penniless, and later tended more towards mysticism and neo-Platonism. The reason-faith controversy continued in a greatly modified form, coming to a climax later in the century with the conflicting opinions of the "Angelic Doctor," the Dominican monk St. Thomas Aquinas (1225–1274), and the "Seraphic Doctor," the Franciscan monk Giovanni di Fidanza, known as St. Bonaventura (1221–1274). In the hypothetical heavenly pecking order, the seraphs rated much higher than the angels, and by the year in which these two men died, official opinion sided with the pious Bonaventura, as opposed to the much more intellectual Aquinas, but it didn't stay that way for long.

In the spirit of Plato and St. Augustine, St. Bonaventura helped materially in building up the Franciscan order. He was more of a mystic than a philosopher, and an anti-intellectual, opposed to the use of reason in theology, indeed, probably afraid of it. Bonventura believed that true knowledge does not come from the impressions that the world makes on the senses—remember Plato's words, recorded in chapter 5: "What is seen in the heavens must be ignored, if we truly want to have our share in astronomy." Genuine knowledge comes instead from the world of spirit, through the soul. Don't try to define or understand God; just become aware of Him and do His bidding. Be careful of relying too much on Aristotle;

remember that he was a heathen (no more, of course, than was Plato).

St. Thomas, on the other hand, used his remarkable talents to reconcile reason with faith, and Aristotle with Christ. Translations of the Greek philosophers into Latin were becoming more readily available, although they came by a devious route. St. Thomas made sure that he got a reliable translation of Aristotle by asking his fellow Dominican William of Moerbeke (near Ghent, East Flanders) to translate directly from the Greek. There had been some direct translations, but William, and presumably some assistants, produced a large number of Latin versions of the works of Aristotle, along with some of the writings by Archimedes, Hero, Hippocrates, Galen, and others. Together with earlier translations from the Arabic and Syriac, by Gerard of Cremona and others, this old knowledge in an old language came as a tremendous shock to the scholars of the thirteenth century of Christendom. Universities were established, or grew out of the old schools of law that had flourished for several centuries in Paris (825) and Bologna (1088). The old school of theology at Oxford developed into a university during the thirteenth century. As noted in chapter 5, in 1209 the "Town versus Gown" problem escalated when a student killed a townswoman and some students were hanged in reprisal. Teaching was suspended, many students and teachers left, some of them moving some seventy-five miles northeast to develop a university at Cambridge.

Aristotle's science has been considered in earlier chapters; it seems to have little to do with the Christian religion, yet it became an integral part of it. The reason is that Aristotle's science was a logical aspect of his philosophy, and Aquinas integrated that philosophy with Christianity. There is an overall design, there is a purpose for everything, and each thing is in the place assigned to it. If it is displaced, it will try to go back to its natural home. Plants and animals are put there for a purpose, ultimately to serve mankind. The final cause of any change is the need to fulfill the purpose of what is changing.

Thomas Aquinas knew his Greek philosophy very well. His comments on cognition begin with:

> Democritos, and with him all the naturalistic philosophers find the cause and means of cognition in the material atoms, which, detaching themselves from objects, impinge on the senses. They do not admit that understanding differs from sensation—Plato on the contrary distinguishes between sensation and intellect. With Democritos, Aristotle admits that the concurrence of the senses is necessary, with Plato he distinguishes sensation from understanding. It seems to me that Plato has wandered from the truth.

And in writing about truth Aquinas adds:

> All that exists has been created for a definite end by the Almighty power of God and after the laws of His intelligence: Now, that intelligence is the source and the supreme law of all truth; then all beings, by the mere fact that they exist, are true in an absolute manner. Beyond the absolute truth which streams from the divine essence, there is a relative truth which is grasped by the action of the understanding. . . ."

Thomas Aquinas believed that the happiness of understanding is the greatest happiness of all, that the highest achievement of the soul would be "that on it should be inscribed the total order of the universe and its causes," and that it should see God.

If ever a man struggled hard to understand, it was Thomas Aquinas. He felt that the Church should reject arguments about when the world was created "lest the Catholic faith should seem to be founded on empty reasoning." He studied and criticized Aristotle's works, and he built a religious philosophy that was branded as a heresy in 1277, but that led to his canonization only 46 years later, thanks to the espousal of his cause by the Dominicans, as opposed to the Franciscans. Later three holy books were sometimes placed on the altar together—the Bible, the Papal Decretals, and Thomas's *Summa Theologica.* Fortunately, papal infallibility was not to be declared until the Vatican Council met on July 18, 1870.

Our interest here, however, is centered on the effects of the writings of St. Thomas Aquinas on scientific thinking. The examples already referred to—the use of reason, studying the order of the universe, the relative truth of understanding, the rejection of arguments about the date of creation—are very promising. Some of his other beliefs were held by other scholars of the time—angels as the highest point of creation (and he describes them in great detail); devils and witches; female babies as male babies who didn't quite make it, their defection probably due to the blowing of a damp south wind at the time of copulation. He blamed Eve, like everyone else had done, because she caused man to disobey God, so that the taint of her sin is borne by all generations. There was a problem though, because at that time the Church had not decided yet that Mary was a virgin—indeed that wasn't official dogma until December 8, 1854. He decided that her sin was justified *a posteriori,* however, because of the holy son that she produced.

Thomas Aquinas's theory of demons was that they cooperate with witches to cause men to have wet dreams, the demon flying off with some of the semen to impregnate some unsuspecting woman far away. He was no scientist, and he was much more interested in what we call Aristotle's philosophy than his science. Unfortunately, Thomism came to be interpreted as endorsing all of Aristotle, including his denial of the vacuum, and the distorted ideas of physics that followed from it. It was a small part of the beliefs of the time, perhaps, but one that had devastating consequences on the development, and more precisely on the developers, of natural philosophy.

The writings of Averroës, now translated, were being taught in Christian society, especially at the University of Paris, and they were a threat to Christian theology. A secular priest named Siger of Brabant (an old duchy of the Netherlands, now a province of Belgium) denied the "first man" theory of creation, and the doctrine that the individual soul lives after death. Astrology had also done its work on his thinking, because he saw all human events as governed by the stars. Since after a period the heavens were supposed to return to their original configuration, he therefore expounded the cyclic theory of the universe, that after some period everything and everybody will repeat what they are doing today. Siger was condemned and imprisoned by the Inquisition in the same year, 1277, that the teachings of Averroës, and some of the teachings of Thomas Aquinas were put on the black list. It is easy to see why the Catholic theologians were so upset—people were teaching at the university that the Christian religion discourages learning, that theology does not contribute to knowledge, that a dead body cannot rise as the same body, now living, and that people should emphasize happiness in this life because it is likely to be the only one that they will ever experience. Heresies!

At the beginning of the fourteenth century a former student returned to the University of Paris to deliver a different message. Johannes Duns Scotus (1265?–1308)) mustered enough subtle arguments in support of the doctrine of the Immaculate Conception to earn the title "Doctor Subtilis." He disagreed with Aquinas, emphasizing the importance of the individual will and de-emphasizing the power of logic in determining religious doctrines. Not that he wasn't capable of logical argument himself, but he saw God's will as the prime mover of the universe, rather than God's intelligence. The Church dogmas were a matter of faith—if you reason about them, you are led to contradictions. There is no way to prove that life continues after death—you just have to believe it. This attitude pleased the Franciscans, who had opposed the teachings of the Dominican Aquinas. Unfortunately, it was pushed too far by his followers, who argued strongly against the new learning that was emerging, and that is the subject of our next chapter. These Scotists, or Duns-men, provided such stupid caviling arguments against these new ideas that they also gave the English language the word "dunce." Duns Scotus didn't really deserve that.

12

Physics, Astronomy, and Persecution

I, Galileo, being in my seventieth year, being a prisoner and on my knees, and before your Eminences, having before my eyes the Holy Gospel, which I touch with my hands, abjure, curse, and detest the error and the heresy of the movement of the earth.

—ADW 1, 142

Liberty, above all, has been the most profound ideal of Judeo-Christianity, liberty of mind, and soul and body.

—Taylor Caldwell, Foreword to *Great Lion of God*

Two bishops, two kings, and a "Doctor Mirabilis" ("admirable doctor") initiated the European scientific revolution in the thirteenth century. One of the bishops was Robert Grosseteste (1175?–1253), first chancellor of the University of Oxford and, later, Bishop of Lincoln and teacher of Roger Bacon*. The other was Albertus Magnus* (1193–1280) of Cologne, Count von Bollstädt, for a few years Bishop of Ratisbon (now Regensburg) in Bavaria, and teacher of Thomas Aquinas. The kings were Frederick II of Hohenstaufen† (1194–1250), King of Sicily, and, for the last thirty-five years of his life, Holy Roman Emperor, and Alfonso X (The Wise) (1221–1284), King of Castile and Léon. The "Doctor Mirabilis" was Roger Bacon. Bishop Grosseteste and King Alfonso concentrated on the physical sciences, astronomy in particular, while Albertus and King Frederick II specialized in the life sciences, zoology and botany. Roger Bacon studied optics and wrote treatises on grammar, logic, mathematics, physics, and alchemy; above all, he argued strongly for the scientific method, as opposed to the reliance on authority, for understanding natural phenomena. This quintet was a galaxy of stars who brought an entirely new look to science in Christian countries and city states after centuries of neglect.

Albertus had a major impact on the world through his student Thomas Aquinas, but he also wrote twenty-six books on zoology and botany and he described nearly a hundred precious stones, very much in the spirit of Aristotle and Theophrastos. He dissented from the common belief that some trees give birth to birds and nourish them with their sap. He was also interested in the physical sciences to some extent, studying the climate and tides, performing chemical/alchemical experiments, but he opposed the idea that metals could be transformed into each other. He believed that the earth was a sphere, and he made an obscure reference to the possibility of people living on the other side of it, a dangerous thought to express unambiguously at the time. He supported the belief, however—indeed the creed—that a comet is a ball of fire thrown by an angry God, and that it is a predictor of diseases, wars, famines, tempests, and other punishments. It was an event of the earth and therefore below the moon, and anyone who disagreed was in deep trouble, as Michael Maestlin was to discover three centuries later. With it all, Albertus, or Albert the Great, deserves his title as a great pioneer of natural science, chemistry, and physical geography. He was beatified by the Roman Catholic Church in 1622, a few years after Galileo's first appearance before the Inquisition.

Frederick II, as Holy Roman Emperor, had his hands full with the Fifth Crusade (1227–1229)—which for a while made him King of Jerusalem as well—and in putting down a revolt by his oldest son. Somehow he found time to study and materially advance the science of zoology. Frederick prepared a detailed treatise on falconry, including anatomical details of falcons and other birds, their migration, and the mechanics of flight. He experimented with the artificial incubation of eggs, and tried to determine how vultures found their prey—did they see it or smell it? He kept a large menagerie and even traveled with it sometimes (prisoners of war were enslaved to perform the dangerous task of transporting the wild animals). But the animals were for study, not sport. However, he used the results of his study of their breeding habits to determine the hunting season, which, as emperor, he could enforce. In the tradition, perhaps, of the poet-caliph of Seville, al-Mutamid, he also wrote poetry. He aided the integration of Islamic thought into the Christian West. He was also excommunicated by the Church three times for political and theological reasons.

His contemporary in Spain, King Alfonso, also contributed a lot to the transfer of Islamic knowledge to the Christian world. He too was busy with his battles—he felt that Muslim scholarship was one thing but continued Muslim presence in Spain was another. Following the campaign of his ancestors against the Moors, he captured Cadiz in 1262 and helped to push them back across the water to Morocco, where many of them had come from originally, and where he felt they belonged. Like Frederick II, however, he was not content with sponsoring translations—he wanted to get involved, especially in astronomy. In 1252, Alfonso published the best planetary tables of the Middle Ages, the *Tabulae Alfonsinae,* based on the old tables from Muslim times prepared by the Toledo Jew Arsechiel, and ultimately going back to Ptolemy. He wrote, with assistance, five *Books on the Knowledge of Astronomy* and much later the moon crater Alphonsus was named in his honor. "Had I been present at the Creation," he is reported as saying, "I could have given the Deity some valuable advice." A great king, with many great qualities, modesty not being among them.

Alfonso's code of laws and other writings founded the Spanish literary language of Castile—in translation here is one of his thoughts on thought: "Thought is born in the mind of man, and it should be wrought not with anger . . . but with reason." He recognized that translations of Arabic writings, particularly those of Averroës, had begun to shake orthodox Christian beliefs and he reported that disbelief in immortality was very common among Christians in Spain, and not just among the scholars. The discovery that there was another religion and that its people were very civilized and its scholars very knowledgeable had come as a real shock. The eastern Muslim empire had been destroyed by Genghis Khan, the Moors were on the retreat in Spain, the two centuries of Crusades were coming to a close, and Marco Polo had left on his travels to return before the end of the century. It was a time not just for theology and scholasticism but for skepticism, rationalism, and science to resume their troubled courses.

In England it was the century of the Magna Carta (1215) and its later modification. Bishop Robert Grosseteste had written a treatise on the *computus,* a set of tables of the time for calculating astronomical events and movable dates. He made some criticisms of the physics of Aristotle, writing about motion and about heat, light, and sound. He studied some properties of mirrors and lenses, had some thoughts about the nature of the rainbow, and used an early form of the inductive method, starting from observation and inducing an explanation, moving from particular cases to a more general view.

After Grosseteste's death, his star pupil, Roger Bacon, commuting between the universities at Oxford and Paris, gained such fame that in 1265 Pope Clement IV wrote for him to prepare and transmit a general summary of the sciences of the time. Under adverse conditions, Roger worked furiously to comply, and three years later sent to the pope what we might call a book, a paper, and a letter—a long manuscript, a shorter one, and a summary. Unfortunately Clement died that year, and Roger never heard from him. Bacon believed very strongly, and made no secret of it, that knowledge of the material world can progress only by experiments backed by mathematical analysis. He rejected any authority that was claimed to be automatically correct and became fiercely critical of many of his contemporaries. He took al-Haytham's work in optics and extended it imaginatively. "From an incredible distance we might read the smallest letters and number grains of dust or sand. . . . So also we might cause the sun, moon, and stars in appearance to descend here below." But he adds, "Thus it is believed that Julius Caesar on the shore of the sea in Gaul, discovered through huge glasses the disposition and sites of the castles and towns of Great Britain." (UA 9, 350)

Thus was the later preliminary work on the telescope by Leonard Digges improved and pushed backwards in time through sixteen centuries; and it wasn't just light that appeared at the image—"bodies could also be so constructed that poisonous beings and influences and infections could be led off whenever men wished." Perhaps the hologram could be said to fulfill this prophecy: "Glasses could also be so constructed that every man could see gold and silver and whatever a man wished; and whoever should hasten to the place of the vision should find nothing."

The indefatigable Roger Bacon had many ideas on a wide range of practical applications of science: "Instruments of navigation can be made without men as rowers, so that the largest ships, river and ocean, may be borne on, with the guidance of one man, with greater speed than if full of men. Also carriages can be made so that without an animal they may be moved with incalculable speed. . . . Also instruments for flying can be made, so that a man may sit in the middle of the instrument, revolving some contrivance by which wings artificially constructed may beat the air in the manner of a bird flying. . . . Instruments can also be made for walking in the sea or rivers, down to the bottom, without bodily peril." (Du 4, 1010)

Bacon experimented in alchemy, knew how to make gunpowder, and, like Albertus Magnus, he was accused of black magic. The Franciscan Order managed to pillory both him and Thomas Aquinas simultaneously, condemning the "admirable doctor" Bacon to prison, and branding as heresy the "angelic doctor" Aquinas's works, as we have already seen. Bacon wrote, "Would that I had not given myself so much trouble for the love of science." He had sent his books to Pope Clement IV in the care of his student, John of London, who was prepared to answer any questions that the pope might have. This may have been the same Englishman named John Peckham (1220?–1292) who later became the Archbishop of Canterbury. Peckham wrote on optics and the eye as well as meteorology. Many of his views on science were learned from Bacon; he found himself in disagreement with Thomas Aquinas, which is not surprising, since he belonged to the Franciscan Order.

The Dominican scholar St. Vincent of Beauvais (1200?–1263) was also accused of magic and of conferring with the devil. He wrote a comprehensive encyclopedia, summarizing uncritically the scientific knowledge of the time, derived from recent translations of Arabic works into Latin, and quoting Albertus Magnus and other Christian authorities. He recognized that many people regarded the miraculous stories of saints as mere fables, but what would you expect from those who didn't even believe in the eternal punishment of hell? In the nineteenth century, a school of historical and religious portrait paintings flourished in Paris at the École des Beaux Arts, and Théobald Chartran (1849–1907) has given us his rendition of St. Vincent instructing the king of France, Louis IX (1214–1270), known as St. Louis. Vincent is standing in his white habit, his black cowl draped down his back and partially covering his bald head, lecturing to a class of one, a very attentive pupil, sitting nearby and staring up at him, thirsty for knowledge. Louis was a pious Christian king, devoted to justice and learning. In 1250, during the Sixth Crusade, he was captured by the Saracens, but he was ransomed four years later and remained king until his death. Meanwhile a Yorkshire man, John of Hollywood (1200?–1256), had crossed the channel to teach mathematics and astronomy in Paris. He was the first to use the recently translated writings of the Arabs on astronomy. His books on the calendar, on Ptolemy's *Almagest,* and on *algorismus*—the study of arithmetic—reached a wide circle of Christian scholars.

Towards the end of the thirteenth century, a few more scientists cropped up in various Christian countries. William, the canon of St. Cloud, in Paris, was the only medieval Christian astronomer to measure the obliquity of the ecliptic—he obtained the value 23°34′. His measurements revealed inaccuracies in the Ptolemaic tables that had been translated into Latin a century or so earlier. He also experimented with the camera obscura, using it to observe the sun, and developed a perpetual calendar. In Germany, Dietrich of Freiberg (d. 1311), the prior of the Warzburg monastery, was a physicist who studied light and colors, including the rainbow, basing his research on the works of Avicenna and Averroës. Three hundred and sixty-two years before Newton's famous experiment, Dietrich studied the dispersion of light by a crystal. Like Roger Bacon, he saw the importance of the experimental method for understanding physical phenomena. In Spain, Arnold of Villanova (1235?–1313) taught philosophy and medicine at Barcelona, moving also to France and Italy, writing books on astrology, theology, and chemistry when he wasn't acting as physician to kings and popes. He interpreted their dreams in terms of the superstitions of the time, and he made it clear while healing them that their wickedness was largely responsible for the evils of the world. For this and his ventures in the sciences he was condemned and imprisoned. Fortunately, Pope

Boniface VIII—pope from 1294 to 1303—rescued him, and he repaid the debt by curing the pope of kidney stones.

In Belgium, Henry of Bate (1246–1310 or later) observed the annular eclipse of the sun on January 31, 1310, but his main contribution was to astrology, translating the astrological works of Ibn Ezra. For an astrologer, he was very scientific, particularly when compared to modern standards. In France, a chemist named Peter, the canon of St. Omer, also flourished during the second half of the thirteenth century, writing a book on colored materials, with recipes for preparing pigments for painting. The Englishman Gilbert impressed his followers so much that they dubbed him "Dr. Desiderablissimus," i.e., the most desirable. In 1250, he became chancellor of the medical school at Montpellier, where he wrote books on taking care of one's health, on herbal medicine, on the works of Hippocrates, and much more. He appears to be the first person in Christendom to recognize that smallpox is contagious; he recommended that travelers drink distilled water, that people on sea voyages should take some fruit with them—and he strongly recommended surgery for some cancers.

From Belgium the physician John of Saint Amand (d. 1300) came to Paris to practise and to write a textbook on Greek medicine as it had been preserved, interpreted, and extended by the Muslim physicians. He recognized the importance of the experimental method in diagnosis and research. A little later, John of Jandun, near Reims, became bishop of Ferrara. He wrote on motion and on nearly all aspects of the known sciences—meteorology, precious stones, metals, animals, the physics and zoology of Aristotle, the works of Averroës and Albertus Magnus, plants and planets, and human souls. Scientific inquiry was, like spring, breaking out all over. With kings, bishops, even popes learning and supporting new ideas in science, it had seemed that a renaissance in learning was at hand. The Church again missed an excellent opportunity to embrace, or at least not to suppress, the inevitable advance of science, against which it was to fight so fiercely during the centuries that followed. Some churches are still fighting it today.

We follow into the next century the works of some scholars at the universities where Roger Bacon had taught, because they really *did* ultimately lead to a revolution, the revolution of Copernicus, Kepler, Galileo, and Newton. One of these men, William of Ockham (a town in Surrey, England) (1285?–1349) studied at Oxford, and was more a philosopher than what we call a scientist, but we remember him here for his statement "Ockham's Razor" that is so essential to science: "*Principia non sunt multiplicanda praeter necessitatem*"—hypotheses are not to be multiplied unnecessarily. The other three scholars, Thomas Bradwardine (1290?–1349) of Merton College, Oxford, Jean Buridan (1300?–1358?), and Nicole Oresme (1323?–

1382) in Paris, decided to make a study of motion, in particular motion on earth. Little had happened on the subject since we left it in chapter 5.

Bradwardine later became Archbishop of Canterbury, but at Oxford he had figured out the difference between what we call kinematics—the description of motion—and dynamics—the causes of motion, thereby resolving some of the problems that had bothered Grosseteste and others. In the early 1330s, Bradwardine and a number of colleagues—the "Merton School"—carried this further by recognizing velocity as a ratio of distance to time, noticing its instantaneous and changing nature, and specifying the kinematical description of uniformly accelerated motion. Although algebra was known to Hindu and Arabian scholars centuries before, it would have to wait until the sixteenth century for François Viète to apply it to problems in physics. With hindsight then, we can say that the Merton School discovered the relation

$$v = u + at$$

for motion with uniform acceleration a. Here u is the initial velocity and v is the velocity at a time t seconds later. They also recognized that the distance traveled (s) equals the average velocity ($\bar{v}$) times the time:

$$s = \bar{v}t$$

Finally, one of them figured out that for uniform acceleration the average velocity is the average of the initial and final velocities:

$$\bar{v} = \tfrac{1}{2}(u+v)$$

Put the three equations together and you have:

$$s = ut + \tfrac{1}{2}at^2$$

Thus, for uniformly accelerated motion from rest (u=0) the distance traveled is proportional to the *square* of the *time*. Also, with a little algebra,

$$v^2 = u^2 + 2as$$

so that, starting from rest, the speed built up is proportional to the *square root* of the *distance* traveled. These equations are taught (or should be) in high-school physics classes, but at the time they were not really obvious. Some 270 years later even the great Galileo assumed in one passage of his writings that for uniformly accelerated motion, which he applied to falling bodies, the velocity increases linearly with the *distance* and (correctly) that the distance increases as the square of the time, although these two statements are inconsistent with each other. It didn't take him long to discover the mistake though.

Bishop Oresme, which he was during the last five years of his life, had learned to draw a graph, plotting velocity against time, and noting that, for uniform acceleration, the curve becomes a straight line and the area under it is the distance traveled—the germ of integral calculus. He also suggested that, instead of saying that the earth is fixed and the stars move, one could say that the stars were fixed and the earth moves, "just as it seems to a man in a moving boat that the trees outside the boat are in motion." We do not know if he had read Lucretius:

> We sail along in a ship, it keeps on moving,
> But seems to be standing still, and what is still
> Seems to be moving past
>
> (L, IV 387–389)

But we do know that for Oresme it was a dangerous thing to say. Whether or not he really believed that the earth moved, he made it clear in his writings that he did *not* believe it.

Wisely, he also waffled on whether there could be people living on the other side of the earth. Six hundred years earlier, this question had been settled unambiguously in the negative by the ninety-first pope, St. Zacharias. Oresme must have known of the horrible fate of Cecco d'Ascoli* in 1327, burned alive in Florence for teaching heresies, including his disbelief in the judgment of the long-departed Zacharias. This execution had occurred a few years before Oresme was born, and in order that all might learn the lesson, the painter Andrea de Cione, known as Orcagna (1308?–1368), was soon commissioned to prepare a fresco at the Campo Santo in Pisa, depicting Cecco suffering the flames of hell. It is hard to believe that the Christian Church could stray so far from the teachings of Christ. Unfortunately, four more centuries of repression, torture, and burnings were yet to come.

One of Oresme's teachers—they were all studying theology—had been Jean Buridan, known, perhaps unjustifiably, as the author of the story of Buridan's ass—the donkey, symmetrically placed between two loads of hay, starving to death because it couldn't decide which one to eat first. Certainly, Buridan did write on free will and decision-making, but our interest here is that he argued against all of the theories about motion suggested by Aristotle and insisted on by his successors. He preferred the ideas about motion that had been proposed by Philoponos, possibly rediscovering them. It was he who introduced the concept of *impetus*—not very well defined, but a quantity we can recognize as momentum. This impetus is a property of the moving object, given to it by the person who throws it, or acquired by it during its fall under gravity. It can be destroyed by a resisting medium, or, for upward motion, by the weight of the projectile. Further—and here was a giant step, but a speculative one—the idea can be applied to celestial objects; the stars were originally given an impetus and have been moving in their "natural manner" (circular paths) ever since. The general belief, dating from Aristotle, that local and celestial motions had nothing to do with each other was questioned for the first time.

Not very sexy stuff, particularly when compared to Boccaccio's bawdy *Decameron* that was published around the same time, and certainly not as world shaking, it would seem, as the great *Divine Comedy* that Dante (1265–1321) had begun to write in the year 1307. The very idea that the planets might be governed by the same laws that determine the motion of an arrow, vaguely perceived by Buridan, was diametrically opposed to conventional belief. The heavens were divine and were governed by different rules; indeed, Dante located them on the seven planets, on the sphere of the fixed stars, and on the empyrean sphere, the highest heaven of all. Lectures were arranged in Milan and Florence to explain the *Divine Comedy*—Boccaccio was the first lecturer in Florence, at an annual salary of one hundred florins. Interest ran high, as people wanted to know what Limbo, Purgatory, and Paradise were like and, in particular, in which location some of their departed friends and enemies had ended up. Dante had mentioned many of the souls in an oblique way, so there was great speculation as to who they all were. The fame of the wonderful poem spread, and with it came the belief that these fantasies were facts. His magnificent vision became a reality, and the last word on astronomy, unchallenged for a hundred years.

Another bishop, Albert of Saxony (1316?–139?) helped to found the University of Vienna and became its first rector. He had studied with Buridan and tried to work out a rule to describe the way that falling objects increase their speed. He did not go along with the crazy idea that the earth was moving, but he did recognize that much of the shape of the earth's crust is due to erosion. He wrote in the spirit of Theophrastos on herbs, precious stones, and minerals, and he taught the new logic and philosophy of William of Ockham. The studies on mechanics were ultimately handed on to Leonardo da Vinci and Geronimo Cardano by the writings of Biagio Pelacani, who received his doctorate from the University of Pavia in 1374. He also wrote commentaries on the works of Aristotle in natural philosophy, including *Meteora,* i.e., "things above." (As early as 450 B.C., Diopithes had directed suspicion against Pericles and Anaxagoras by proposing a decree requiring public accusation of those who neglected religion or proposed new doctrines about *Meteora.*)

In England, a very courageous man, John Wycliffe (1320?–1384) helped to limit the control that the church had over secular matters while he attacked the worldliness of the church in general, and of the friars in particular. He also was a Fellow of Merton College, Oxford; in 1377 he was accused of heresy by the reigning pope, Gregory XI, and summoned to appear before the bishop of London to answer the charges. The hearing was disrupted by the crowd that was well aware of the way that the friars behaved, and public opinion was so strong that Wycliffe was acquitted at another trial the following year. He then preached about a religion that is personal and lies within, rather than one that relies on formalism alone. He denied the right of priests to offer absolution; and he denied, too, the doctrine of transubstantiation, and for this and other heresies he was forced (1382) into "early retirement," relinquishing his teaching position at Oxford, and retiring to his rectory at Lutterworth. There, until his death two years later, he worked on his translation of the Bible into the English of the time—the time of Chaucer. The translation and editing were completed by his assistant, John Purvey, and published in 1388.

Wycliffe has been called "the morningstar of the reformation"; he was a good and great man. At the Council of Constance in 1415 the dead Wycliffe was denounced as a heretic and it was ordered that his remains be removed from the consecrated churchyard at Lutterworth. The next pope, Martin V, informed Bishop Richard Fleming of Lincoln that there had been too much delay and the sentence must be carried out. The grave was opened, Wycliffe's bones were burned, and his ashes cast into the stream called Swift that passes by the church. A later clergyman, Thomas Fuller (1608–1661), wrote that this took the ashes "into the Avon, Avon into the Severn, Severn into the narrow seas—then into the main ocean, and thus the ashes of Wycliffe are the emblems of his doctrine which is now dispersed all the world over." (CB, 1043)

It was a German clergyman and natural philosopher, Nicholas of Cusa (1401–1464), who was the first person to openly reject the classical theological model of the cosmos. As suggested two thousand years earlier by members of the school of Pythagoras, he thought that the earth might be moving—rotating on its axis—and he realized that for a hypothetical observer on the sun the earth would appear to be moving in an orbit. He also worked on improving the Alphonsine astronomical tables, and suggested that pot-plants should be weighed carefully to see how their weights were changing as they grew. He also studied the body of canon law and pronounced some of its papal decrees as counterfeit. He made many other contributions, not the least being a proposal to Pope Pius II to formulate a program of reform for the Church and its decadent dignitaries. Nobody but Pius seemed to be interested. (Only a few years earlier, the eastern Christian empire, still centered in Constantinople, had been overwhelmed by the Muslim Ottomans under Sultan Muhammad II. The decimation of Constantinople by the Crusaders two hundred and fifty years earlier still rankled, but Emperor Constantine XI had hoped for help from his fellow Christians in Rome, under Pope Nicholas V. It never came. That was practically the end of intellectual studies in Constantinople for a long time.)

During his short life (1436–1476) another German, Johann Müller, known as *Regiomontanus,* set up an observatory at Nuremberg, recording in 1472 what would come to be known as Halley's comet. Müller questioned the legacy of Ptolemy, wondered about motion of the planets around the sun, and saw that more precise observations were required. His astronomical almanac for the years 1475–1506 was used by Columbus and other navigators, and Pope Sixtus IV invited him to Rome to assist with the reform of the calendar. Nothing more was heard of the early ideas about motion proposed by the Oxford and Paris schools until Leonardo da Vinci (1452–1519) addressed the question, along with practically all other questions about nature that could be conceived at the time.

How could a man like Leonardo, almost an exact contemporary of Christopher Columbus, suddenly appear out of nowhere, it seemed, and make such magnificent contributions to art, science, medicine, and engineering? His science alone was worth more than that of any other man before him outside of the best of Athens, Baghdad, and Cordova. At his death his young companion, Francesco Melzi, wrote, "it is not in the power of nature to create another."

He read the works of Buridan and stated, correctly, that the speed of a body falling vertically is proportional to the time of fall from rest, but he also stated, incorrectly, that it was proportional to the distance through which it had fallen. At that time there was no way to find out by experiment. At one point he stated that "everybody will follow its path in a straight line as long as the nature of the violence done by its motive force persists in it," but, like Aristotle, he could not extrapolate this apparent suggestion of Newton's first law to include motion in a vacuum. He appeared to understand projectile motion intuitively, since a number of his drawings of military bombards show curved paths, in some cases almost parabolic, for the explosive objects those primitive cannons were designed to project. Leonardo contributed ideas on devices like pulleys, levers, cranes, on the transmission of sound through water, on hydrostatics, hydraulic engineering, and canal construction. He observed the resonance of a lute when a note from another lute is sounded nearby, and of course he had a deep knowledge of the eye, and of light and shade. He stated that "mechanics is the paradise of the mathematical sciences, for by its means one comes to the fruit of mathematics," meaning its application. Mathematics was his worst subject, but only because it had so much competition. The work of Theophrastos on botany was well known to him, and he recognized that the rings in the cross-section of a tree revealed its age, their widths denoting the annual rainfalls. Leonardo dissected, he said, thirty human cadavers, and with his remarkable artistic skills he produced many detailed anatomical drawings. He even poured wax into the heart of a dead bull in order to understand its shape! He recognized that the blood returning to the heart is not the same as the blood leaving. He was a one-man renaissance of learning and art.

As we have noted, the percentage of truly creative geniuses in the population probably doesn't vary so much from one time and place to another, but what they are able to accomplish depends critically on the spirit and attitude of their immediate contemporaries, especially of the "Big Boss," if there is one. It is fortunate that Leonardo lived at a time when the Roman Popes were especially interested in strengthening their secular powers, giving correspondingly less attention to theology. More than that, nearly all of them were very supportive of art and architecture, although mostly of the work of others rather than that of Leonardo. Sixtus IV had the Sistine Chapel built, and naturally it was named for him; Innocent VIII sponsored the Mantegna frescoes in a Vatican chapel, but ran short of money and had to sell some cardinal hats to make ends meet. Alexander VI was too busy with wars and with his son and daughter, Caesar and Lucretia Borgia, to do much else than watch over the far-flung Church, which he did with diligence. However, he had forgotten or purposely ignored Acts 8:20 in which Peter condemned Simon for thinking that "the gift of God may be purchased with wealth." Alexander and his successors raised a lot of money by simony, enough to stir Martin Luther into action in the coming decades. And during Alexander's papacy, Nicholas Copernicus spent the year 1500 in Rome, giving some lectures in mathematics.

Pope Julius II brought many artists and architects to Rome, including Michaelangelo and Raphael, he gave the world a new Saint Peter's, and he commissioned the wonderful painting on the ceiling of the Sistine chapel that presents an image of the creation of our earliest ancestor by a large God in the image of a man. Following Julius II came Pope Leo X† (1475–1521), listed as one of our "Helpful Despots," but also referred to in chapter 8 as a candidate for the title of Beast of the Apocalypse. His support of the arts, and his encouragement of translations and of the archaeology of ancient Rome was accompanied by the decadence of his administration, and his profligate waste of money that left Rome bankrupt. In the last year of his life he conferred the title of "Defender of the Faith" on England's Henry the Eighth because Henry had attacked Martin Luther in print. It was also the year in which the excommunicated Luther was banned by the Diet of Worms, having denied the supremacy of the pope, and having publicly burned his notice of excommunication.

In that same year also, Hernando Cortes captured Mexico City, and Ferdinand Magellan was killed on the small island of Mactan in the Philippines. When Juan Sebastian del Cano and his men completed the voyage around the world that Magellan had led from Spain, the thousand-year-old belief that there were no inhabitants on the other side of the world was confronted with some new information. St. Augustine had pointed out that "scripture speaks of no such descendants of Adam," and in any case the Almighty would prevent people from living there because they wouldn't be able to see Christ arriving from heaven at the time of his Second Coming. As noted, it had been made an official dogma in the eighth century that such people did not exist, and an entrenched belief like that can't be destroyed suddenly by the tales of a bunch of sailors. It took close to two hundred years for the fact, supported by much more data, to be accepted by everyone, but around that time it had caused enough doubts for the Spanish theologian Tostatus to argue that "the apostles were commended to go into all the world and to preach the gospel to every creature; they did not go to any such part of the world as the antipodes; they did not preach to any creatures there; therefore no antipodes exist." With the return of the survivors of Magellan's expedition, however, the ancient edict of Pope Zacharias was quietly laid to rest as a credo of the Church.

Five years later the troops of Charles V, Holy Roman Emperor and King of Spain, ravaged Rome's inhabitants and its treasures and imprisoned Pope Clement VII. He and Charles soon made up, however, but by the time of the pope's death in 1534 England and Denmark had broken away from the Catholic Church of Rome. Charles V saw to it that Spain, France and Italy would stay in the Catholic fold and that the pope would not sanction the divorce of Henry VIII from Catherine of Aragon, who happened to be Charles's aunt.

As Protestantism began to develop in the more northerly parts of Europe, the choice of which variety of Christian faith to adopt in any area was determined by its ruler. When one ruler died or was deposed, his replacement often opted for another creed. The inhabitants were then faced with the law of the jungle—to adapt, emigrate, or die—the same law that people had faced in the wars between Christians and Muslims in Spain and the Near East. In a period of thirty years, the English were Roman Catholics, then Henry VIII-type Catholics, severed from Rome, then Protestants (Edward VI), then Roman Catholics again (Mary), and finally

Protestants yet again under Elizabeth I. It wasn't a question of theology, it was a question of political power, an undeclared civil war, with each faction in charge persecuting all rivals. The Christian-Muslim battle remains with us today, especially in Beirut and Armenia-Azerbaijan; the Catholic-Protestant battle continues in Northern Island.

The common worship of God, and now the common worship of him through Jesus, was powerless to prevent leaders from inflicting unspeakable evils on anyone suspected of having any idea that was sympathetic to any other creed. Martin Luther had translated the Bible into German, and when William Tyndale* (1492?–1536) wanted to translate it into English during the reign of Luther's avowed enemy, Henry VIII, he got no help from the Bishop of London, but he eventually managed to get his translation published in Germany. The shocking story of the destruction of these Bibles by English bishops, and of Tyndale's imprisonment and tortured death forms a dark blot on the history of England and Belgium.

Others who translated the Bible into English fared according to the sometimes-fluctuating decrees of the ruling monarchs. A year before Tyndale's death, Miles Coverdale (1488?–1569) published in Zurich the first English translation of the whole Bible, including the apocrypha. Within a very few years his *Cranmer's Bible* (after the Archbishop of Canterbury, Thomas Cranmer, 1489–1556) was shown to Henry VIII, who, surprisingly, ordered that copies be placed in all churches. Not so fortunate was John Rogers (1500?–1555) who worked to improve the translations by Tyndale and Coverdale. Rogers spoke against the pope at a time when the Roman-Catholic Queen Mary had succeeded her father, Henry, and he was sentenced to death. "His wife, with nine small children and one at the breast, followed him to the stake" (from the *Martyrdom of John Rogers* February 4, 1535). Archbishop Cranmer was also burned at the stake along with many other Protestants and Henry–Catholics during the short but terrible reign of Bloody Mary. All in the name of Christ.

In 1542, Charles V encouraged Pope Paul III to establish an Inquisition in Rome, at first to root out any converts to Luther's teachings (or the newer challenge of John Calvin [1509–1564]), but later to persecute all who deviated from orthodox beliefs. It was the year before Nicholaus Copernicus was to see a copy of his revolutionary book, just before he died. Copernicus, well aware of ideas from the long past, eventually came to believe that it was not enough to patch up the theory of Ptolemy—an entirely different approach was needed, and he provided one. Copernicus had delayed publication of his work almost to his last hour, partly because he feared persecution from the Church but perhaps mostly because he knew that his theory wasn't quite correct, and he wouldn't be able to defend the consequences of all of his calculations. More than that, he knew that he would not be able to take the ridicule that would multiply itself as people became aware of his insane idea that the earth was moving. There had been some examples of that already, although he had been secretive about his work. The stories of his struggle with the problem and with himself, and of the personalities and problems of those who followed him—Tycho Brahe (1546–1601) and Johannes Kepler (1571–1630)—have been told magnificently by Arthur Koestler in *The Sleepwalkers.* Here we restrict ourselves to a few technical details—would that the struggle to understand were simply a matter of overcoming "a few technical details."

It was mentioned in chapter 9 that, along with the works of other Greek astronomers, Aristarchos's treatise on the solar system had been translated into Arabic during the latter part of the ninth century and the beginning of the tenth. In 1488 a translation into Latin of this translation from the Greek was published in Venice. The idea that the sun might be at the center, with the earth and other planets circling it, had been tossed around by others before this translation became available, and Copernicus's contemporary, the poet-philosopher from Ferrara, Celio Calcagni, produced a manuscript that had the fixed sun and stars, together with the moving earth, as its unsupported thesis. It was Copernicus who backed up the idea with detailed calculations, but since it was based on circular motions, it needed as much patching up as Ptolemy's standard theory in order to bring it into agreement with the data. After all, it was essentially the same idea as that of Aristarchos, and we have seen how that was rejected, partly because it didn't have all the wheels within wheels that the Procrustean bed of circular motion required. So Copernicus decorated his model with the planets attached to small circles (epicycles), the centers of which moved on other circles (deferents) that were centered on the sun rather than the earth. He rejected the equants of Ptolemy as being too artificial, but the job of getting each planet to move in the right path, and to arrive at the right point of it at the right time was just as difficult with deferents centered on the sun as it was when they were centered on the earth. Of course the basic circular motion was much simpler, and the model showed clearly why Venus and Mercury always appear close to the sun. In addition, it exhibited a regularity, later to be made more precise by Kepler, that the periods of the planets increased monotonically with the radii of their orbits.

Five years after Copernicus died, there was born in Nola, near Naples, a baby who grew up to be a restless philosopher and an avid supporter of the theories of Copernicus, Nicholas of Cusa, and Lucretius. Giordano Bruno* (1548–1600) was a Dominican priest, but after four years he was accused of heresy and fled, wandering throughout Italy before trying the Calvinist atmosphere of Geneva. Calvin had died only twelve years earlier. Bruno found that the tradition of his intellect, acuteness, and courage did not make up for that of his intolerance, despotism, and rigor, and he moved to Toulouse to study for the M.A. He went on to Paris, England, and Germany, each for a few years, stirring up interest and hostility wherever he went. In 1589, he found himself excommunicated by a superintendent of the Lutheran Church. Eventually Bruno decided to return to Italy—this time to Venice—but the Inquisition caught up with him in 1593 and transferred him to a prison in Rome. Seven years later he was delivered to the secular magistrate with the command "That he be dealt with as mercifully as possible, and punished without effusion of blood." He replied, "Your sentence strikes more terror into your hearts than into mine," and on February 17, 1600, he was burned at the stake in Rome's Campo dei Fiori, the Field of the Flowers.

Bruno had imagined that there could be planets encircling other stars, and he saw the earth and the other planets as living organisms. He believed in an infinite God, whose universe was to be studied; he thought Aristotle's view of it should not be accepted without challenge and enquiry. He had been influenced by the early neo-Platonists, Plotinus in particular, and he embraced a spiritual pantheism that regarded the whole universe as a personal God. He was searching for an absolute principle from which the mysteries of mind and matter could be explained. His cause was not helped by his ill-concealed contempt for the Church and its priesthood. Perhaps it was ultimately helped by it though, because Bruno's life and death initiated the defeat, though not even now the final demise, of medieval thinking about the world.

During this time there had arisen a serious objection to the model that Copernicus had proposed. It was an objection that couldn't be overcome by adding an epicycle or tinkering with the value of a parameter. If the earth moves around the sun, why don't the stars appear to change their positions during the year? The improved instruments that Frederick II, king of Denmark and Norway, had financed for Tycho Brahe allowed him to measure angles with an accuracy of considerably better than a tenth of a degree. Brahe would therefore have been able to detect this "parallax" effect if the stars were less than a hundred billion miles away, a distance that it was impossible to imagine. Either the stars are more than about a thousand times as far away as the sun, or the earth doesn't move. Brahe chose the latter alternative, with the earth at rest and the sun moving around the earth and the other planets moving around the sun. For a description of the solar system it was almost equivalent to Copernicus's model, but it avoided the parallax problem and added the comforting feeling that, despite all of these theories, we are at rest at the center of the universe.

It was a geoheliocentric system, in the spirit of the model first postulated by Heracleides nearly two thousand years earlier, and referred to in chapter

5. It was also like the model of Felix Capella, who was close in space and time to St. Augustine, but not in interests. Erigena too had a similar idea, suggesting that Mars and Jupiter might be circling the sun, as noted in chapter 11. Brahe saw it as a compromise between the extreme models of Ptolemy and Copernicus, one that held the earth fixed, but for observational rather than theological reasons. (A serious attempt to observe this stellar parallax was made in 1727 by James Bradley [1693–1762] but the effect was not measured until 1838 by F. W. Bessel [1784–1846] using improved equipment and thereby directly determining the distance to the star 61 Cygni. Bradley, however, measured the motion of the earth indirectly from the effect that this motion has on the angles at which light reaches us from the stars.)

Tycho Brahe also threw new doubts on the validity of the Aristotelian belief of perfection and immutability in the skies by his observation of a nova in 1572, and by his measurements that showed that the comet of 1577 was far above the moon in a region that had been believed never to change.

Convinced that Copernicus was basically right, and equally convinced of the divine mathematical beauty of the universe, Kepler struggled with the data for years and finally, to his great elation, thought that he had found the answer: "Circumscribe a twelve-sided regular solid about the orbit of the earth; the sphere stretched about this will be that of Mars. Let the orbit of Mars be circumscribed by a four-sided solid. The sphere which is described about this will be that of Jupiter. Let Jupiter's orbit be circumscribed by a cube. The sphere described about this will be that of Saturn. Now place a twenty-sided figure in the orbit of the earth. The sphere inscribed in this will be that of Venus. In Venus's orbit place an octahedron. The sphere inscribed in this will be that of Mercury. There you have the basis for the number of planets."

That there are only five regular polyhedra was known to Theaitetos (c. 415–369 B.C.), a contemporary of Plato. Inclusion of the earth as a planet in addition to the five planets known at the time made six spheres between which the five regular figures could be placed, thereby determining the radii of the orbits and the number of planets.

Fortunately, Kepler did not stop at this. The dedication of Tycho Brahe to the need for accurate astronomical observations was strongly motivated by his secret use of these data in astrology. Kepler eventually became an assistant to Brahe and thereby gained access to this jealously guarded information. Because of its proximity and what turned out to be the eccentricity of its orbit, Mars had been the most intractable of the planets, and Kepler spent the first six years of the seventeenth century trying calculation after calculation to fit the observed motion of Mars. He returned to the equant of Ptolemy which Copernicus had spurned, and he gave a bigger role to the sun, in accordance with his philosophy, by letting the planes of the various orbits intersect the sun rather than (as Copernicus had imagined it) the earth, but an error of 8′ of arc persisted, beyond the accuracy of Brahe's measurements.

Kepler faithfully recorded his errors and triumphs, and although he made a number of mistakes that he failed to catch, he calculated the motion of the earth as seen from Mars, postulated an egg-shape for the orbits, and tried other shapes. He was drawn to the elliptic shape because he understood its geometry, which he used as a basis for understanding other shapes, but he failed to recognize the elliptical nature of the orbits until a clue appeared from the data on Mars.

It had become apparent to Kepler that the orbit of Mars was an oval, and that the difference of the two axes divided by the length of the larger one was 0.00429. As a planet moves around the sun, the angle between the line joining it to the sun and the line joining it to the center of the orbit changes. For Mars the maximum value of this angle, as determined from Brahe's measurements and Kepler's analysis, was 5° 18′, and the secant of this angle is 1.00429. From the multitude of numbers that his studies had embedded in his mind, he spotted this coincidence, which, if the orbit is not too far from being a circle, is a consequence of the properties of an ellipse. It still took some time for Kepler to recognize that the orbit *was* an ellipse, indeed he put this coincidence aside because he had another idea to test. Maybe the orbit was an ellipse! Eventually, he got it all together, and now we can say that when that maximum angle is expressed in radians it has the value 0.0926, and that if the motion is an ellipse, the eccentricity of the ellipse must have that same value. (The eccentricity, e, is defined as the square root of the difference of the squares of the lengths, a and b, of the semi-axes, divided by the length of the larger one: $e = \sqrt{a^2 - b^2} \,/\, a$.)

The magic number 0.00429 noted by Kepler is directly related to the eccentricity, being one half of its square: $\frac{1}{2}(0.0926)^2 = 0.00429$ for ellipses that are close to being circles. Although they were close to circles, they were, nonetheless, ellipses in their own right, and the shadows of Plato and others who had insisted that heavenly motions must be circular began to slowly fade away. Of course you have to be very wary about coincidences like this. Kepler was able to show that *if* Mars moved in an ellipse, then these two numbers had to be the same, and that led him to suppose that the other planets also moved in ellipses and to test the consequences of that assumption. Without this sort of understanding the equality of two numbers may have as much significance as for those discussed as numerology in chapter 8.

Why had Ptolemy's clockwork system describing the motions of the planets survived for so many years? For all sorts of psychological, religious, and political reasons of course, and it had been corrected from time to time whenever someone noticed that it had been too far off the mark. While still in his teens, Tycho Brahe had noticed that even the corrected form of Ptolemy's tables, which had been offered by Alfonso X and published only three hundred years earlier, were already off by a month or so. That was intolerable, but it was an error of only 0.003 percent. For all other measurements of the time, and many made since, that sort of accuracy was beyond the reach of the best scientists, but if you knew what you were doing you had only to look up and see that the planets weren't where they were supposed to be. Nevertheless, there must be something correct behind all this theory; otherwise, after a lapse of a few hundred years, there would be no connection at all between prediction and observation. How did Ptolemy and his predecessors incorporate the consequences of two of Kepler's Laws (each planet moves in an ellipse with the sun at one focus, and the line from a planet to the sun sweeps out equal areas in equal times) without having the least idea of them?

We saw in chapter 5 that two circular motions can easily be combined to make an elliptical motion. One condition that would ensure this is that the planet moves around one circle (the epicycle) with constant angular velocity while the center of that epicycle moves on another circle (the deferent) with the same angular velocity, but in the opposite sense. The sum and difference of the radii of the circles would then define the semi-axes of the ellipse. If the radii are equal, the path is a straight line. That arrangement, however, would have the planet moving around the ellipse with constant angular velocity, in violation of the second of the laws quoted above, or, as Ptolemy had seen it, of the data. Ptolemy had discovered that if it were assumed that the angular velocity of a planet was constant, not about the center but about a particular point (the center of the equant) displaced from the center, much better agreement with observation resulted. This is his theory of equants, which, as we have seen, Kepler tried and Copernicus had rejected as a fake. And so it was, but why did it work? An ellipse has two foci, and it turns out that when a planet moves along its elliptical path according to Kepler's laws, its angular velocity about the *other* focus is very close to being constant. It is remarkable that Ptolemy was able to find this point, not knowing that it was the empty focus of the ellipse along which the planet moved. The error here is proportional to the square of the eccentricity of the orbit, so that an eccentricity of one percent, as for Mars, would mean an error of only one-hundredth of a percent. If a point some distance from this empty focus is chosen as the center of the equant, the error would be proportional to the eccentricity itself, in this case one percent, and completely unacceptable. For Mars, that empty focus is approximately twenty-six million miles from the center of the sun. If its eccentricity had been very much greater, Kepler's numerical "coincidence" would not have occurred, and Ptolemy may not have developed his theory of equants.

So much for the regular celestial motions, but what about the apparently irregular occurrences like eclipses and comets? There is a legend, unfortunately only a legend, that the wise philosopher Thales predicted that an eclipse of the sun would occur on the date that we now call May 28, 585 B.C. When it happened, the Medes and the Lydians, who had been fighting each other for ages, declared a peace, ensuring it by a marriage arrangement between the families of their leaders. (We've lost something from all this science. Solar eclipses don't seem to have that effect these days.) Eclipses of the moon were recognized early as being due to the shadow of the earth. The Pythagoreans and Anaxagoras recognized this and made inferences from the fact that the shadow's edge was circular. Philolaos, however, thought that the eclipses might be due to the counter-earth that he had introduced into cosmology.

One way to test the reliability of an ancient historian is to check on any record of eclipses that he may have made. Herodotos (486–426? B.C.) claimed that an eclipse had happened in 480, just before the battle of Salamis, but that was not correct. Thucydides (471?–400 B.C.), on the other hand, told of the dates (August 3, 431) of a partial solar eclipse, of an annular solar eclipse (March 21, 424), and of a lunar eclipse (August 27, 413)—all of which have been verified. Of course these occurred during his adult life; the eclipse mentioned by Herodotos was thought to have occurred when he was a small boy, and it was already a legend by the time he grew up. It doesn't take very long for a non-event to become a generally accepted happening.

Unusual events make people afraid, especially if they appear to be signs of the anger of a god. Plutarch tells how Pericles (499–429 B.C.) prepared a large fleet to attack Epidauros:

> . . . it happened that the sun was eclipsed, and it grew dark on a sudden, to the affright of all, for this was looked upon as extremely ominous. Pericles, therefore, perceiving the steersman seized with fear and at a loss what to do, took his cloak and held it up before the man's face, and screening him with it so that he could not see, asked him if he imagined there was any great hurt, or the sign of any great hurt in this, and he answered no. "Why," said he, "and what does that differ from this only that what has caused that darkness there is something greater than a cloak?"

A terrible plague had struck, the expedition fared badly, and the Athenians relieved Pericles of his command. The eclipse *must* have had something to do with it; at least, so it would have seemed.

Aristotle also noted how the shadow of the earth on the moon shows that the earth is round, a fact verified by moving north or south and observing that some other stars appear while some that were familiar were no longer visible. Archimedes's orrery, mentioned in chapter 5, was sufficiently accurate to predict eclipses of both the sun and moon.

The deaths of ancient Greek heroes were believed to be accompanied by a darkness over the earth, presumably due to an eclipse. For the Romans, darkness was thought to have descended at the deaths of Romulus and of Julius Caesar. In Christian countries, darkness is believed to have engulfed the world during the time of the crucifixion (Luke 23:44), but neither Seneca nor Pliny the Elder, alive at the time, tell of it.

St. Augustine recognized that some philosophers were able to foretell the occurrence of eclipses, but he reproached them:

> For with their understanding and wit, which Thou bestowedst on them, they . . . foretold, many years before, eclipses of those luminaries, the sun and the moon—what day and hour, and how many digits—nor did their calculation fail—they foresee a failure of the sun's light which shall be, so long before, but see not their own . . . (they do not) slay their own soaring imaginations as "fowls of the air" nor their own diving curiosities (wherewith, like "the fishes of the sea" they wander over the unknown paths of the abyss) nor their own luxuriousness as "beasts of the field" that Thou, Lord "a consuming fire" mayest burn up those dead cares of theirs and recreate themselves immortally.

The war between science and Christian theology had begun.

In the eighth century A.D., Bhavabhúti wrote about Màdhava and Màlati, the Romeo and Juliet of India. As the heroine Màlati is about to be sacrificed by the priest and priestess of the goddess Chamundà, Màdhava rescues her and says, among other things:

> Blest was the chance
> That snatched my love from the uplifted swords
> Like the pale moon from Ràhu's ravenous jaws.

Ràhu was the dragon who was believed to cause eclipses by swallowing the moon.

There are countless legends like that in different countries. They are primitive, but no more primitive perhaps than Increase Mather's statement nearly a thousand years later that the comet that appeared around the time of the death of President Chauncey of Harvard in 1672 was a sign that nature was grieving over his departure. But perhaps the good Mr. Mather was engaging in hyperbole, although other statements he made indicate that he was serious. If eclipses could cause some distress, comets could lead to panic. The very early belief that comets were balls of fire thrown by an angry god was uninfluenced by Christianity, except for the change from god to God. Albertus Magnus, one of the great men who sparked the thirteenth century, certainly believed that was what comets were. A comet, a heavenly light, or a special star often appeared at the birth of anyone who was special enough to have a legend built up about him—Abraham, Moses, Lao-Tse, Asclepios, Buddha, Christ, and others. Julius Caesar lived and died with portents. As his wife Calpurnia warned him not to go to the Senate on that fateful day, Shakespeare gave her these words to say:

> When beggars die there are no comets seen;
> The heavens themselves blaze forth the death of princes.

to which Caesar replies:

> Cowards die many times before their deaths;
> The valiant never taste of death but once.
>
> (*Julius Caesar,* Act 2, Sc. 2., lines 30–33)

Comets, as we have seen, were officially declared to lie below the moon in the space of mortals and of sin and corruption. If you want proof of a comet's significance, remember the comet of 1066 and what a terrible year that was for England.

During the early years of the sixteenth century the fear of comets grew, stimulated by the general belief in superstition and by exhortations from the pulpits. The comets were condensations of man's sins; their variously shaped tails were specific messages of warning from God. They brought pestilence and famine; they predicted or accompanied important events. It was impious to study anything about them except their divine meaning. Peter Apian (1495–1552), however, kept track of them and noted that their tails pointed away from the sun. Blaise de Vigenère (1523–1596), a French humanist and cryptographer, pointed out in 1578 that many important monarchs had been born and had died without the advent of a comet. Thomas Erastus (1529–1583), a Swiss physician, argued that burning purifies, so that if comets were balls of fire thrown by an angry God, they could hardly start a plague. Once more we think of Kepler's teacher, Michael Maestlin (1550–1631), who made extensive measurements of the Great Comet of 1577 and decided that it was *above* the moon. Tycho Brahe had observed the same comet with superior instruments, but not yet the telescope, and the results agreed. Sure of his measurements, but scared, Maestlin published his conclusions with a theological argument that it is evident that he did not believe, but that kept him out of trouble.

A Lutheran pastor, David Fabricius (1564–1617), became interested in astronomy, built his own instruments, and published a report on the comet of 1607. He was a friend of both Brahe and Kepler; indeed, his measurements of the motion of Mars were used, along with those of Brahe, in Kepler's calculations, although Fabricius did not like the elliptical motion that Kepler deduced from them. His son Johannes, born in 1587, grew up to become a physician, but he also studied the motion of sunspots and apparently learned about the sun's rotation from his observations. He reported on this in 1611. Tragically, Johannes died at the age of twenty-eight, and his father was murdered by an angry parishioner two years later.

Since this chapter is confined to events that occurred up to the year 1642, the year of the death of Galileo and the birth of Newton, we shall return in chapter 15 to later developments and conclude this study with a summary of the status of that other problem—motion on earth. Some of the works of Archimedes, first translated into Latin in 1269 (by William of Moerbeke), and an Aristotelian discourse on mechanics were republished by Niccolò Tartaglia (1500?–1557) in Latin (1543) and later in Italian. Tartaglia did much more than publish earlier manuscripts. He criticized the mechanics of Aristotle, and he was the first person to learn about the motion of objects by performing experiments. Without the theoretical understanding of ballistics that came later, but with sound practical experience, Tartaglia discovered that the maximum range of a cannon is obtained by pointing it at 45° to the horizontal. He declared this discovery to be a military secret and backed up his findings with a report on an experiment with the largest cannon he could get his hands on—a two-thousand-kilogram giant. About that time Domingo de Soto suggested that falling bodies move with constant acceleration and applied the earlier results of the Oxford and Paris schools to the problem, but there appears to be no record of any experiment he may have performed.

In 1554, Giambattista Benedetti (1530–1590), a student of Tartaglia, published a criticism of Aristotle's work on motion, supporting and extending Buridan's idea of "impetus." In particular, he insisted, correctly, that impetus, i.e., momentum, defines the tendency for an object to move in a straight line. He saw that "natural" or force-free motion is in a straight line, not a circle, and he also pointed out that when a stone is released from a sling it starts to move in a straight line, but gravity immediately intervenes and curves the path. Others had thought that it would move in a straight line until it had run out of impetus and then drop straight down, seeking its position on the earth.

Geronimo Cardano* (1501–1576) contributed the statement, "two balls of the same material falling in air arrive at a plane at the same instant" but again there is no evidence of any experiment that he performed. He may have read about John Philoponos. Cardano was a remarkable character, and three hundred years later the Scottish astronomer John Nichol (1804–1859) wrote about him in the *Appleton Cyclopaedia of Biography* with characteristic nineteenth century English prose:

> One of our true "curiosities of literature," born in Pavia in 1501, said to have caused his own death in 1576, that he might not, by living longer, falsify his prediction of that event! . . . Of great industry, undoubted originality and power, and extensive acquirements, his fame yet rests for the most part on his pure charlatanerie. As a moral entity, if indeed the term can with decency be applied to him, he was also a man of contradiction: he loved knowledge, sought apparently for truth, and experienced high aspirations; nevertheless, he never shrank from deceit and falsehood; his practical life full of disorder; his scientific faith worth nothing—he stole from Tartaglia and published as his own the famous rule for the solution of cubic equations. . . .

"Cardan's Rule," which should be called "Tartaglia's Rule," can be expressed most simply as the solutions of the cubic equation:

$$y^3 + 3ay + 2b = 0$$

The solutions are $y = (-b+c)^{1/3} + (-b-c)^{1/3}$ where $c = (b^2+a^3)^{1/2}$.

Cardano was also a physician and an astrologer. In 1552 he was summoned to Scotland to attend an ailing John Hamilton (1511–1572), the Archbishop of St. Andrews. John Nichol grudgingly admits that his prescriptions effected a "celebrated cure." For persecuting Protestants and assisting Mary, Queen of Scots, and for murdering her enemy (and brother) James Stuart Murray (1531–1570), the archbishop was later strung up in his pontifical vestments and hanged by his kinsman James Hamilton. The true story is that the murder of Murray had nothing to do with the Catholic-Protestant conflict—James Hamilton himself had shot Murray for seducing his wife.

You cannot believe every story that you hear, of course, although that one is more reliable than the often-repeated story that Galileo climbed the Tower of Pisa and dropped off two balls to see if they would hit the ground at the same time. However, in 1586 Simon Stevenus (1548–1620) and his assistant Jan de Groot, in the Netherlands, did drop two lumps of lead, one much heavier than the other, through a height of ten meters. They verified by these experiments that the two pieces reached the ground at the same time. In addition, an enormous amount of empirical evidence on projected and falling objects had accumulated over the centuries, but the information was anecdotal and unsystematic. It was against such a background, though he may not have been aware of all of it, that Galileo* (1564–1642) finally released the study of falling objects from its Aristotelian framework.

One argument Galileo advanced concerned a thought-experiment of a stone dropped from the top of a tower, not so much to argue about *when* it would land as to *where* it would land. Supporters of Aristotle claimed that, if the earth were rotating from west to east, the tower would move while the stone was falling, so that the stone (headed toward the center of the earth by its "natural" motion) would land some distance west of the tower, contrary to experience. Galileo expanded on ideas anticipated by Lucretius and first developed by Oresme to defeat this argument. He applied his analysis to a cannon firing east as opposed to a cannon firing west, to a moving boat, to a hunter shooting a moving target, and to other situations in order to demonstrate that the motion of the earth, or of the boat, or of the gun, would be communicated to anything at rest on it. (As late as World War I at least one commander of a British bombing squadron had to have it demonstrated to him that a bomb released from a moving aeroplane does not fall straight down.)

Understanding of the motion of falling objects had been delayed partly by a traditional reluctance to experiment, and partly by an inability to measure time accurately. We know now that after only two seconds a solid object dropped from rest is moving at more than 40 mph, a speed that the technology of the time could not measure. Galileo therefore directed his attention to distance as a function of the time, measuring the latter by a water clock or by pulse beats, and measuring distances in "braccia," or cubits (20 ± 2 inches). Earlier he had noted the regularity of the pendulum, but he had not applied it to the measurement of time.

To slow the process so that it could be more easily measured, Galileo states that he took a piece of wooden moulding about 20 feet long, cut, lined, and polished a straight channel in it, placed it in a sloping position, and rolled a round bronze ball down it. Repeated measurements verified that the distance increased with the square of the time. The relation was verified for different slopes, the constant of proportionality increasing with the slope. He extrapolated on the one hand to free fall, and on the other to smaller slopes until, for the horizontal case, the acceleration was zero. "Horizontal" meant for him, as for us, a nearly spherical surface with center at the center of the earth. Unaccelerated motion was, therefore, to him motion in a circle, not in the straight line recognized earlier by Benedetti, and later by Descartes, and, more precisely, by Newton.

During Galileo's lifetime experiments *were* conducted from the top of the tower of Pisa, first by Giorgio Coresio in 1612 to show that different weights do not reach the ground at exactly the same time, and later, the year before Galileo died, by Vincenzo Renieri who reported in a letter to Galileo (March 13, 1641): "From the summit of the Campanile of the Cathedral (at Pisa) between the ball of lead and the ball of wood there occur at least three cubits of difference." These appear to be the first quantitative measurements of the effects of air resistance.

At the age of twenty-five Galileo had been appointed as professor at the University of Pisa. Three years later he moved to Padua, staying there until the fateful year 1610, when at Florence he became First Mathematician and Philosopher to Cosimo II, Grand Duke of Tuscany. He had continued his studies of the motion of objects on earth and discovered that a projectile moves along a parabola, although air resistance can greatly modify that. He was able to sort out the utter confusion of the centuries caused by air resistance and its interference with the effects of gravity. Galileo understood the basic independence of the horizontal and vertical components of motion and came to the remarkable conclusion that, if a solid object is fired horizontally, and at the same moment another is dropped vertically from the same spot, they will reach the ground at the same instant, provided that air resistance is negligible. He then turned his eyes and his mind to the heavens, as if his upsetting of Aristotle's notions about earthly motions wasn't enough.

Toward the end of the sixteenth century, Roger Bacon's prediction that "from an incredible distance we might read the smallest letters" was beginning to come true after a lapse of three hundred years. In 1571, the English mathematician Leonard Digges experimented with convex and concave lenses. In 1608, two Dutch scientists, Hans Lipperchey (1587–1609) and Jacobus Metius (1580–1628) independently invented the refracting telescope. Galileo heard of these developments and next year built his own instruments. He invited some of the leaders of Venice to climb to the upper level of St. Mark's Cathedral to look through his telescope and see ships that were fifty miles away. They were very impressed and recognized the military value of the discovery. Early in 1610, when he offered his professional and other colleagues at Padua the first sight of the moons of Jupiter through his more powerful model, many refused. Those who did take a look were convinced that they had seen the handiwork of the Devil himself. The academic authorities recognized that this could be interpreted as evidence against Aristotle and in favor of the Copernican theory, and their outcry against Copernicus, muted while his idea was seen to be a mere hypothesis, broke out and condemned the telescope and the man who had caused it to produce these diabolical apparitions.

Up to this point there had been relatively little opposition to Galileo—in fact Jesuit scholars supported him and carried out their own observations of the satellites of Jupiter. In 1610, Galileo looked at the moon through his telescope or, as Milton put it soon afterward:

> —the moon, whose orb
> Through optic glass the Tuscan artist views
> At evening, from the top of Fesolé
> Or in Valdarno to descry new lands
> Rivers, or mountains, in her spotty globe.
>
> (*Paradise Lost,* Book I, 287–291)

The moon was not supposed to be perfect anyhow, and the discovery did not interfere with the belief that the earth was the center of the universe. Both Galileo and a number of Jesuit scholars observed the phases of Venus, and that certainly meant that Venus was circling the sun, as Tycho Brahe and a string of others throughout history had maintained.

Unfortunately, Galileo thought that he had a monopoly on all the discoveries of his time in physics and astronomy. He embraced the ideas of the long-deceased Copernicus—ideas that even their author had recognized to be shaky—and ignored the more recent work of Kepler. In 1612, when he received a report of the discovery of sunspots by the Jesuit scholar Chrisoph Scheiner (1575–1650), Galileo claimed priority for the discovery, but was unable to back up his claim. Some Jesuits naturally turned against him for that. He made more enemies among the academic Aristotelians with his obvious contempt for their failure to accept the law of Archimedes for floating bodies. Whenever the occasion demanded it, and more often when it did not, Galileo wrote letters in support of the old theory of Copernicus with the earth moving as just another planet.

Since the authors of the biblical books had naturally assumed that the earth was fixed at the center of all things, the unproven motion of the earth around a fixed sun conflicted with the Bible. Even that conflict was not very serious and it has now been conveniently forgotten. It was based mainly on Psalm (19:4–6):

> He has set a tent for the sun, which comes
> forth like a bridegroom leaving his chamber,
> and like a strong man runs its course with joy.
> Its rising is from the end of the heavens and
> its circuit to the end of them.

The famous event recorded in Joshua (10:13) also seemed to mean that if the sun "stood still" that day it must have moved both before and since: "The sun stayed in the midst of heaven and did not hasten to go down before a whole day" (New Oxford translation).

The sad events that followed have been reported eloquently and often, and a summary appears in the *In Memoriam* section of this book. On February 26, 1616, the Roman Catholic Inquisition ordered Cardinal Bellarmine to command Galileo, in person, not to teach or discuss these new ideas. It also required that the book by Copernicus have removed from it any mention of the factual nature of the earth's motion. Galileo agreed, but returned to Florence to write his masterpiece, the dialogue between Salviati and Sagredo, on the one hand, supporting Galileo's position, and a naive person aptly named Simplicio, who tried to justify the old ideas. The fact that Simplicio was given some lines from Pope Urban VIII to speak fanned the flame, and after several appearances before the Papal Inquisition, Galileo was imprisoned, threatened with torture, and required to swear that he would denounce anyone who supported the heresy of the motion of the earth. This "heresy" and all that went with it was banned from all Catholic colleges and schools, and theologians had a field day writing tracts that condemned these ideas and, with a multitude of biblical quotations, "proved" them to be wrong.

The Protestants were not far behind in their condemnation of this scientific genius and pioneer. More than two hundred years later some Lutheran clerics were still preaching the validity of the theory of Ptolemy. In fact, they ranted against Copernicus, Galileo, Newton, and the astronomers who followed them—the truth, after all, is to be found in the Bible and nowhere else. Fortunately, on this issue, those who take their science from a literal interpretation of the Bible have now retreated and have turned their attention to other areas that we discuss later.*

*The story of Galileo has been told in a detailed and scholarly way by Professor Stillman Drake: *Galileo* (Oxford Univ. Press, 1980) and *Telescopes, Tides and Tactics* (Univ. of Chicago Press, 1983), as well as in his translation of the famous dialogue, listed under *Select Bibliography*. Everyone should read at least one of these books—lest we forget.

13

Witches, Devils, and Lunatics

Thou shalt not suffer a witch to live

—Exodus 22:18

Then the demons came out of the man and entered the swine, and the herd rushed down the steep bank into the lake and were drowned

—Luke 8:33

The Exodus damnation of witches was repeated in Leviticus 20:27 and Deuteronomy 18:10 and ultimately became enshrined in the beliefs and superstitions of Christian churchmen and their followers. It was thought to be a clue, given directly by God, to an understanding of so many otherwise inexplicable calamities, ranging from the idiosyncrasies of the weather, to the occurrence of physical and mental diseases. Not only that; it sanctioned an outlet for frustrated men to vent their collective spleen on countless defenseless women, though that was hardly the official, or even the conscious, reason for its continued sanction by the churches. Witches were very real; after all, hadn't Saul consulted the Witch of Endor as related in I Samuel 28:7-20? Not that everyone believed that witches were responsible for the evils for which they were blamed. The poor wretches themselves knew better until, persecuted and tortured out of their minds, they often confessed to crimes that they could not possibly have committed, their intolerable pains then being released by execution. There were others too who spoke against witchcraft.

Late in the sixth century more than a hundred thousand Lombards —("long beards")—crossed the Alps to settle in the valley of the River Po, which stretches across northern Italy. They set up an enlightened form of government according to which their king was *elected.* In 643 they followed this up with the enactment of laws of tolerance—freedom of worship, even for the Arians, and protection of the poor. Despite the growing belief throughout Christendom in the evils that witches were supposed to invoke, the Lombards expressed an early official disbelief in witchcraft, even making the belief an object of ridicule. Very progressive thinking for the time, but unfortunately other Christians were not prepared to accept these Christian attitudes. There was, of course, St. Agobard, a voice crying in the wilderness of the ninth century, as thoroughly opposed to the prosecution of witches as he was to other forms of superstition. He believed that storms were the results of natural causes and, as we have noted, spoke up against the filthy nature of the Christian cities, the abominable ordeal by fire, and the use of graven images in worship. On the other hand, the enlightened Jewish scholar Maimonides accepted the literal meaning of the quote from Exodus that appears at the beginning of this chapter. He also believed that death should be the punishment of any Jew who broke the rules of the Sabbath, or who did not believe in the Jewish Law or who adopted any other heresy.

The Visigoths had invaded Italy in the fifth century and spread their control to Spain until they were beaten by the Muslims in 711. Some of their laws outlived them or, more accurately, survived after the Visigoths had been absorbed into other groups. Their laws recognized that witches could be in league with the devil and his many demons, and that they could cause storms. Compared to the terrible purges that occurred in later centuries, however, the punishment was relatively humane: two hundred lashes and a shaving of the head. King Cnut of England also recognized the power of witches and made laws to define their punishment, particularly for the deaths they supposedly caused through their magic.

In Germany in the early part of the eleventh century, Empress (now Saint) Kunigund, wife of Henry II, heard that her reputation as a faithful wife was being questioned. According to legend, she established her innocence and confuted the rumor-mongers by walking barefoot over red hot irons. That type of courage, repeated by so many people since then, was not regarded as witchcraft—it was seen as a work of God, not of Satan. Two centuries or more later, Thomas Aquinas wrote a long dissertation—nearly a hundred pages—on angels and devils, noting how the latter could cooperate with witches to give them the power of the "evil eye," by which they could inflict injury on others merely by looking at them. In 1275, the year after Aquinas died, the first French trial for witchcraft led to the execution of Angèle de la Barthe, found guilty of copulating with the devil. Someone had watched her perform a fertility rite and reported it to the Christian authorities, whose sexual fantasies were thereby stimulated.

Early in the fifteenth century, it looked as though, after the Battle of Agincourt, England was about to capture the whole of France. Masters of northern France, they moved towards Orléans, laying siege to it on October 12, 1428. Jeanne, the young daughter of a French farmer, Jacques d'Arc, turned the tide against them, but she would be accused of witchcraft and ultimately be tried for heresy. From the age of thirteen she heard a voice, which she interpreted as God's, telling her that she could deliver France from the English. When this voice was supplemented by visits from St. Michael, St. Margaret, and St. Catherine, complete with jewels and haloes, she was convinced that her mission was divine. No matter that the English thought that God was on *their* side, no matter that the saints she saw were variations on images she had probably seen in the local church, no matter that the voice she heard could be interpreted as a product of her subconscious mind—the fact remained that she believed, and "We walk by faith, not by sight" (2 Cor. 5:7). Joan of Arc's faith really moved mountains. Her overwhelming conviction that she was a messenger of God won her an interview with the Dauphin, later Charles VII—but was she inspired by angels or by devils? He had his doubts, but he gave her the benefit of them. It was a fortunate decision—through her efforts he would be crowned within a year.

On April 29, 1429, Joan reached Orléans and rode in triumph through its streets with an army of knights. They had brought provisions to a starving garrison; more important, perhaps, they had brought the belief in a miracle, the defeat of their oppressors. The English got the word, too, and began to wonder if God had switched sides; and after their defeat came, they regarded Joan as a witch. Despite an arrow that lodged in her chest, and another that pierced her thigh, she continued to inspire the French to pursue the English, overreaching herself at times until, a little more than a year later, she was captured. After some negotiations she was handed over to Pierre Cauchon, Bishop of Beauvais (?–1442), for trial by the Inquisition. In that year also, a follower of Joan, by name Pierrone, was found guilty of witchcraft and burned at the stake.

In order to protect itself, the Church had decreed that access to God lay only through its clergy, and a claim to direct divine inspiration such as Joan's was therefore a heresy punishable by death. The edict had been deemed necessary because of the enormous number of false claims to supernatural powers that witches and others were alleged to make—how could you be sure that they weren't voices from the devil unless communications were syphoned through the Church? The prophets of the Old Testament would have been in real trouble during the fifteenth century! Joan acknowledged the supreme authority of the Church in matters of faith, but these voices were from God, she fervently believed, and she would answer only to Him for what she had done in response to them. She was "tried" by the court, but English soldiers surrounded the place. If the Inquisition wouldn't condemn her, the English would. On May 31, 1431, she was found guilty and burned, but twenty-four years later the verdict was declared unjust. By that time the work that she had begun was completed; the English were driven out of practically all of France. In 1920, she became a saint of the Church. Perhaps there is still hope for Bruno and Galileo.

For some time there had been stories of Satanism, the worship of Satan in defiance of Christianity, with its Black Mass, a travesty of the Christian Mass. There were also stories of the Sabbat, or Witches Sabbath, a midnight orgy supposedly held by witches and demons. William Howitt, in his *History of the Supernatural* (1863) tells us:

> Witches, sorcerers and sorceresses are people who deny God and renounce him and his grace; who have made a league with the Devil; have given themselves up to him body and soul; who attend his assemblies and sabbaths, and receive from him poison powder and, as his subjects, receive command from him to injure and destroy men and animals; who, through devilish arts, stir up storms, damage the corn, the meadows and the fields and confound the powers of nature.

If you really believe that, and most people did, it is no wonder that witch-hunting was more than a popular gruesome pastime —it was thought to be necessary for survival. In 1430, the "Good Duke Humphrey" (1391–1447), Duke of Gloucester, married his mistress Eleanor Cobham and brought witchcraft to the British royal family. She hired Margery Jourdemain to make a waxen image of the king (Henry VI), and thereby cause evil magic in an attempt to destroy him. Eleanor wanted to be queen, and Humphrey liked the idea, too. In 1441, the women were accused of witchery and Margery was executed. Four years later Humphrey too was arrested; he died while in custody.

In 1437, and again in 1445, Pope Eugene IV exhorted the inquisitors to use greater diligence in rooting out these evils. Remember the prophecy made to King Ahab (I Kings 20:42) after he released the Baal-worshipper, Bar-hadad, King of Edom: "Thus saith the Lord. Because you have let go out of your hand a man whom I appointed to utter destruction, therefore your life shall go for his life and your people for his people." In 1484, the year that he was elected, Pope Innocent VIII issued a bull on the subject of witches. The faithful were forbidden to confer with them in any way; they were blamed for plagues and storms and accused of sexual unions with devils. The Inquisition was now commanded to increase its witch-hunting activities, leaving no way untried for the detection and apprehension of such nefarious creatures. The pope's predecessor in office, Sixtus IV, had been forced to condemn the Carmelite nuns of Bologna because they had acted like witches in seeking knowledge from devils. One hundred and fifty years earlier, Dante had placed witches in Hell, but not in its most repulsive realm; they were classed only with liars, seducers, and flatterers:

> Whence in the second circle have their nest,
> Dissimulation, witchcraft, flatteries,
> Theft, falsehood, simony, all who seduce
> To lust, or set their honesty at pawn,
> With such vile scum as these.
>
> —*Inferno* Canto XI (HCl 20, 46)

With the beginning of the sixteenth century, however, no punishment for witches would be harsh enough, and the attitude continued to harden.

The bull of Pope Innocent VIII stimulated the Dominican inquisitors, Jacob Sprenger and Hendricus Kramer, to write and publish the book *Malleus Maleficarum* (the *Hammer of Witches*), a guide to the detection, apprehension, and destruction of witches. Watch out for a woman boiling a brew in a cauldron; she may be preparing a prescription given to her by the devil, or she may be preparing to eat a child that she has sacrificed. Watch out for animals, because witches can transform themselves into beasts, or an animal near a woman might be someone she has transformed. Some men too can perform these acts of evil, but it is mostly women, because they are more sensual and are known to be instruments of the Devil anyhow. The hunt was on! During the first twenty years of the sixteenth century the fires burned, and in Como and Brescia alone in Northern Italy women met that agonizing death at an average rate of one every two weeks. The grisly work was performed by secular officials to whom the Church handed over their convicted witches. When some officials refused to carry out the task, they were threatened by Leo X with excommunication, and the concomitant eternal punishment. The same threat had worked in a good cause when St. Ambrose had threatened Theodosius I a millennium earlier and, more than another millennium earlier still, Elijah had humbled Ahab with more earthbound threats (I Kings 21:29). The shocking purge of witches continued for two more centuries, the hundreds of victims becoming thousands, and then tens of thousands or, according to some, hundreds of thousands or even millions. When superstition and religion join forces, reason is overwhelmed and human kindness perishes with it. Goethe wrote that all traces of Christianity became extinct in Rome.

Like practically everyone else, Martin Luther believed that invisible devils were everywhere. His older contemporary, Niccolò Machiavelli, not one to put up with nonsense, believed that "the air is peopled with spirits" and that events on earth are heralded by signs from the heavens. Luther and Calvin, impressed by the comment of St. Paul—"O foolish Galatians, who has bewitched you from your faith. . . ?" (Gal. 3:1)—encouraged witch hunts. Despite Luther's comment—"I would have no compassion on these witches; I would burn them all."—he was less enthusiastic about it than Calvin was, and fewer witches were burned in Luther's Wittenberg at that time than in Calvin's Geneva. The Spanish Inquisitors, on the other hand, did not concentrate on looking for witches; they had other targets in mind. So also did Christian II, "The Cruel," king of Denmark and Norway from 1513 to 1523. He ended the death penalty for witchcraft but massacred the Swedes without mercy.

The Christian Church, rapidly becoming the Christian Churches, struck out at all threats, real or imaginary, during the uncertainties of the Reformation. The *In Memoriam* section at the end of this book shows a sad concentration in the sixteenth and seventeenth centuries—scholars had become a threat and had to be disciplined. Because of their time-honored religious hygiene, the Jewish people suffered less than the Christians from the plagues that often devastated civilization. Since plagues in general, and those of 1519 and 1522 in particular, were thought to be works of the devil, it followed that the Jews, along with the witches, must be in league with Satan, adding yet another excuse for persecuting and murdering members of this long-suffering race. In 1520, Cornelius Heinrich Agrippa von Nettesheim* (1486–1535) tried to save a woman who was on trial before the Inquisition for witchcraft. He was imprisoned and exiled from France and hounded by the Dominicans for the rest of his life.

While Luther was bringing the Reformation to Germany, Huldreich Zwingli (1484–1531), with his different style, was bringing it to Switzerland. He focussed at first on speaking out against the sale of indulgences by a Franciscan friar who had come to Switzerland as a member of the pope's sales force. He then changed the emphasis on the Mass, often replacing it with a sermon, as Luther was doing. He refused to date "original sin" back to Adam and Eve, or to believe in Purgatory or the power of priests to forgive sins. He acquired the cooperation of the Zurich Council, and in 1529, when a Protestant minister was burned at the stake in Catholic Schwyz, some twenty-five miles south of Zurich, Zwingli persuaded the Council to declare war. For a while, only an uneasy peace—a cold war—ensued.

A few months later, Zwingli and Luther met and a division of Protestants among themselves became apparent before Protestantism had really developed. Zwingli saw the Eucharist as a symbol; he did not believe that it was really the body and blood of Christ that was being administered. Luther quoted to him Mark 14:22 and Luke 22:19, "This is my body . . ." and they parted, agreeing to disagree. Two years later the fragile peace with the Catholics was shattered and Zwingli was killed during the battle between Catholics and Protestants on the road between Zurich and Schwyz. A comet appeared that year, and Zwingli had insisted that it foretold an impending calamity. We do not know how much this belief influenced him, but we do know that Paracelsus, fighting the scientific and medical authorities as Zwingli was fighting the ecclesiastical, dedicated his *Explanation of the Comet* to him.

A few years earlier, Johannes Denck and Ludwig Hetzer provided the first translation of the Old Testament prophets into German. Inspired by Zwingli's stand against the Catholic Church, they and others began the Anabaptist movement. They were soon persecuted for denying the divinity of Christ, for showing contempt for civic government, and for arguing that the faithful should be baptized during their adult life, when they can consciously make the decision to do so. Zwingli tried without success to convince them to baptize their babies and to cooperate with the government, and so another Protestant split originated. Hetzer was imprisoned for heresy and beheaded on February 4, 1529. The Anabaptists were persecuted by Catholics and Protestants alike, partly because they took the law into their own hands, and assumed dictatorial control over the city of Münster until they were routed by a Catholic army, with nobody spared. Members of the sect in other cities decided to abjure the use of force thenceforth.

When a group of people embraces a particular faith or a particular article of a faith, that belief is clung to with a passion that is strengthened among the survivors of persecutions. The early Christians persecuted by the Romans, the Nestorians who lost the battle of orthodoxy, the Protestants splitting among themselves and, both earlier and later, the Jews, all spread their version of religion by emigrating for their beliefs. Menno Simons (1492–1559) organized the conservative Anabaptists in Holland and Germany, giving the world the peaceful Mennonites. Other Protestant denominations, notably the Baptists, had their origins in the theological turmoil of the first half of the sixteenth century. The Methodists' beliefs originated a little later with Jacob Harmensen (1560–1609), also called Arminius, a Dutch theologian. His writings were studied very "methodically" (as their Oxford colleagues put it) by John and Charles Wesley a little more than a century later.

Each of the sects was a threat to the others and to the original Christian Church, and the Church was a threat to them. In that environment of mistrust it was natural to look out for your own safety, to suspect any idea that could be interpreted as opposed to your cherished beliefs, and to look for people to blame for anything that you didn't like and didn't understand. It is for this reason that a short discussion of these theological conflicts has been introduced here, because witch-hunting, the persecution of scientists, and the mistreatment of the insane rose to their maximum levels of degradation during the spread of these conflicting religious dogmas. In this as in other strife intolerance stems from insecurity.

Of course there had been a large measure of insecurity long before the Reformation, with men falsely accused as well as women. The Cistercian monk of Cologne, Caesarius of Heisterbach (1180–1240?) had written that many men had entered into a Faustian pact with the devil. He didn't call it that, because Dr. Johann Faust came later (1480?–1540). Faust began to exhibit his magic and soothsaying in 1506, acquiring thereby a great deal of fame and the blessing of Cologne's archbishop. Later in the century the story of his pact with the devil was published. As usual, legends develop about unusual people, particularly after they have died, and it was believed that Mephistopheles had carried him away. A few years later, the English dramatist Christopher Marlowe (1564–1593) wrote *The Tragedy of Doctor Faustus*, and after a further two hundred years Johann Wolfgang von Goethe (1749–1832) began Part I of his masterpiece, *The Tragedy of Faust*, with Faust alone sitting at his desk:

> I have alas! Philosophy
> Medicine, Jurisprudence too,
> And to my cost Theology,
> With ardent labour studied through
> And here I stand, with all my lore
> Poor fool, no wiser than before.
>
> . . . .
>
> Up! Forth into the distant land!
> Is not this book of mystery
> By Nostradamus' proper hand,
> An all-sufficient guide? Thou'lt see
> The courses of the stars unroll'd
> When nature doth her thoughts unfold
> Spirits! I feel you hov'ring near;
> Make answer, if my voice ye hear!

(Translated by Anna Swanick UA 21, 59)

Faust's contemporary, the Italian philosopher Pomponatius (1462–1525), a professor at Padua, questioned (1513) the power of devils; he was a genuine iconoclast who also questioned the arguments of Aristotle, Thomas Aquinas, and others. He survived by announcing that these heretical thoughts were part of his *philosophy*, but as a *Christian* he said he had no doubts at all. Next year three hundred more witches died in Como. William Roper, attorney-general in the reign of Henry VIII, wrote a biography of his father-in-law, Sir Thomas More (1478–1530), author of *Utopia.* In it he quotes More's advice to his children: "The devil, seeing a man idle, slothful, and without resistance ready to receive his temptations, waxeth so hardy that he will not fail still to continue with him, until to his purpose he hath brought him." Roper reports that one time his wife, More's daughter, had the "sweating sickness" and physicians despaired for her life. More went to his chapel to pray and came back with the suggestion to give her a "glister" (an enema). The physicians hadn't thought of that, and it cured her. Once again, quiet meditation helped to retrieve an idea from the subconscious or to provide access to a message from God, depending on your point of view, and her faith completed the healing.

While belief in devils can be used to scare children into good behavior, centuries ago it led to hideous cruelties. Back in 621 B.C. the lawmaker of Athens, Draco, had prescribed death for nearly all offenses, and draconian penalties were reinvented by the Christians more than two thousand years later to combat the influences of these imagined powers of evil. In the year in which Sir Thomas More was decapitated—the sentence of hanging being commuted by Henry VIII—Michael Caddo, in Geneva, was also tortured and executed. Caddo had been found guilty of obtaining from the devil a mysterious cerate, or powerful ointment, and spreading it on the walls of the city in order to bring a plague. Henry continued his infamous reign by introducing, in 1541, laws that prescribed the death penalty for practices that, it was alleged and believed, were performed by witches. When his daughter, Elizabeth I, came to the throne, witchcraft was re-enacted as a capital crime. A very unfortunate precedent was established in the courts of law: uncorroborated evidence was admitted. In 1566 Agnes Waterhouse, aged sixty-four, was accused of using her cat to cause acts of violence, even murder. Her daughter, also accused, and scared for her own fate, testified against her own mother. Agnes was found guilty and hanged, although there was no real evidence against her. Later (1582) in the Age of Elizabeth another woman, Ursley Kempe, had some illegitimate children who conspired to testify against her in court. Ursula was a midwife and, at times, a wet nurse, and she was offered leniency if she would admit to witchery that would implicate others. She did so, but the lying magistrate ordered her execution anyhow.

In England, however, the intensity of the witch-hunts was considerably less than in Scotland, where it is reported that more than one woman in every hundred was burned as a witch during the last forty years of the sixteenth century. One of the first to ridicule the belief in witchcraft was a Belgian physician, Johann Wier (1515–1588), an apprentice of Cornelius Agrippa. His book *De Praestigiis Daemonum—The Illusion of Demons*—was published in 1563 and therefore written with great caution. He even stated that he believed in witchcraft but that he was doubtful if witches could perform all of the miraculous acts that were ascribed to them. It was Satan himself, he argued, who put these delusions into the minds of women and made them confess under torture to this witchery. Protestants and Catholic clerics joined in denouncing him and his Satanic ideas. In England, Reginald Scot (1538?–1599) followed with his practical guide for the cultivation of hops and, more to our present concern, his *The Discoverie of Witchcraft*, published in 1584.

It was the year that Francis Throckmorton, an ardent Catholic, was executed for plotting with the Spanish ambassador and with Mary, former Queen of Scotland, to place Mary on the throne instead of Elizabeth, and to make England a Catholic nation. Mary's son, James VI (1566–1625), was now King of Scotland, and in 1589 he arranged to marry Anne, daughter of the king of Denmark. The royal ship transporting them was nearly wrecked in an unusually strong storm. James was convinced that someone had plotted against him and had worked with the devil to produce that storm. Four suspects were tortured into confessing, and one of them, John Fian, was subjected to the most revolting and excruciating tortures —his legs crushed in the "boots," wedges driven under his fingernails, and much more—before being burned to death. This is the same James VI who later became James I of England, but before that happened he authorized witch trials in North Berwick not far from the border with England. One of the first to be tried was a lay healer, Geillis Duncan, whose employer decided that her healing powers arose from a pact she had made with the devil. She was arrested, tortured, and forced to name others. She died in 1590. In the next year, Eufame Macalyane visited the lay healer, Agnes Sampson, to get some relief from the pain of delivery of her twin sons. She was burned alive on Castle Hill in Edinburgh. In 1592 Agnes Sampson was forced to confess to a plot against the king and queen, and she was strangled and burned. In England the eighty-year-old Alice Samuel was brought to trial. She worked for a family, and the three young children had made her life utter hell for four years. Many times they scratched her face until blood flowed, since they believed that if you did that to a witch nobody would bewitch you. They testified against her at her 1593 trial, and two days later she was hanged.

In France, the respected political economist Jean Bodin (1530–1596) had emphasized the importance of studying history as a background for understanding political trends. He also believed in astrology, demonology, and the application of numerology to the welfare of the state. Along with this, he was a firm believer that witchcraft was the source of many evils, and he called for the strongest of punishments for witches. His words were heeded, for his *Demonomanie,* published in 1580, was very persuasive. Nicholas Remy (1515–1590) terrorized and persecuted the women of Lorraine, condemning nearly a thousand of them as witches during the last fifteen years of his life. Peter Binsfield, Assistant Bishop of Trèves (present Trier, in the Saar region of Germany), was responsible for more than six thousand executions. As an incentive for confession, he offered to those accused of witchcraft the option of being strangled before being burned if, under torture, they would admit to their alleged crimes. Underneath it all, there was apparently a spark of humanity in this pious bishop.

At that time, Dietrich Flade* (?–1589) was rector of the University of Trèves and Cornelius Loos* (?–1595?), a Dutch Catholic theologian, was one of the professors. Flade was an eminent legal scholar of law, and chief judge of the electoral court. (In the "Golden Bull" of Charles IV in 1356, Trèves had been chosen as one of the seven areas whose archbishop or other leader could vote on the choice of a Holy Roman Emperor.) Witches and magicians were brought before Flade, and he routinely sentenced them until he began to realize that they were not guilty. He could no longer believe that they rode on broomsticks, and were responsible for diseases, hexes, and storms, and he courageously hesitated to condemn them. The powerful Archbishop of Trèves had him arrested and accused him of having sold himself to Satan, to which, under extreme torture, he confessed. He too was strangled and burned.

Professor Loos also came to doubt all of these stories about witches and his doubts troubled him, because he was a devout Catholic, very much opposed to the Protestant movement. He wrote a mild criticism of witch-hunting and of belief in supernatural sorcery, but the ecclesiastical authorities stopped publication. His unpublished manuscript, *True and False Magic*, was discovered late in the nineteenth century in the Jesuit library at Trèves. Loos was imprisoned in 1593, brought out of jail, and forced to recant on his knees before the church leaders. He spent the next two years or so under surveillance, in and out of prison, until the plague relieved him of the further persecution that had been promised to him. In Wittenberg, where Martin Luther had initiated the Reformation in 1517, Benedikt Carpzov (1565–1624), chancellor of Saxony, showed that Lutherans could be as tough on witches as anyone else. The German jurist and satirical poet Johann Fischart (1546?–1590) known as Mentzer, translated Jean Bodin's book so that all Germans might learn about the evils that witches were supposed to bring upon the world.

The situation did not get any better as the new century dawned. Installed as King of England in 1603, James VI, now James I, tried to reinstitute the divine right of kings. Fortunately, he didn't get away with that, but he did succeed in stimulating more terrible witch trials by writing his book of horrors, *Demonologie*. He pointed out in this disgraceful book that witches could control the emotions of men and women, inducing love or hate at will; that they could kill by destroying a wax model of a person—putting a fatal "hex" on them, to use a modern American word; and that not only witches, but anyone else who dealt with them should be killed. His *Counterblaste to Tobacco* made more sense and, if listened to, would have saved many more lives than he terminated. We remember him chiefly because the *Authorized Version of the Bible* was published during his reign, and named for him. We may also remember that he ordered copies of Reginald Scot's great book against the persecution of witches to be burned by the hangman. That was as close as he could get to hanging the author himself, since Scot had died a few years earlier.

One infamous trial during James I's reign involved Anne Redfearne, who had been initiated into witchcraft by her mother. A man by the name of Robert Nutter was found murdered, and because he had tried to rape Anne and she was seen as a witch, it was natural that she should be arrested. Much to the outrage of the local people, she was found to be "not guilty," so another charge was laid against her—murdering Nutter's father, who had been dead for twenty years. This time she was found guilty and, to the joy of the crowd, she was executed in the year 1612.

There were so many accusations that it became necessary to try a number of women at once. Elizabeth Southern was a defendant at one of the early mass-trials, and she really was a witch, or so she claimed. For a number of years, she and another woman had been engaged in a feud, outdoing each other in trying to bewitch each other. Elizabeth was very proud of her witchcraft, but she died in 1613 while in prison awaiting trial. Perhaps her enemy had hexed her and won the final round of their long battle.

The disease of witch-hunting spread throughout Europe, and later it was to infect America. In Spain, a Basque woman, Maria de Zozoya, was tried with the group that she led, and they were all burned alive. In Southern Germany, Katherine Guldenmann was born in 1547 and brought up by an aunt who was later burned as a witch. We know more about Katherine than the others because of the meticulous notes that were kept by her oldest son, Johannes. On May 15, 1571, Katherine married Heinrich Kepler, an adventurer, mercenary soldier, and tavern keeper. He ultimately disappeared, apparently wanted for some capital crime. According to his own testimony, Johannes was conceived at 4:31 A.M. on the morning of

May 16, seven and a half months before he was born. He cited his weakness at birth as evidence of its premature nature, and the horoscope he built for himself made his conception legal. Legal or not, from this unlikely couple came a revolution in the understanding of astronomy.

From the beginning, Frau Kepler was by no means a peaceful person, and her quarrelsome attitude got worse as she grew older. She knew how to make magic potions from the herbs that she collected. In 1615 she quarreled with another old woman who, to get even perhaps, complained that one of Katherine's witches' brews had made her sick. The word got around, and others suddenly remembered feeling ill after drinking one of Katherine's poisonous concoctions. Then there was the story of the woman who had died after drinking one, and hadn't Katherine tried to get her father's skull in order to cast it in silver and make a drinking goblet, a present to Johannes from mother? Besides, she made the cattle sick by merely touching them. Katherine was therefore arrested and chained. She was accused of witchcraft at a time when several witches were being burned each year in villages like Weil der Stadt, Kepler's birthplace, and in neighboring Leonberg where Katherine now lived. The family moved her to Linz some two hundred or more miles east, in Austria, but in 1620 she returned and was arrested again. Johannes came to her aid—he had just published the third law of planetary motion—and litigation continued for more than a year. Finally, the Faculty of Law at the neighboring University of Tübingen ruled in the case—she must be required to confess under the threat of torture. Here Katherine showed her true strength. Confronted by the executioner and the sadistic tools of his trade, and provided with graphic descriptions of the horrors that were impending, she refused to confess. She was released but could not go back to Leonberg for fear of being lynched. She died a few months later.

During the sixteenth and into the seventeenth centuries, many people besides Michael Caddo were accused of smearing a devilish poisonous substance on the walls of houses in order to bring the plague. Alessandro Manzoni's masterpiece *I Promessi Sposi*—or *The Betrothed* as we call it—is centered on Lombardy between 1628 and 1631 and deals in detail with the plague that devastated Milan in 1630. Manzoni tells how "anointers" had been tried and found guilty in Palermo, Geneva (as we have seen), Casale Monferrato, Padua, Turin, and finally in Milan. "Here one, there many unhappy creatures were tried and condemned to the most atrocious punishment, as guilty of having propagated the plague by means of powders, ointments, witchcraft." One morning in June 1630, a woman looked out of her window and saw a man writing on a piece of paper and moving his hand across the walls. She spread the word and he was arrested, even though he was Sr. Piazza, Commissioner of the Tribunal of Health. He was tortured and forced to implicate an innocent barber named Mora. The two of them were paraded through the streets of Milan and subjected to unbelievable tortures inflicted on them by the angry mob. They were then confronted with the official torture until death. Manzoni showed that the laws that allowed the use of the rack and other instruments of torture were evil, but that the real evil lay with the judges who, anxious for a conviction, exceeded the law by abandoning all proper legal procedures. One can hear the lawmakers saying, "Racks don't torture people; people torture people," and leaving the racks to work their evil another time.

All of these tales of witchcraft and evil magic were believed so widely that it was practically impossible to gainsay them. In 1646, Sir Thomas Browne (1604–1682), the very orthodox physician at Norwick, published his widely read *Religio Medici*, in which this comment appears: "For my part, I ever believed, and do now know, that there are Witches: they that doubt of these are . . . a sort, not of Infidels, but Atheists." More than thirty editions of this, and translations into five other European languages, spread the word. His actions spoke even louder, at least to two unfortunate women who were condemned to death on his testimony as an "expert witness" as a physician.

Then the scourge came to America. In *The Liberties of the Massachusetts Collonie in New England, 1641,* it is written, "If any man or woman be a witch (that is hath or consulteth with a familiar spirit) they shall be put to death." In 1650, Margaret Jones, a medical practitioner accused of witchcraft by male physicians, was the first woman executed for witchcraft in America. The witch hunt continued throughout the century, particularly in Salem, where, for example, Goody Glover was executed in 1698. Some children that she had scolded came down with strange diseases; it was clear that she had bewitched them. She was found to have some dolls, obviously used for hexing, and she couldn't recite the Lord's Prayer in English. "Guilty!" About the same time a black slave named Tituba was arrested as a witch on the testimony of three young girls of the household. She was found not guilty and sold to whomever could pay the expenses of her imprisonment. All told, more than two hundred were imprisoned, another fifty-five tortured, and twenty were executed.

Slowly, the tide turned in Europe, and after the usual delay, America would follow Europe's lead. Burning for witchcraft ended in Geneva in 1652, as people began to wonder about a creed that preached the love of Jesus and yet practiced these officially legal murders. Sweden was slow to start mass witch-hunting. In 1669 Gertrude Svensen was accused of kidnaping children and handing them over to the devil. She implicated others under torture, and all seventy of them, including fifteen children, were burned without any real evidence. Illicit sexual activity became the charge in 1670 in Scotland for Jane Weir (incest with her brother), and in France for Madelein de Demandola (intercourse with a priest). It would be more than another century before the last beheading of a witch in Scotland (1782). A few years earlier (1775), Anna Maria Schwägel was the last woman to be tried and executed for witchcraft in Germany, where the hunting had been particularly brutal and intense.

During the seventeenth and eighteenth centuries there were people who condemned these atrocious witch hunts and had the courage to speak and write their convictions. There were also those who continued to believe that witches should be exterminated. Thomas Hobbes (1588–1679) thought that witches should be punished, not for the evil acts they were alleged to perform but for *pretending* to do impossible things. In *Of Man*, the first part of his *Leviathan*, published in 1651, he wrote, "The prediction of witches, that pretended conference with the dead, which is called necromancy, conjury and witchcraft . . . is but juggling and confederate knavery." Richard Baxter (1615–1691) was an enlightened Puritan and a chaplain in Oliver Cromwell's army. He preached against the excesses of both Cromwell and the Royalists during England's Civil War, and tried to persuade the authorities of the Anglican Church to accept moderate dissenters. But he was no peacemaker when it came to the subject of sin, describing a cruel God bent on vengeance with his eternal tortures of Hell. How good it was, Baxter believed, how completely in accord with the will of God, that witches were being burned in Salem, Massachusetts.

Over in Holland, on the other hand, Balthasar Bekker* (1634–1698), a Dutch theologian, published his book *The Bewitched World* in 1591, making it clear that he did not believe all of these stories about witches and devils. He insisted on applying reason to the interpretation of the Scriptures and argued that, among other things, reasoning leads to the conclusion that the devil is myth rather than reality. Bekker's Protestant brethren expelled him from his pulpit and he was lucky to escape with his life. In Leipzig, Professor Christian Thomasius (1655–1728) followed in Bekker's footsteps with his disbelief in magical powers, and in the existence of a real devil. Some of his writings were attributed to an unnamed student as part of his dissertation, but the subterfuge availed Thomasius but little, especially when he stated that there was nothing wrong about a Lutheran marrying a Calvinist. His arrest was ordered, but he escaped to the University of Halle where he made a fundamental contribution to the growth in Germany of the new knowledge that was being stimulated and discovered by Gottfried Wilhelm Leibniz.

After the Restoration and Charles II's return to the throne (with Oliver Cromwell not only dead and buried but disinterred and hung on a gallows), the American Congregational clergyman Increase Mather (1639–

1723) stopped preaching in England and returned home. For nearly sixty years thereafter, he was pastor of the Second Church in Boston, and for sixteen of those years he was also President of Harvard College. He spent the years 1688–1692 in England, working for a better charter for Massachusetts with Sir William Phipps, who then came over to govern the colony during the next two years. Increase and Sir William saw that the witch-hunting mania was subsiding in England, partly from disgust, perhaps, and partly from doubt about the alleged miraculous powers of witches. On returning to Boston, Increase Mather published his *Cases of Conscience Concerning Evil Spirits*, and the new governor granted pardons to some women who had been condemned to death by the court. This action particularly irritated Lieutenant Governor William Stoughton (1630?–1701) who had presided over the Salem witch trials; he never presided over another one. It also ran against the convictions of Cotton Mather (1663–1728), son of Increase and assistant pastor under his father for nearly forty years. Cotton supported the idea of trials for witchcraft and executions of those found guilty.

In 1898 there appeared a short historical novel, *Ye Lyttle Salem Maide* (Lamson, Wolffe and Co.), by Pauline Bradford Mackie. It includes a description of the trial for witchery of its heroine, Deliverance Wentworth. It is historical fiction, but not historical fantasy, and several prominent people of the late seventeenth century appear in it—Stoughton, for example, and Cotton Mather "a young man of ascetic face and austere bearing, clothed in black velvet with neckbands and tabs of fine ribbon."

The seven judges wore "crimson velvet gowns and curled white horsehair wigs" except for one who had a black skull-cap. It was Judge Samuel Sewall (1652–1730) who by 1692 would condemn nineteen persons to be executed for witchcraft. Five years later he "rose in his pew in the Old South Church in Boston Town, acknowledging and bewailing his great offence." So quickly did attitudes change in the "Massachusetts Collonie" in the 1690s.

Cotton Mather stepped forward to address the judges. Pauline Mackie puts these words in his mouth, some of them taken from his writings.

"Especially against New England is Satan waging war, because of its greater godliness. For the same reason it has been observed that demons, having much spite against God's house, do seek to demolish churches during thunderstorms. Of this you have had terrible experience in the incident of this prisoner. You know how hundreds of poor people have been seized with supernatural torture, many scalded with invisible brimstone, some with pins stuck in them. . . . Yea with mine own eyes have I seen poor children made to fly like geese, but just their toes touching now and then upon the ground, sometimes not once in twenty feet, their arms flapping like wings."

Cotton's own book *Wonders of the Invisible World*, published in 1693, gave an account of the trials of witches, with emphasis on the effects of these evil spirits when they invade human bodies. He was confronted by a Boston merchant, Robert Calef, who wrote critical letters to him and his father and then had the temerity to publish *More Wonders of the Invisible World*, a direct attack on Cotton's book. Increase Mather, now President of Harvard College, ordered copies to be burned in the college yard in the year 1700. Robert Calef died nineteen years later, very unpopular for a while, but the tide was turning in his favor.

Back in England, John Wesley (1703–1791) was less than sympathetic to witches, as we have noted. He believed in good and bad magic, in miracles, and the intervention of ghosts in human lives. A case of "good" magic that he quoted was the story of the Catholic girl who lost her sight when she read the Catholic book of the mass, but regained it whenever she picked up the New Testament. Perhaps for him it was an allegory. Wesley was a strict and pious man, and although as late as 1768 he stated that "the giving up of witchcraft is in effect the giving up of the Bible," he did not seek to punish witches by torture.

Times had changed anyhow—way back in 1735, an act of parliament had removed witchcraft from the list of punishable crimes, but the idea died slowly. In 1765 the eminent lawyer Sir William Blackstone (1723–1780) affirmed the existence of witchery, and eight years later the divines of the Associated Presbytery in Scotland expressed their concern that people generally no longer believed in witches. They made it clear that they believed in the evil magic of witchcraft anyhow by passing a resolution that confirmed their conviction. Finally, in 1781, the benevolent dictator, Joseph II (1741–1790), Holy Roman Emperor and King of Germany, issued the *Edict of Toleration*. Throughout his realm witchcraft and magic were no longer punishable.

Joseph struck a blow for knowledge when, in spite of a unique "summit meeting" with Pope Pius VI, he closed monasteries that did not support studies, operate schools, or care for the sick. Fourteen students at the University of Innsbruck denounced one of their teachers who had told them that the world was more than six thousand years old, and the case ultimately came before the emperor himself. His judgment: "The fourteen students should be dismissed, because heads as poor as theirs cannot profit from education." That was harsh—the students were not necessarily "poorer" than others, but they had been subjected, perhaps from childhood, to a common intellectual strait-jacket. That was more than two hundred years ago, before anything was known about using radioactivity to determine when past events had occurred, or about the evolution of mankind from animals.

While witches were supposed to have cooperated with the devil, there were others, it was believed, who were possessed by a devil against their will. The devil, or his associated demons, imps, succubi or incubi could sneak into the body through an open mouth, or could already have been sitting on some food that was eaten. Making the sign of the cross, however, would frighten them away. If they did get inside, there is no telling what they might do—except that it would be all bad, mostly bringing the scourge of mental and physical illnesses. If the "devil" is thought of as equally invisible bacteria and viruses, it is easy to see how infectious diseases could be accounted for in this way. And didn't witches transfer sicknesses from one person to another, just as "carriers" do today? Isn't it wise to keep your mouth shut when in a place where you are exposed to infection? Don't doctors, dentists, and nurses wear those medical masks in order to stop these invisible "devils" from getting inside their bodies, and to keep their own devils from escaping? Sometimes the supernatural devils that people believed in were naturally occurring germs.

The paradigm of the devils—the generally accepted belief that invisible devils were responsible for all of the evils that afflicted mankind—was already flourishing in the time of Jesus, and many people believe that what was thought to be true then is true for all time, because it is "in the Bible." One group that was badly hurt by this attitude included those who were suffering from mental illnesses. As we have seen their treatment was intended to disgust the resident demon so that he would leave, to torture him by torturing his host, to feed abominable substances to him, to speak the most hostile and obscene words to him, and to make life intolerable for him by making it intolerable for the poor victims of the illness. For *eighteen hundred years* that was the standard Christian treatment of the insane. Originally, it was thought that the lucid intervals that appear from time to time in some afflicted persons were controlled by the phases of the moon. The Latin word *luna* for "moon" gave us the word "lunatic," used widely into our twentieth century, often as a term of opprobrium.

Back in the fourth and fifth centuries B.C., Hippocrates had recognized insanity as a disease of the brain. Plato's attitude, as expressed in the *Republic*, was to keep insane people out of sight so that society could pretend that they did not exist: "If anyone is insane, let him not be seen openly in the city, but let the relatives of such a person watch over him at home in the best manner they know of, and if they are negligent let them pay a fine." As noted in chapter 7, Asclepiades, in the second century before Christ, as well as Themison and Celsus, Jesus's contemporaries, and later Caelius Aurelianus and Paul of Aegina had all recognized the need for humane treatment of the mentally ill. This was echoed in the ninth and early tenth centuries by a German monk and chronicler, Regino of Prüm, who regarded insanity as a disease rather than the effect of internally confined demons. He was Abbot of Prüm from 892 to 899, moving to St. Martin's at Trier (Trèves) until his death in 915. That was an important position, but his words were not heeded, at least on this subject.

In Constantinople, Michael Constantine Psellus (1018?–1079?) revived the study of Plato, and also wrote an unfortunate book, *De Operatione Daemonum*. He wrote on many other subjects too—history, agriculture, physics, astronomy, medicine, alchemy—and that enhanced his reputation, but his book telling how demons operate seems to have gathered much more attention than the rest. It was what people wanted to hear. Why do demons enter the body? Answer: because they are cold and want to enjoy the warmth of a human body. (Perhaps even that was not as warm as their native habitat!) This sort of thinking, mixed with arguments from the Scriptures and Plato and infused with quotes of the babbling of some demented souls, crystallized the belief in the devil theory of insanity. As with other ideas that the authorities wanted to emphasize, the writings were supplemented with images that everyone could understand. Devils carved in stone adorned many cathedrals, and a saint is portrayed in one of the windows of Chartres Cathedral in the act of exorcizing a devil from a lunatic. A holy Latin adjuration is issuing from the saint's mouth while a horrid horned demon is emerging from the mouth of the person possessed. People became so scared that a devil would enter their bodies when they were asleep, or were awake but not fully alert, that mass hysteria set in from time to time. Distraught mobs gathered at holy shrines to cry for help. Nuns in a German convent started to bite each other, and the practise spread to convents in Holland and Italy. In France, the less dangerous habit of mewing like a cat spread through the convents, and a convent of Ursuline nuns was invaded by demons.

We have seen how Paracelsus (1493?–1541) challenged the age-long assumptions in science and medicine, as Luther, Zwingli, and Calvin were opposing the old assumptions of Christian theology. An imbalance of Galen's "humors" was not the cause of illness; instead, one should look to the chemicals in the body and prescribe chemicals as remedies. "Possession by devils" is not really that at all; it has nothing to do with either devils or saints, and it can be cured by prescriptions and proper care. Paracelsus moved from one place to another throughout Europe, openly contemptuous of his fellow physicians, and running into both fame and trouble wherever he went. Many of these doctors had never seen a patient until they got out of medical school, their heads filled with medieval medical theories. Like his contemporary, Johann Faust, he became a legend during his lifetime; had he too sold himself to Satan? Was he, in fact, Dr. Faust?

Paracelsus was a precursor of psychoanalysts in asking how nightmares are possible, and in recognizing them as sexual fantasies. His theory was based on the demons of the time, in particular the lascivious male incubi and the (mostly) female succubi, but he was searching for a real understanding, not just repeating the ancient inanities. He recognized that casting a spell on a part of someone's body comes not directly from the sorcerer to the limb in question, but that it must come via that someone's mind. "It is not the curse or the blessing that works, but the idea. The imagination produces the effect." In order to treat a schizophrenic Paracelsus would make an image in wax and convince the patient that this was his other personality. He would then destroy that personality by burning the image. He had a remarkably advanced view of magic, spells, incantations, hexes, and the like—they are neither good nor bad in themselves; rather it is the intention of the person who works them that counts. Is he or she trying to bring evil or to bring good? In England, however, a contemporary of Paracelsus, Sir (later Saint) Thomas More, who was quoted earlier in this chapter, supported the practise of scaring demons out of the mentally disturbed. That treatment, or the alternative torture of preventing the patients from sleeping, was enough to turn a perfectly sane human being into a raving maniac. Later in that sixteenth century, Calvin's successor, Theodorus Beza, asserted that the idea that insanity is a natural illness is "refuted both by sacred and profane history."

Very early in the next century, the English poet and translator, Richard Carew (1555–1620) wrote in his *Survey of Cornwell*:

> In our forefathers' daies—there were many browsenning places for curing of mad men, and amongst the rest, one . . . called Saint Nunnespoole which . . . gave name to the church. . . . The watter running from Saint Nunne's well fell into a square and close walled plot, which might be filled at what depth they listed. Upon this wall was the franticke person set to stand, his backe towards the poole, and from thence with a sudden blow in the brest, tumbled headlong into the pond; where a strong fellowe . . . tooke him and tossed him up and downe, alongst and athwart the water, untill the patient, by foregoing his strength, had somewhat forgot his fury. Then was hee conveyed to the Church, and certain Masses sung over him; upon which handling, if his right wits returned, Saint Nunne had the thanks; but if there appeared small amendment he was browsenned againe and againe, while there remayned in him any hope of life, for recovery.
>
> (ADW, 2, 130)

A year or two before the publication of the *Survey of Cornwall*, Shakespeare's *As You Like It* had its first stage presentation. In Act 3, Scene 2, Rosalind, disguised in male attire, tells Orlando: "Love is merely a madness, and, I tell you, deserves as much a dark house and a whip as madmen do."

In 1632, London's Bethlem ("Bedlam") Hospital for the Insane received a medical man as governor, in some recognition of the medical nature of mental problems. Founded in 1247 as a priory of the Order of the Star of Bethlem, the hospital began to admit "lunatics" in 1403, an earlier law of Edward II vesting their property in the crown. By the eighteenth century, however, the administration charged an entrance fee for the public to look and laugh at the chained lunatics, a human zoo. It was even more disgusting than when it had been opened as a lunatic asylum—enough to bring the word "bedlam" into the language. The English physicians Thomas Willis (1621–1675) and Thomas Sydenham (1624–1689) described various forms of nervous affliction by direct observations. Sydenham strongly recommended bleeding in the treatment of "mania," prescribing that "young subjects, if of a sanguine habit, are to be bled to the extent of nine ounces on two or three occasions, with three days between each bleeding." A caution was given not to exceed this amount of bleeding "otherwise idiocy and not recovery will result." The bleeding was followed by a course of purgative pills. The humors of Galen—blood, phlegm, and black and yellow bile—were still part of the belief systems of leading physicians. In France, Jean Baptiste Denis (1645–1701) tried blood transfusions to cure insanity. When he tried it on a love-sick youth, with a fatal result, he was sued for malpractice.

There followed a time when long-established beliefs in religion and philosophy were being challenged. New and more liberal attitudes were developing, the validity of the theory of witchery was seriously questioned in some regions, but practical changes in the treatment of the insane were slow in coming. In 1751, however, the Society of Friends founded a small hospital in Pennsylvania to provide more humane treatment for a few mentally disturbed people, and to offer it as a good work and therefore acceptable to God. A similar hospital was built in Virginia in the 1770s, and in the following years similar hospitals appeared in the other colonies.

An Englishman, John Howard (1726–1790) taught his countrymen a great deal about prisons and asylums. He inherited a fortune and lived "for the good of others"—*Vivit propter alios* was inscribed on his tombstone in Kherson, on the mouth of the River Dnieper, presently Soviet Russia. Howard sailed for Portugal to see if he could be of any assistance following a recent earthquake, but a French privateer seized the ship on which he was traveling and he was imprisoned. The experience led to his interest in prison reform; when released, he pressured the British Parliament on the issue. His wife died, leaving him with a very young son whom he raised alone on his estate in Bedfordshire. Unfortunately, the son suffered from a "hopeless derangement" and had to be placed in an asylum, stimulating his activist father to press for reform in mental hospitals as well.

Howard traveled throughout England, Europe, and the Near East learning as much as he could about prisons, asylums, and the incidence of the plague. He reported that in Constantinople he had found an insane asylum that was much superior to the famous St. Luke's Hospital in London. Way back in the seventh century, the Greek physician and surgeon Paul of Aegina had written a medical encyclopedia, had been encouraged by the conquering Caliph Omar, and, as noted in chapter 10, had prescribed merciful treatments for the insane. This attitude still prevailed in Constantinople when John Howard visited it near the end of the eighteenth century. And he was not the first European to observe it; several hundred years earlier some monks who had traveled to Muslim territory returned with a similar report. Howard was on his way to Russia to continue his campaign for better conditions, and to learn first hand how the Russians treated their convicts and insane. Unfortunately he was taken ill en route and died in Kherson.

France was fortunate to have a series of iconoclastic writers challenging established beliefs of the day. It began with the skepticism of the essays of Michel Eyquem Montaigne (1533–1592). René Descartes (1596–1650) followed with his antischolastic philosophy and modern scientific thought. Pierre Bayle (1647–1706) in his 1697 *Dictionnaire Historique et Critique* analyzed and criticized the accepted paradigms of history and philosophy, arguing that it was useless to use reason to try to prove the existence of God, or to determine his nature. Bayle went back to the Greek philosopher and skeptic Pyrrho (365?–275) who argued that many philosophical disputes could lead to no certain conclusion and that the best way to approach them is to say, "I know nothing about it, and abstain," turning instead to questions that *could* be treated rationally. The eighteenth century then brought François Marie Arouet, known as Voltaire (1694–1778), "the incomparable satirist," "the apostle of unbelief," "a bad-hearted man," "a defender of victims of religious intolerance"—depending on whose comments you read. Dr. Durant ended his book on *The Age of Voltaire* with the statement, "When we cease to honor Voltaire we shall be unworthy of freedom." Voltaire was the centerpiece of the Age of Enlightenment, the century in which long-established beliefs in Christendom were finally receiving serious challenges.

It wasn't all happening in France, of course. Old theological beliefs about science were not just being challenged; they were in retreat, following the work of Isaac Newton and his associates in England, Christian Huygens in Holland, and Baron von Leibniz in Germany. But in other areas France led the way with Denis Diderot (1713–1784), the encyclopedist and philosopher, and Jean Jacques Rousseau (1712–1778) who wrote articles for Diderot's encyclopedia, quarreled with him, attacked Voltaire, quarreled with David Hume, and spent the last decade of his life partly insane. Which brings us to Baron Paul Henri Dietrich d'Holbach (1723–1789), the French materialistic philosopher who also contributed to the encyclopedia and insisted that mind must be considered as part of the body, since mental disorders can be cured only by attacking their physical causes. If mind and soul were regarded as separate eternal entities—mental aberrations, he argued, would never be understood. And "to say that the soul will feel, think, enjoy, and suffer after the death of the body is to pretend that a clock shivered into a thousand pieces will continue to strike the hour."

D'Holbach (originally von Halbach—he was born German) was wealthy and he supported both science and the arts. Twice each week distinguished guests came to his home to discuss philosophy and raise arguments against the Christian church—Diderot; Jean le Rond d'Alembert, the mathematician and philosopher; Claude Helvétius, whose work *De L'Esprit* was condemned by the Sorbonne and publicly burned; George de Buffon*, naturalist, forced to recant his early glimpse of evolution; Rousseau, practically the only one of the guests who wasn't an atheist; and several others.

The French Parliament condemned a number of books, and ordered that their authors be arrested and punished. Two of d'Holbach's books were on the list: *The System of Nature* and *Christianity Unveiled.* He escaped punishment, however, because he had had the foresight to print them under the name of a gentleman who had been dead for ten years, and his friends weren't about to give him away. He was consistently an atheist, religiously devoted to improving man's lot in the world, fed up with the leadership of kings and priests, turning instead to philosophers and other scholars. Socrates had said that he was prosecuted "for inventing new gods, and for not believing in the old ones." When a strongly held belief is supported by faith alone it cannot accept a challenge and often responds by hostility towards the challenger.

The mad George III (1738–1820)—king of Great Britain and Ireland (at least officially) for sixty years—lived across the Channel at the time of these French encyclopedists. His first major mental breakdown occurred in 1765, five years after he had been crowned, and a series of them occurred in 1788, 1801, 1804, 1810, and 1811; he went blind and never recovered from his last seizure. With Frederick North (known as Lord North) as his prime minister (1770–1782), George was well enough to function, but his ruinous policy alienated and then lost the American colonies. Great indignation was aroused throughout the British Isles by the treatment that he received when the mania overtook him. The treatment of his case was left largely in the hands of Dr. Francis Willis (1718?–1807) because, alone among the physicians, he asserted that he could cure the king at his asylum for lunatics in Greatford, Lincolnshire. The royal patient was treated with very little respect; if he got dangerous he was knocked down and put in a straitjacket. On the other hand, treatment was generally kind and the patients were allowed a great deal of freedom. George was given a lot of Peruvian bark from the genus Cinchona that contained quinine and other fever-lowering alkaloids. Dr. Willis's reputation and fortune rose to great heights when a "sane" George was delivered back to his stressful throne.

In France, physicians began to recognize that the treatment of the insane was causing the poor patients to deteriorate, living in constant fear, many of them in chains, all in filthy conditions. A basic change finally began when Philippe Pinel (1745–1826) was appointed as the directing physician of the large asylum at Bicêtre, in the Gentilly commune of Paris, and later of another asylum at nearby Salpetrière. He wrote tracts on *Mental Alienation, Clinical Medicine,* and *Nosography* (the classification of diseases), and he was a very popular lecturer. His chief contributions, however, lay in his actions rather than in his words. Pinel introduced an element for treating patients that Christian theology had never contemplated, despite the teachings of Christ: kindness. Following the French Revolution the new government gave him permission to remove the patients' chains and let them emerge from their horrible cells to sunshine, fresh air, and both physical and mental exercise.

Dr. Durant called it "one of the many triumphs of secular humanitarianism in the most agnostic of centuries." The English statesman and biographer Viscount Morley of Blackburn (1838–1923) had written in his 1809 biography *Diderot and the Encyclopaedists*, "It is certainly true as a historical fact that the rational treatment of insane persons and the rational view of certain kinds of crime were due to men like Pinel, trained in the materialistic school of the eighteenth century. And it was clearly impossible that the great and humane reforms in this field could have taken place before the decisive decay of theology."

Once again it was the Society of Friends (or Quakers) who pioneered humane treatment for the insane, this time in England. In the same years that Pinel was beginning to implement his ideas in France, a Quaker merchant, William Tuke (1732–1822) independently started a similar treatment for some patients in York. His wife suggested that the word "asylum" be replaced by "retreat," and care of the insane became a family tradition. Their grandson Samuel (1784–1857) continued this humane treatment, and great grandson Daniel Hack (1827–1895) became a London specialist in mental diseases, and co-author of *A Manual of Psychological Medicine* and other studies of insanity. Samuel improved on his great-grandfather's methods and had a big influence on mental hospitals, particularly in Canada. Many physicians had not listened to William Tuke, and terrible conditions continued to prevail in some of the asylums. William himself had been strongly criticized in the *Edinburgh Review* (April 1803) and no support of his humanitarianism was forthcoming from the Anglican clergy, or for that matter from John and Charles Wesley and their descendants.

Pinel's work in France was continued by his successor and student Jean Étienne Dominique Esquirol (1772–1840) who in 1838 published what could be called the first textbook in psychiatry, at least the first nontheological text. Esquirol backed up his arguments with statistics, and was appalled that in France, too, the old inhuman practices continued in some of the hospitals.

Since that time there have been many other pioneers in the field of psychiatry. They are too numerous to mention except for Sigmund Freud (1856–1939) and Carl Gustav Jung (1875–1961), whose works are very well documented, along with Alfred Adler (1870–1937) with his theory of the inferiority complex. Psychiatry, hypnotism, prayer, and meditation probe our unconscious minds in their various ways. The unconscious is a vast and magnificent neural network with limited access and, as the sum of our past experiences, it compels, amazes, and surprises us. It is our god and our devil within, the still small voice of conscience as well as the lust of rape and murder. We do not yet know enough about how to control it or to profit from it. It has taken human beings a long time to advance beyond the thinking described in chapter 1. At that stage, everything that was not understood—practically everything that happened—was attributed to good and evil spirits. Rejection of the gods-and-devils hypothesis is now leading us to better understanding, and lifting us up one small notch on the ladder of evolution.

1. PERICLES

2. THUCYDIDES

3. CHARLEMAGNE

4. LOUIS IX (left) and ST. VINCENT (right)

Some rulers, though wielding great power, had the vision to use their power to further the quest for understanding rather than to suppress it. In the ancient world, Pericles (1) was idealized by the historian Thucydides (2) as an inspirational leader who urged his people to live by reason rather than superstition. Pericles supported scholars and artists, and made Athens a center of learning and culture.

In the Middle Ages, Charlemagne (3) supported the scholar Alcuin, collected books from all over the realm, established a school at his capital, and became a student himself. Similarly, Louis IX (4, left) of France was devoted to learning and appointed St. Vincent of Beauvais (4, right) as his teacher. Beauvais wrote a comprehensive encyclopedia, summarizing uncritically the scientific knowledge of his time, much of which was derived from recent translations of Arabic works into Latin.

1. COPERNICUS

2. MARTIN LUTHER

3. HENRY VIII

4. ERASMUS

But not all scholars were so fortunate as to live at a time that was tolerant of learning and free inquiry. Copernicus (1), fearing persecution from the Inquisition, would not allow his great treatise on the solar system to be published until after his death. He lived in the age of the Reformation and Counter-Reformation, when the Catholic Church was torturing or executing those who deviated in any way from traditional beliefs, especially those who followed the iconoclastic teachings of the renegade priest Martin Luther (2).

One of Copernicus's contemporaries was King Henry VIII of England (3). He was called the "Defender of the Faith" by Pope Leo X in 1521 for attacking Martin Luther in print. Not long after, he himself was excommunicated by the Church for divorcing his first wife, Catherine of Aragon, and marrying Anne Boleyn.

But perhaps worse than the official oppression of church or state was the oppression of mass ignorance that held the general public captive to superstitions. Many were victimized during the age of witch burnings and many independent thinkers were accused of being in league with the devil. Only a few, like the wily and satirical Erasmus of Rotterdam (4), were able to escape with their lives. Erasmus supported the circulation of the Bible in the vernacular languages and he offered a number of insightful critiques of various biblical passages, showing that they were spurious. In 1540, at the command of Pope Paul IV, his collected works were publicly burned. The Inquisition would probably have liked to do the same to him, but he was fortunate enough to die a natural death.

14

Scriptures, Weather, and Diseases

What can be a grosser superstition than the theory of literal inspiration?

—Frederick Temple (1821–1902), Archbishop of Canterbury

To him who feels the hailstones patter about his ears, the whole hemisphere appears to be in storm and tempest.

—Michel Eyquem de Montaigne (1533–1592)

Strangely, almost ironically, we owe what is claimed to be the most accurate version of the Bible to the Assyrians, whom we left in chapter 6, with their cruel zoo and their emphasis on military might. We read in I Chronicles (5:26): "And the God of Israel stirred up the spirit of Tilgath-pileser, king of Assyria, and he carried them away, even the tribe of Reuben and the tribe of Gad and the half tribe of Manasseh, and brought them to Halah and Habor and to the river Gozan, cities of Media. And they still dwell there to this day." Media, the home of the Medes, was the region of present north Iraq and northwest Iran, spilling over into Armenia. The Assyrians had conquered vast territories, including this one, but in 612 B.C. the Medes took their revenge. Teaming up with Nabopolassar of Chaldea, Cyaxares, King of the Medes, destroyed Nineveh, the Assyrian capital. (He also conquered Armenia and involved himself in the long war with Lydia, referred to in chapter 12—a war halted by an eclipse.) After the Medes triumphed the Assyrians kept a low profile for many centuries, preserving the Aramaic language that was closely related to Hebrew and sometimes called "Syriac." During the first centuries of the Christian era, many of these people were converted to Christianity, and Armenia is Christian to this day. These Christians were often persecuted, but since they broke off relations with the Roman/Byzantine Church, they were tolerated by Persian invaders, who were Rome's enemies. When Constantine was converted to Christianity in the year 313, however, the Christians were seen as supporters of the Holy Roman Empire, and they were persecuted until the Muslims conquered the Persians a little more than three hundred years later. As we have seen, they were then tolerated for a while, provided they paid their taxes.

All that time, these Christians preserved their written heritage—the Bible in the Aramaic language. They remembered the words of Revelation 22:19, "If any man shall take away from the words of the book of this prophecy, God shall take away his portion from the tree of life. . . ."—almost the last words in the Bible. They were meticulous in preserving manuscripts and writing accurate copies, to the point where in 1957 the Catholicos Patriarch of the East claimed that "the Church of the East received the Scriptures from the hands of the blessed Apostles themselves in the Aramaic original, the language spoken by our Lord Jesus Christ himself, and the Peshitta is the text of the Church of the East which has come down from the Biblical times without any change or revision," as quoted by Dr. Lamsa in the introduction to his translation of the Peshitta (i.e., "true") text into English. We may have some reservations about receiving this "from the hands of the blessed Apostles," but, because of the care bestowed on it, and because it was not necessary to translate it, that text has remained unaltered over the centuries.

Other versions of the Bible have come by a variety of routes. We have noted earlier that the Septuagint translation of the Old Testament, from Hebrew and Aramaic to Greek, upset orthodox Jews, appalled to have the Hebrew text so profaned. It became a legend that this affront to the Almighty caused a darkness for three days over the face of the earth, though this alleged event lacks supporting evidence. When Christianity was preached to the Greeks, this ultimately became the official Old Testament of the Greek Orthodox Church. The Jewish martyr Rabbi Akiba ben Joseph (50?–132), chief teacher at the rabbinical school at Jaffa, contributed toward establishing the canon of the Hebrew Old Testament—a compilation of the Pentateuch, the Prophets, and the *Hagiographia,* i.e., "written by divine inspiration." Sadly, for his support of the Hebrew uprising the Romans tortured him unmercifully, flaying him alive. One of his disciples, Aquila of Pontus, provided another translation of the Old Testament into Greek, and in the third century Origen wrote the *Hexapla.* As the name implies, this gave six versions of the Old Testament side by side—the Hebrew text, the Hebrew in Greek letters, the Septuagint, and the translations into Greek by Aquila and two others.

Toward the end of the fourth century, the irascible St. Jerome produced the first Latin Bible—the Vulgate. In translating the Old Testament into

Latin, he used the Greek Septuagint and the other translations given by Origen, although when there was disagreement between them he went back to the original Hebrew. For the New Testament, he used earlier Latin translations and Greek manuscripts. The Vulgate soon became the official Bible. It was basically a translation of a translation, a fact that was forgotten centuries later when, as we have seen, good men who translated it into their native languages were severely persecuted. After the pioneering, brave, and tragic efforts of Wycliffe, Tyndale, Rogers, and others to produce a Bible that English-speaking people could read, there have been many English versions. With all of its errors of triple translation, the King James version has remained the official Bible for many churches. People like tradition, and though by some standards close to four centuries is not a very long tradition, it is a tradition nevertheless.

In chapter 9 we noted the mistranslation of Matthew 22:9—"compel" them to come into the marriage feast was later replaced by "invite." Another example, from Matthew (19:24), arises from the confusing similarity of the words for "rope" and "camel" in the Peshitta text: "It is easier for a *rope* to go through the eye of a needle" rather than a "camel." Harder to understand is the similarity between the word *awaley* for "ungodly" and the word *eweley* for "babies," and they look even more similar in the original writing than in these transliterated forms. For example in Psalm 144:7, "deliver me out of great waters, from the hand of strange children" and also in the eleventh verse of the same Psalm, the prayer is really for deliverance from the *wicked.* One dot under the word distinguishes "*you* have deceived" from "*I* have deceived." Thus in Jeremiah (4:10) the reproach "Ah, Lord God! Surely thou hast greatly deceived this people and Jerusalem" in the King James version should really be the confession "O Lord God, surely *I* have greatly deceived. . . ." The Aramaic word usually translated as "sword" in Matthew (10:34) is an idiom for "division"—"I come not to bring peace but a division." There are many other examples given by Dr. Lamsa in his article "Words Resembling One Another" as part of the introduction to his translation of the Aramaic text (*Holy Bible,* 1986).

It was bad enough that some theologians had propagated unorthodox, and therefore heretical ideas; bad enough that witches were cooperating with the devil, and that the satanic spirits did not respond as they should to devout prayers and Latin swear-words. On top of this, the scientists, influenced by the Muslims, were getting out of hand, trying to overthrow the official model of the universe and criticizing the teachings of the great Aristotle, teachings that had been integrated with Christian theology by the revered Saint Thomas Aquinas. But when men criticized the Bible and threw doubt on the authenticity of some of its passages, that was really too much. It would seem that the stability that comes from tradition, whether in Latin, English, or any other language, is not enough. It is necessary to have stability of *interpretation.* Neither the words nor the meaning of the text can be changed without threatening the whole theology.

Exegesis, or Biblical interpretation, began with the first Jewish scribe who decided not to limit his work simply to transcribing the Torah. We will not examine the details of the many Jewish and Christian writings that followed, some very profound, others afflicted with the sophistries of the gematria discussed in chapter 8. We will, however, look briefly at the conflicts that arose from the struggle of Christian theologians to understand what they regarded, and still regard, as the Word of God. Origen again pioneered the way by arguing that the Bible should be understood at three separate levels—the literal, the moral, and the allegorical. To this was later added the "spiritual," "mystical," or "anagogical." This defined the "Four Senses of the Bible," dovetailing with other "fours"—seasons, compass points, gospels, etc. Thus whenever the word "Jerusalem" appears in the Bible, it *literally* means "a city in Palestine," it *morally* means "the soul who believes," it *allegorically* means "the Church," and it *anagogically* means "the heavenly eternal Jerusalem." While there were these hidden meanings, however, the literal meaning of the Pentateuch must be recognized as written or dictated by God, being a precise statement of historical facts as recorded by Moses.

That belief was universal in Christendom until Abraham ben Meir ibn Ezra (c. 1092–1167) wrote his commentaries on the Bible. In chapter 8 we have noted some of his work, and the poem to him by Robert Browning. Here we may add that he appears to have been the first scholar to question the common belief that these first five books of the Old Testament were written by Moses and handed down unchanged over the centuries. Clearly the last chapter of Deuteronomy was added by someone else, since it describes the death of Moses, but some other recorded events are known to have occurred later. Rabbi ben Ezra, as Browning calls him, was very wise—he attributed some of his critical remarks to another rabbi who had died earlier. Ibn Ezra, however, went along with the traditional view of the "Song of Solomon"—that it was an allegory of the love of God.

We have also referred to another great Jewish scholar, Maimonides (1135–1204), who interpreted the Adam and Eve story as an allegory. In his view, Adam was the active spirit, Eve the passive, and the serpent was imagination. In his *Guide to the Perplexed* Maimonides recognized the various meanings attributed to biblical words and passages and emphasized that reason should be applied by the believer who is having some doubts. We should not take literally the parts of the Old Testament that were written for our uneducated ancestors. This use of allegory got out of hand with his immediate successors, some of whom saw the whole of the Old Testament as an allegory. Historical figures in the Bible became legendary, the Jewish rituals merely symbolical—leading to a serious split among Jews of the early thirteenth century. One of them, Rabbi Solomon, was so upset by these allegories that he denounced them before the Dominican Inquisition, arguing that they were heretical for Christians as well as for Jews. In 1242, the Talmud was publicly burned in Paris, two years after the death of Rabbi David Kimchi, who had defended Maimonides' teachings and acted as arbitrator and healer of the split between the two Jewish factions.

Two centuries later, the Latin scholar Lorenzo Valla (1407–1457) showed that the Apostles' Creed was created several centuries after the apostles had died, and that the works that people had believed for centuries were written by the Dionysius whom St. Paul had converted were in fact composed much later. There was also the story of Abgar V (Ukkama) ruling from Edessa over his domain of northwest Mesopotamia. This Abgar (there were twenty-eight others) lived at the time of Jesus, and is supposed to have written to him and received a reply. Abgar, with all his wealth and power, was a leper and was allegedly cured by one of the disciples who visited him. Valla showed that this too was a report that would not withstand scholarly enquiry. He was a fanatic about Latin—he would have liked to abolish Italian and return to that language that he loved so fervently. He spent much of his effort in showing what inferior Latin others had written, including that of the Vulgate Bible of St. Jerome. He also wrote a diatribe against chastity, and made it clear that he practised what he preached. He avoided trouble from the ecclesiastics partly by doing something that made them even more angry at him. There was a document, eleven hundred years old supposedly, that was believed to have been written by the emperor Constantine to the pope of the time, Sylvester I, granting him secular control over western Europe. Nicholas of Cusa, Valla's contemporary, argued against its validity, and Valla mustered his immense knowledge of Latin to prove beyond any doubt that it was a forgery. Alfonso V, King of Aragon, Sicily, and Naples, was naturally delighted with Valla's work, and when Valla was brought before the inquisitors, Alfonso used his considerable clout to persuade them to release him.

The illegitimate son of a Mr. Gerard was born in Rotterdam on October 28, 1467. He didn't like the name "Gerard," which means "strong with the spear," and changed it to the Latin "Desiderius"—"loved" or "desired"—and to emphasize the point, he added "Erasmus" from the Greek for "worthy of love." In the same way that the Greek/Latin "ballistic missile" means "a thrown thing that is thrown," so Desiderius Erasmus means "a loved person who is loved." (It could be argued that he had not really departed from his father's name, for in early societies a good spear-thrower was very much loved and appreciated.) But he wasn't always loved, at least by those of the monastic orders upon which he launched his satirical attacks.

Erasmus also made enemies by supporting the circulation of the Bible in the vernacular languages of Europe, though he himself was more at home with Latin than with his native language. In 1516 he published the Greek Testament and a new Latin translation. That didn't bother people so much as the fact that in studying ancient manuscripts he had discovered—or rediscovered, for others had noted it—that the verse I John 5:8 did not appear in any manuscript before the end of the fifth century, when it had been added by an unknown writer. Erasmus therefore omitted this text, "And there are three to bear witness, the Spirit and the water and the blood, and these three are one," because he saw that it was spurious. Martin Luther agreed with him and omitted it from *his* translation, but people don't like changes in their holy books, and the verse was reinstated, providing "biblical" support for the doctrine of the Trinity. This does raise a question though. If the Peshitta text dates back earlier than the late fifth century, as is claimed, how is it that this verse appears in it, duly translated by Dr. Lamsa?

Erasmus also discovered that some of the epistles attributed to St. Paul were written by others, and for these and other heresies his work was condemned by the University of Paris and other august bodies. The first edition of the collected works of Erasmus (Basel, 1540, 9 volumes) was burnt by the command of Giovanni Pietro Caraffa (1476–1559) while he reigned as Pope Paul IV during the last four years of his life. In this role he also reorganized the Inquisition, and became unpopular because of his nepotism. Erasmus was beyond this Paul's reach because he had died in 1536. Again Martin Luther, while convinced of the infallibility of the Scriptures, was equally convinced that the origins of some of the books of the Bible were not what traditional beliefs held. He recognized that St. Paul's epistle to the Hebrews was not written by him, and he removed the Epistle of St. James from his New Testament as being unworthy of inclusion.

Michael Servetus* (1511–1553), mentioned at the end of chapter 10, was arrested and brought to trial before the French Inquisition partly for expressing the same kind of doubts, that Judea was the "land of milk and honey" described in Exodus 3:8 and 33:3—an image continued by the hymn of St. Bernard of Cluny: "Jerusalem, the Golden/With milk and honey blest." He was accused of other heresies, but somehow managed to escape to Switzerland. The metaphor of "falling out of the frying pan into the fire" is tragically accurate here, because Switzerland was Calvin's territory. Calvin had him imprisoned and burned as a heretic for trying to make a liar out of Moses. A French contemporary of Servetus, Sebastianus Castellio* (1515–1563), deviated from the Calvinist doctrine, and, reasonably enough, Calvin expelled him from his rectorship at Geneva. However, Castellio had his own interpretation of the Song of Solomon—he doubted that the love songs were describing Christ's passion for his Church, or Jehovah's love for Israel. Calvin and Beza hounded him to an early grave. Born a few years after Castellio, the Spanish poet Luis de Leon* (1527–1591) also wrote about his interpretation of the Song of Songs and other books from the Old Testament. The Spanish Inquisition imprisoned him for five years and forced him to rewrite his commentary according to official beliefs.

Christian and Jewish orthodoxies alike resented any interpretation of the Scriptures that differed from established beliefs, particularly an interpretation that differed very widely. In the next century, Uriel Acosta* (1591–1647) offended the Jewish congregation in Amsterdam by asserting that the Mosaic Law was of human origin, and that the immortality of the soul was not implied by the Old Testament. Acosta had converted from Catholicism and moved from Portugal to escape the Inquisition, but he was scourged and forced to recant by his adopted Judaist colleagues.

More well-known is the philosopher Baruch Spinoza* (1632–1677), a son of Portuguese Jews who was born in Amsterdam. He wrote about three kinds of knowledge:

(1) hearsay and vague impressions, received passively,
(2) comparison of various phenomena and the establishment of relationships between them,
(3) the discovery of some Absolute Principle.

He wrote much more, too, but here we note his attitude to biblical research, for he dismissed (1) as worthless and saw (2) as useless unless perhaps it could help achieve (3). That meant that one could not claim that any writing, biblical or otherwise, was infallible; it is better to teach people decent conduct than to make them recite a creed. He was excommunicated by the rabbis, to the shame of his wealthy parents who were pillars of the synagogue. Christians, too, were appalled at the teachings of this great philosopher and joined in his public vilification.

Another distinguished man, Richard Simon (1638–1712), wrote *Critical History of the Old Testament* (1678) and *Critical History of the Text of the New Testament* (1683). The first of these so outraged Jacques Bossuet (1627–1704), tutor to the Dauphin, and later Bishop of Meaux, that he obtained an order to stop publication and to burn all copies already printed. Simon was immediately removed from his membership of the French Congregation of the Oratory, a Roman Catholic religious society that had been formed in 1611. The analysis of the Old Testament that Simon had made led him also to the conclusion that Moses could not have written the first five books of it. He also attacked the time-honored belief that Hebrew was the language which God spoke, and that He had given this language to Adam and Eve, later confusing everyone with the Tower of Babel, as we have seen. Simon was followed by a Swiss theological critic, Jean Le Clerc (1657–1736), who moved to Amsterdam in search of an atmosphere more sympathetic to his studies of the Old Testament. He also recognized these arguments about the authorship of Genesis and advanced more criticism of other books, only to be roundly condemned by the theologians of the day.

So it continued until more recent times. Thomas Woolston* (1670–1733)—or Woelston—was a minister of the Church of England but was too sympathetic to deism to be allowed to continue in that role. Following the revelations of Copernicus, Kepler, and Woolston's contemporary Isaac Newton, together with the philosophy of Hobbes and Locke, an English deist movement developed during the seventeenth and eighteenth centuries. The common thread of the diverse philosophies of these deists was to develop a religion based on reason. Woolston was prosecuted for publishing his *Six Discourses on Miracles* in which he questioned the reality of the New Testament miracles. He was fined, and imprisoned until his death three years later. Other deists of the time also ran afoul of the law. The Irish free-thinker John Toland (1670–1722) published his *Christianity not Mysterious* in 1695, designed to show that "there is nothing in the Gospel contrary to reason." This raised a storm of criticism in London, so he returned to Ireland, only to find that the Irish Parliament had ordered that his work be burned by the hangman. He returned to England before the decree could be extended to include himself.

In Germany, Gotthold Lessing (1729–1781), the dramatist and critic, also contributed to biblical criticism when he published *Fragments of an Anonymous Writer.* Toward the end of his life, Lessing was keeper of the ducal library at Wolfenbüttel, Brunswick, and these fragments were allegedly discovered there. It was suspected that they were really written by Lessing himself, although they have been attributed to the theologian Hermann Reimarus (1694–1765) who also had not claimed their authorship. It was dangerous to be an author of material such as these "fragments" that attacked the absolute authenticity of the Scriptures. They also asserted that the foundation of Christianity could not lie only in the Gospels, because these were contradictory, and even spurious in parts, and should be the subject for further scientific enquiry. No, Lessing stated, the true foundation of Christianity is in the hearts and souls of men, not in the fallible written word.

This was followed a few years later by the studies of Johann von Herder (1744–1803). In his *Spirit of Hebrew Poetry* he argued that the Song of Solomon is a beautiful oriental love poem rather than an allegory of the love of God. This was hard to accept—what would a love poem be doing in the Bible? Wasn't sex wicked, and weren't those erotic parts, if thought of that way, downright sinful? Hadn't they been put to sacred music a couple of centuries earlier by the composer of masses, hymns, and magnificats Giovanni Pierluigi da Palestrina (1526?–1594) extolling the union of soul and body—*not* of body and body? Love poem indeed! The very idea was blasphemous. Not only that, Herder had asserted that the Psalms were written by different authors at different times. He was offered the professorship of theology at Göttingen, but the condition was that they should first investigate his orthodoxy. Knowing he would flunk that test, he moved to a safe haven at Weimar.

In 1792 the Roman Catholic Scotsman, Alexander Geddes (1737–1802) provided a new translation of the Old Testament. He also argued that the Pentateuch had more than one author, and that it could not have been written before the time of David (1000 B.C.), some 300 years after the death of Moses. Both Catholic and Protestant authorities disowned Geddes for this, but the days of burning heretics at the stake had finally gone. Less sadistic methods, however, were developed and used into the nineteenth century. Partly because of Christianity, and partly despite Christianity, the laws changed and no longer allowed the death penalty for heresy. John William Colenso* (1814–1883), appointed Bishop of Natal, South Africa at the age of thirty-nine, translated the New Testament into Zulu, and supported the natives in their suppression by the Boers. He allowed polygamous Zulus to join the Church without giving up all but one of their wives, and he announced that he did not believe in the eternal punishment of hell. For that matter, Colenso did not believe that Moses wrote the Pentateuch; in his mind it was a later forgery. After ten years of leading his See, Colenso was deposed by the Bishop of Cape Town and branded as a heathen by British publications.

In Germany, Ernst Hengstenberg (1802–1869), a Protestant theologian, strongly opposed all of this rationalism and modern biblical criticism. He was backed by the incompetent King of Prussia, Frederick William IV (1795–1861), another monarch who went insane. The word was penetrating theological circles, however, and other scholars studied the Bible and came to the same conclusions as had Castellio, Simon, Herder, Geddes, and a line of theologians dating back to Rabbi ben Ezra. In 1869, Abraham Kuenen of Holland published *The Religion of Israel* in which he noted the conclusions drawn by others, emphasized that the Old Testament is a mixture of history and legend, and that there is nothing supernatural about the predictions that appear in it.

In England generally, and at Oxford University in particular, scholars were slow to pick up on these developments. When Edward Everett (1794–1865), former President of Harvard and U.S. Senator, appeared at Oxford to receive an honorary degree the students were extremely hostile to him because they judged him to be too liberal in his religious convictions. For the same reason, Benjamin Jowett (1817–1893), an Oxford professor, was charged with heresy but in 1860 was acquitted. Attitudes changed in England towards the end of the century, and Dr. Jowett eventually (1882) became Oxford's vice-chancellor. One sin that Jowett had committed was to join with six other very worthy gentlemen of England in publishing *Essays and Reviews*. They described the ideas of the German and Dutch scholars and added some of their own, including doubt about the belief in eternal punishment. The book was condemned as atheistic by "Soapy Sam," Bishop Samuel Wilberforce, and other bishops and archbishops in England and Ireland joined in with their condemnations. Two of the authors were prosecuted and suspended from their ministeries for a year. Another, Dr. Frederick Temple (1821–1902), headmaster of Rugby, was afraid that the publicity would reflect adversely on the school, but refused to dissociate himself from the conflict.

Scholarly ideas about religious texts and about sciences that threw doubt on those texts also had an uphill battle with William Ewart Gladstone (1809–1898), member of parliament for over sixty years and prime minister of Britain for fourteen of them. Gladstone attempted to reconcile the two different stories in Genesis and the scriptural and scientific theories of creation. He fought hard for orthodoxy and, busy man that he was, prepared the legal plea against Bishop Colenso. In chapter 16 we consider some of his ideas about the theory of evolution. There were other people who continued late into the nineteenth century to insist that Christianity stood or fell on the acceptance or rejection of the literal meaning of the Old Testament. Their spiritual heirs are with us into the late twentieth century, and undoubtedly will continue to advance their narrow doctrine into the twenty-first century and beyond. The basic reason is that Jesus and Peter referred to miraculous events that occurred to Lot's wife (Luke 17:32) and to Noah (I Peter 3:20; 2 Peter 2:5) and these events must therefore have happened as recorded. That, plus the fact that to many people the Old Testament is as sacrosanct as the New.

One of the many messages of the Old Testament is that God can control the weather, and the New Testament tells us that Jesus could do it too. The Isaiahs invoked God to bring storms, whirlwinds, and other natural destructive weather on the "drunkards of Ephraim," (28:2) on "Ariel, the city where David dwelt" (29:1), and on the Assyrians (30:30). Amos prayed for the same fate to befall the Ammonites, Jonah was punished with a storm because he disobeyed the Lord, and the God of Moses brought tempests and other disasters to the Egyptians. The author of the eighty-third Psalm in verse 15 prays, "Persecute them [i.e., thy enemies] with the tempest and make them afraid of thy storm." In I Samuel 2:10 we read,

"The Lord shall defeat his adversaries, out of heaven shall he thunder against them," and in the same book (12:17) the theme persists: "I will call to the Lord and he shall send thunder and rain." Psalm 77:18 emphasizes the same warning, "The voice of thy thunder was in the heavens; the lightnings lightened the world; the earth trembled and shook."

It is no wonder that when Christians adopted these Hebrew writings as part of their sacred text superstitions masquerading as Christianity would dominate all thinking about the weather for more than a millennium and a half. A century after the first Isaiah, the first of Aeolian lyric poets, Alcaeus of Lesbos, wrote:

> The rain of Zeus descends, and from high heaven
> A storm is driven
> And on the running water brooks the cold
> Lays icy hold
>
> *Defying the Storm* (UA 3, 145)

As time developed the Hebrew God took over the role formerly assigned to Zeus, or Jupiter, or Odin, with angels replacing their assistants Aeolus, Vulcan, Hugian, Muninn, and a host of others. Natural disasters persisted despite Christian belief, however, and the Christian image of God as a more caring power caused men to shift the blame to devils. Not that God couldn't and wouldn't send warnings to people who got out of line. The heathen got what they deserved, but He wouldn't do that to good Christians, would He?

One reason that gods were invented was to try to explain natural phenomena external to ourselves—all the way from creation to thunderstorms—and the puzzling phenomena that occur inside our own minds. We have seen how, partly on biblical authority, devils were blamed for everything from insanity and the evils allegedly performed by witches to the emptying of the prostate during sleep. Perhaps these devils were those pagan gods—they were powerful enough and, since they were pagan, they were automatically evil. Psalm 96:5 tells us now that the gods of the nations are idols, but the Hebrew word for "idols" was translated in the Vulgate as "devils."

The belief was strengthened that malicious invisible beings were everywhere, ready to pop into your mouth and drive you insane, as we have seen; also ready to produce calamities of weather, disease, and famine. We could call this belief in the ubiquitous devils "the paradigm of the demons," a paradigm being a commonly accepted model or pattern. It was God who sent the lightning though; if you were struck by lightning it was clearly on account of your sin, the nature of which was related to the part of your body that was burned. The word was spread that a number of people had been struck in the crotch; one of them had been a priest at Trèves—that made the gossips happy.

The way to avoid all of these evils was therefore very clear. Commit no sin, or at least confess it, so that God wouldn't be angry, and use the symbols of the Church to frighten the devils away. Hymns, holy water, and the sign of the cross, of course, but more potent perhaps could be the relics of saints, sometimes carried in processions, and the ringing of church bells. Some saints had their specialty, like Saint Taurin, to bring rain, or Saint Piat, to stop it. The most powerful was Saint Barbara, the virgin martyr from the third century, who was invoked to bring protection against storms. Another technique was to build a big bonfire to frighten away the demons of thunder, lightning, and hail. The medicine men of North American tribes would have been right at home. Pope Urban V (1310–1370) was not the only pope to bless the *Agnus Dei*—Lamb of God—a small waxen image that was regarded as a wonderful disperser of thunder.

When churches and cathedrals were struck by lightning, it was obviously the work of the devil. It must be the bells that he was aiming for, because towers and belfries were hit so often. If the bells were consecrated, washed, and oiled, with the sign of the cross traced invisibly upon them, their ringing would repel the evil demons of the storms, it was believed. Inscriptions on the bells bear testimony to their purpose. In Switzerland:

> An dem Tüfel will ich mich rächen
> Mit der hilf gotz alle bösen wetter zubrechen
> (On the devil my spite I'll vent
> And, God helping, bad weather prevent)

Other examples have been given by Dr. White, all with the message that the triple purpose of the bell is to calm the weather, repel the demons, and summon men to worship.

Occasionally someone questioned the traditional beliefs about the weather and tried to develop a more rational basis of understanding. Honorius (?–1140) was a professor of theology and metaphysics at Autun, Saône-et-Loire, France. He understood that rain comes from the clouds rather than from some enormous and invisible tank up in the sky, and he separated meteors from meteorology by recognizing that thunder did not come from stones being dropped from heaven. The spirit of the time was opposed to this sort of thinking, however, and early in the next century Caesarius of Heisterbach (1180?–1240?), writing in his Cistercian monastic cell in Cologne, produced *VIII Libri Miraculorum*—eight books of miracles.

The books are full of anecdotes illustrating the miraculous intervention of the Almighty in preventing disasters or in diverting them to save the holy. Despite the studies of Ibn al-Haytham and Roger Bacon, the rainbow posed a problem because God had told Noah that "the bow *shall be* in the clouds" (Genesis 9:16). That meant that this was the first rainbow, created specially by God as a sign that there would not be another Deluge like that one. Cardinal d'Ailly (1350–1420), the "Hammer of Heretics," some of whose contributions we discussed earlier, had some idea of the optics involved and recognized the relative position of cloud and observer needed in order for this beautiful phenomenon to be seen. He therefore argued that before the Flood there had been no rainbows because God had prevented the clouds from getting into the right position.

Martin Luther didn't believe that bells would drive away any devils, although he believed that the devils were the cause of adverse environmental conditions, with God contributing to the discomfort and peril of those who denied Him. Sermons and books on the subject multiplied, warning the flock about the terrible things that God would do to them if they were blaspheming, or breaking the Sabbath, or performing some other wickedness. It was almost a return to the Old Testament image of Jehovah. God's aim was excellent and if he hit you with one of his thunderbolts it was intended, and if you survived you had better mend your ways. Less specific disasters were a warning for not paying tithes to the clergy, or for not preserving the churches in good repair, at least according to the Protestant pastor Georg Nuber, as preached from his pulpit in Swabia (part of Bavaria). Increase Mather also preached that God and his angels were sometimes responsible for hurricanes and tornadoes, but the bad angels, the fallen angels of the devil, also had their hand in these disasters. By the time that Cotton Mather developed his full preaching potential, however, the word from Europe was that these established beliefs were in doubt. The mechanics of Newton had led to a fundamental understanding of motion, both of the planets and of objects on earth. The motion of air in the atmosphere was too complicated for the theory to predict in detail—it still is—but there was arising the conviction among those who knew about these matters that the weather was a natural phenomenon. Four years before the death of Newton, Cotton Mather preached a sermon about a huge storm, his ideas about it departing from the old religious/superstitious beliefs.

Benjamin Franklin's (1706–1790) experiments with electricity in the middle of the eighteenth century led him to invent the lightning rod, a conducting wire running down from the top of a building in order to offer surges of electrons from lightning flashes to find a safe path to the ground. Franklin was smart enough to have it project above the top of the building and to end in a sharp point. Electrical fields become concentrated around sharp points, making it more likely for electrical discharges to occur between the point and a thundercloud. King George III of England heard about it, but he wasn't about to be told by a Yankee how to do it, so he ordered lightning rods with *rounded heads* to be installed. They didn't work.

As we have seen, despite a great deal of propaganda to the contrary, the use of holy water, exorcisms, church bells, and the torture of witches hadn't worked either. Since the witches were believed to be in league with demons, they should bear responsibility for the terrible lightning damage that had fallen on some churches, and for the deaths from lightning of so many bell-ringers. Perhaps though it wasn't the devils—perhaps man was being punished for his sins in true Old Testament fashion. If that were the case, it would be impious to put a secular lightning rod on a church steeple in an attempt to thwart the will of God.

Slowly, however, it became apparent that this invention worked. Lightning rods were installed, mostly in Boston at first where they were praised by those who saw that the buildings on which they were placed were not damaged again by lightning. They were blamed by others for being impious constructions, and for being responsible for the earthquake of 1755. The word spread to Europe, the earthquake theory was discounted, and many ancient churches, particularly in the mountains, are still standing, with their spires undamaged, because of Franklin's metallic electrical by-pass. The tower of St. Mark's Cathedral in Venice had been damaged by lightning some ten times, including several occasions when it had to be rebuilt. The angel on the top, the relics inside, and the magnificent bells had failed to save it. It has not suffered further damage from lightning since one of Franklin's "heretical rods" was installed in 1766. Once again, science had helped religion to free itself from superstition. The symbols of Christianity have helped many people in their personal lives, but they can't control the weather, or violate the laws of nature in any other way.

They can't cure any diseases that aren't related to mental health either, and not even many of those that are, although it is tempting to believe that miracles can intervene to save us from illnesses and death. Of course it can be argued that *all* diseases are related to mental health, so that the statement doesn't mean anything. It is certain that one's attitude to a disease can play a big part in its cure. Faith that you will be healed helps greatly, whether it is faith in Jesus Christ, or faith in your physician, or faith in someone who has convinced you of the curative properties of some magic potion or procedure. The power of the mind over the body is not really understood, and one of many reasons for this is the power of the mind to fool itself. The Scriptures and other literature relate many anecdotes that feature apparently miraculous cures. The story of Naaman, a general of the Aramean army, and his cure from leprosy (2 Kings 5:1–19) has been told in countless Christian Sunday schools and churches. Not only did Elisha tell him to bathe in the Jordan seven times (a magic number); he also prophesied to Gehazi, who had conned Naaman out of a couple of talents and some clothes: "The leprosy of Naaman shall cleave to you and your servants for ever," and the record tells us, "he went out from his presence a leper as white as snow," (2 Kings 5:27). (One form of leprosy produces large white patches on the skin.) If you believe that, you would certainly believe that both the River Jordan and the bones of Elisha would possess supernatural healing powers. With a different background you would believe in the man/god Asclepios and his daughter Hygeia, goddess of health. That belief works too, provided that it is backed up with practical hygiene.

Periodically, Europe was ravished by plagues, among them the bubonic plague, transmitted from infected rodents to people by the fleas that had feasted on both. The great surgeon and physician Guy de Chauliac (1300?–1368), mentioned earlier, recommended isolation as the best way to avoid this Black Death, saving the life of Pope Clement VI by suggesting that he lock himself up in a room during the rampant phase of the plague of 1348. From this came the idea of preventing people from entering a city, or at least isolating them for the biblical number of forty days—*quaranta* in Italian—to see if they carried the disease. Conditions were so disgustingly unhygienic that it is a wonder that mankind has survived. It was not much consolation, when so many were dying horrible deaths, to believe that they were being punished for their sins. The scholar Erasmus spent some years in England at the end of the fifteenth century and early in the sixteenth. At that period the dreaded sweating sickness was killing people within hours of its first symptoms, and Erasmus saw that it flourished and spread because of the filthy conditions.

In the year 1530, Girolamo Fracastoro (1483–1553), an Italian physician from Verona, published a poem in Latin: *Syphilus sive Morbus Gallicos—Syphilis, or the French Disease.* The disease, relatively new it seemed, was described, its venereal nature and therefore method of propagation recognized, and its name determined. But Fracastoro was interested in many more phenomena, both medical and astronomical. He studied epidemics, typhus, and the plague in particular, and reasoned that contagion was due to very small spores that passed in the air from one person to another. He also had the idea of putting two lenses together to form a telescope, and wrote a treatise on it but apparently never built one. And at that time Paracelsus, to whom we have referred on several occasions, was challenging the writings of Galen and all of his latter-day disciples.

In France, Ambroise Paré (1517?–1590) began his professional life as an apprentice to a barber, moving on to become a surgeon to royalty. The harquebus, forerunner of the musket, was doing a lot of damage to the troops—bullets didn't leave clean wounds like swords and axes did. It had been discovered that boiling oil or cauterization would prevent what we call infection of wounds, but the side effects were disastrous. Paré's unit ran out of boiling oil, so he decided to dress the wounds with turpentine, yolks of eggs, and oil of roses. The next day those who had been treated in this way were feeling much better than those who had received the harsh treatment. Thus was discovered the ability of the body to heal itself if it is given a chance. As Paré put it: "I dressed his wounds; God healed him."

John Caius (1510–1573), physician to the English royalty and Master as well as patron of Gonville and Caius (pronounced "keys") College Cambridge, wrote two books on the sweating sickness that Erasmus had attributed to unhygienic conditions. In his 1552 book, titled *A Boke or Counseill against the Disease commonly called the Sweate or Sweatyng Sicknesse,* John Caius also recommended cleanliness, but for many years afterwards there was little progress.

In Italy, the renaissance in medicine had begun at the height of the renaissance in art; Fracastoro was a contemporary of Michelangelo and the Flemish physician Andreas Vesalius (1514–1564) was four years older than Tintoretto. The latter came to Pavia after studying anatomy in Paris under Jacques Dubois (1478–1555) who was sometimes known as Sylvius. (The canal connecting the third and fourth ventricles of the brain was named "the aqueduct of Sylvius".) Dubois used human cadavers for dissection, but he still held the writings of Galen in great respect, not realizing that Galen had dissected animals. Vesalius continued his studies of human anatomy, but he was puzzled when his observations were in disagreement with time-honored beliefs. So great was the authority of Galen that the first book on human anatomy that Vesalius published (1538) included shapes and structures that no human had ever possessed, copied from Galen. Fortunately, a few years later Vesalius dissected an ape and learned that his observations now agreed with the great authority. The religious, social, and political taboos against dissecting humans had misled anatomists for thirteen hundred years.

Medicine and art came together as Vesalius continued his studies of the anatomy of humans, unfettered by the tradition of Galen, and assisted by his friend, the painter Jan Stevenszoon van Calcar (1499?–1550). The first accurate atlas of human anatomy, *De Humani Corporis Fabrica,* was published in 1543, Calcar designing and preparing the wood-cuts for the illustrations. Vesalius had been courageous in dissecting human corpses, against the beliefs and ecclesiastical edicts of the time, and courageous too in exposing errors carried from physician to student thousands of times. Anatomically men are so similar to monkeys that for many centuries the best medical men could not tell them apart until Vesalius pointed out the differences. It was too early for the theory of evolution to emerge; we examine that in chapter 16.

Twenty years after his book was published, Vesalius made a pilgrimage to the Holy Land; some say it was undertaken as a condition for commuting the death sentence pronounced on him by the Inquisition for dissecting the human body. Certainly the abdication of his sponsor Charles V in 1555 didn't help him, although Philip II kept him on as royal physician. He had made a lot of enemies by his criticism of Galen, by his dissections, and because of his great reputation. Another source relates that his enemies accused him of dissecting a Spanish nobleman whose heart was still beating, bringing a charge of murder against him. The clergy, the medical faculty, and the relatives of the deceased publicized the accusation, and Vesalius left on the pilgrimage from which he did not return. On the way back, possibly to assume a professorship at Padua, he was shipwrecked and stranded on the Ionian island of Zante (present Zakynthos) where he died of cold and hunger.

The professorship that some historians assert he was planning to assume had been vacated by the premature death of his former pupil Gabriel Fallopius (1523?–1562). We know of him chiefly for his discovery of what are now called the Fallopian tubes, but most of his work was centered on the head—the structure of the ear system, of the tear ducts, of the canal traversed by the facial nerve (aqueduct of Fallopius), and much more. One of his pupils, Hieronymous Fabricius ab Aquapendente (1537–1619), was in fact appointed to the prestigious professorship at Padua soon after Fallopius died, and taught there for nearly half a century. He initiated the science of embryology, publishing his book on the development of the fetus in women and in some animals. The book appeared in 1600, the year in which Giordano Bruno was burned for his ideas, and a younger fellow-professor of Fabricius at Padua, Galileo Galilei, was testing the thermometer that he had invented.

In his turn Fabricius also had a distinguished student at Padua, the Englishman William Harvey (1578–1657), a graduate of Caius College. Returning to England, Harvey was appointed professor of anatomy and surgery to the Royal College of Physicians. In 1628, his book officially announced in detail his discovery of the circulation of the blood through the arteries and veins of the body, not just the part of the circulation in which the blood is propelled through the lungs to restore its oxygen content. It took twenty years to persuade his colleagues that the heart provides the power for circulating the blood, sending it through the arteries and somehow into the veins, that act like little valves, directing the blood back to the heart. For medicine, the nature of the opposition to new ideas was beginning to change; the struggle to understand was becoming centered on convincing one's professional peers by sound arguments and demonstrations rather than on avoiding the wrath of churchmen. This was not true of all disciplines. Five years after Harvey's book was published, Galileo was imprisoned by command of the Pope.

Galileo's microscope (1624) later provided the missing piece of the puzzle of the circulation of the blood. Marcello Malpighi (1628–1694) was the first anatomist to use the microscope, and with this instrument he discovered the capillaries, postulated but unseen by Harvey, through which the blood reaches the veins from the arteries. This observation was verified and extended by Anton von Leeuwenhoek (1632–1723), who also discovered microorganisms and red blood corpuscles, and who for the first time observed spermatozoa. In chapter 9 it was noted that Francesco Redi demonstrated in 1668 that the sudden generation of maggots in decaying meat does not occur unless flies are allowed to lay their eggs on the stuff. Leeuwenhoek was only a few years younger than Redi, and with his microscope he was able to describe the reproduction of insects, finally in 1683 putting an end to the long-held "theory" of spontaneous generation and originating a new science—bacteriology. It would be two hundred years, however, before Louis Pasteur (1822–1895) would show that even bacteria are not spontaneously generated, and Robert Koch (1843–1910) would make a real science out of the study of them. The struggle to understand diseases was delayed for that long period by the collective judgment of the experts that bacteria were not very interesting or important.

The English physician Thomas Sydenham (1624–1689) had the good fortune to miss a lot of the medical training that his colleagues were subjected to, learning instead and prescribing at the bedsides of countless diseased patients. He followed the course that each disease took, from its first symptom to eventual death, or for some fortunate patients, renewed health. Sydenham insisted on clinical observation rather than theory, and studied epidemics in terms of their places and their times of the year. A little more than two thousand years after Hippocrates, he established medicine on a sound pragmatic basis. Before the development of medical schools in America, a few doctors took on apprentices who also learned a lot of their medicine by observing patients rather than hearing lectures on Galen. It had taken all that time to begin the overthrow of this particular Authority; yet, as noted in chapter 7, Galen himself had vowed "to trust no statements until, as far as possible, I have tested them for myself."

"Jail fever," or typhus, showed no mercy, nor did typhoid, cholera, yellow fever, or smallpox. By the middle of the seventeenth century influenza had crossed the Atlantic and multiplied in the colonies, and yellow fever was beginning to take hold; diphtheria, smallpox, and other diseases were to follow soon after. The plague that ravished London in 1665 caused the colonists to quarantine British ships. In England, the curing of the "king's evil," scrofula, was supposed to occur by touching the hand of the king, particularly during the years 1660 to 1685 when Charles II was on the throne. The Stuart kings and their public-relations people put out the story that they, and they alone, had inherited the divine right to heal people by the "king's touch," as opposed to those pretenders of the House of Hanover—not quite what some politicians claim today, but in the same spirit.

The concept and practice of inoculation was slow to catch on in the West. It seemed to be the height of madness to take some pus from a person who had smallpox and apply it to a scratch on the skin of a person who had never had the disease. That appeared to be the most certain way anyone could imagine to transmit the fatal disease. Most people to whom this happened got the disease all right, but in a very mild form that gave them immunity from future infection. This inoculation was known many years earlier to the Chinese physicians, and the information spread to the Caucasus Mountains, between the Black and Caspian seas. The practice was adopted by the Circassian people of the area to protect the beauty of their daughters from the scars of smallpox. This people had a reputation for great beauty—being tall, with dark eyes, oval faces, and chestnut colored hair. Voltaire wrote about them:

> The Circassians are poor, and their daughters are beautiful, and, indeed, it is in them that they chiefly trade. They furnish with beauties the seraglios of the Turkish Sultan, of the Persian Sufi, and of all those who are wealthy enough to purchase and maintain such precious merchandise. These maidens are very honorably and virtuously instructed to fondle and caress men; are taught dances of a very polite and effeminate kind; and how to heighten by the most voluptuous artifices the pleasures of their disdainful masters for whom they are designed.
>
> (*Letters on the English* XI)

But not if the beauty was scarred with smallpox. The Circassians therefore adopted the practice of inoculating their children during the first few months after birth. It worked very well, very few lives were lost, and the balance of trade became more favorable. The Turks picked up the custom, and inoculation became common among their children immediately after weaning. The Lady Mary Wortley Montague (1690–1762), wife of the British ambassador to Constantinople, was so impressed by this that she arranged for her six-year-old son to receive the same treatment. Voltaire continues, "The chaplain represented to his lady, but to no purpose, that this was an un-Christian operation, and therefore that it could succeed with none but infidels." The boy survived and his mother worked hard to introduce the practice to England after she returned there in 1718. Voltaire did not note that one way that word of this got to America was through an article in the *Transactions of the Royal Society of London.* Cotton Mather read it and told Zabdiel Boylston (1679–1766) who began trying the procedure in 1721. In that and the following year a serious epidemic hit Boston. Two percent of the inoculated patients died, and sixteen percent of the others, but the pulpits resounded with condemnations of Dr. Boylston. The good Christians bombed his house and that of Cotton Mather; the lawmakers prepared to ban the procedure, but the statistics slowly made converts.

George Washington (1732–1799) ordered his troops to receive inoculations, although some ministers continued to warn their congregations that it was not right to interfere in this way with the will of providence. That particular debate took on a new twist when inoculation was replaced by vaccination, primarily by the efforts of Edward Jenner (1749–1823). For the last fifty years of his life, Jenner was a physician at Berkeley, in Gloucestershire, England, a few miles inland from the mouth of the River Severn. The country folk who were his patients were worried a little about the pustular eruptive disease on the udders of their cows, but they believed an "old wives' tale" that if you caught that disease from a cow you wouldn't catch smallpox. Jenner took this lead seriously instead of laughing at it, and inoculated a boy with cowpox that a milk-maid had caught, later giving the boy a dose of smallpox. In the present century he would have landed in jail for experimenting on a human like that, but rules and available facilities were very different then.

Jenner's communication to the *Transactions of the Royal Society of London* was rejected—his colleagues thought that he had not really established as a fact that cowpox would bring immunization against smallpox. He therefore published his results in the form of a small book which appeared in 1798, and following that the Royal Jennerian Institution was formed, with him as its head. Jenner spent the rest of his life spreading the word about his miraculous discovery, but it was not easy. As the physician James M'Connechy wrote in the 1861 edition of the *Appleton Cyclopaedia of Biography:*

"It is a melancholy to be obliged to state that Jenner's life was embittered by the controversies to which his discovery had led and that an amiable, a virtuous, and an accomplished man was disturbed by petty squabbles to which his nature was utterly abhorrent." More than a century after Jenner's discovery, some English clergymen were still convincing their flocks not to have their babies vaccinated, and I know this is true because I was one of those babies.

Another route to good health—hygiene—was also blocked for years despite the exhortations of some wise men to do something about it. We have noted how this occurred during the fifteenth and sixteenth centuries, but it continued beyond that. Isaac Newton's physician Richard Mead (1673–1754) wrote about the incidence of epidemics and the contagious nature of some diseases. He too emphasized the importance of cleanliness, as did Sir John Pringle (1707–1782), physician to George III. Sir John specialized in diseases that flourished when men lived close to each other, as in jails, or in the army, or in hospitals. He insisted on more sanitary conditions and was able to improve the health of the army somewhat, but civilians tended to cling to their old ways, and it was hard to influence them. There was one influence, however, the *influenza coelestia,* the "influence of heaven," and Pringle helped to bring the Italian word into the English language.

Comparative anatomy was also developing at this time, as a result of the studies of the Scottish anatomist John Hunter (1728–1793), with whom Edward Jenner had studied for a couple of years (1770–1772). Hunter kept a menagerie in his home consisting of innumerable insects, fish, birds, and animals. They ranged in size from bees to a bull, and he would have been a prime target for antivivisectionists. From experiments with dogs and observation of his own ruptured Achilles tendon, he discovered surgical methods for treating club feet. Hunter moved from that to a pioneering study of the natural history of the human teeth, being the first to study them in a scientific manner. His service with the army in Portugal led to his *Treatise on Blood, Inflammation and Gunshot Wounds;* his accidental self-inoculation with syphilis led to a treatise on venereal disease. Hunter contributed much to surgery, physiology, and dentistry as well as comparative anatomy, and he even recognized the danger to his heart caused by his bad temper but was unable to overcome it. On October 16, 1793, a colleague contradicted him; Hunter became furious, collapsed, and died.

His older brother William (1718–1783) was the first professor of anatomy at the Royal Academy, where his lectures transformed the teaching of the subject. William Hunter is best known for his contributions to obstetrics, and for his book *The Anatomy of the Human Gravid Uterus,* consisting of thirty-four plates engraved by leading artists of the time, with explanations in English and Latin. He did not live to see the publication of the full text that was to accompany it; his nephew Sir Matthew Baillie edited his notes and published them in 1794.

In Switzerland, Albrecht von Haller (1708–1777) studied medicine and other biological sciences and so impressed George II of England (1683–1760) that George appointed him Professor of Medicine, Surgery, Anatomy, and Botany at the University of Göttingen. (George was also Elector of Hanover while he was King of Great Britain and Ireland.) Albrecht had been a precocious young man. It was reported that, at the age of nine, he had composed for his own use a Chaldaic grammar, a Hebrew and

Greek lexicon, and a historical dictionary of more than 2000 articles. At the age of twenty-one he published a book of lyrics, *Die Alpen,* recognizing the glory of the countryside and the mountainous Alps and arguing that city life led people away from beauty and therefore from God. It was Albrecht who clarified the connection between the nerves and the muscles, traced the nerves to the brain or spinal cord, damaged specific parts of the brains of animals to determine what functions were inhibited, studied details of the anatomy of the brain, the heart, and the unborn fetus—and wrote several philosophical romances. Under him physiology blossomed as a branch of science.

With all of this expansion of knowledge about human anatomy and physiology, and the light that these disciplines cast on surgery, it became more important to find effective anaesthetics as opposed to holding patients down while they were being cut up.

The German physician Franz Mesmer (1734–1815) should be mentioned in this connection even though he was a quack. He believed—or at least *said* he believed—that the whole universe is pervaded by an invisible element that penetrates all bodies, and that our nervous systems are attuned to this influence and respond to it periodically. Many scientists and religious leaders have believed in the existence of an invisible all-pervasive fluid or spirit, and we discuss this "paradigm of the ghost" in the next chapter. Mesmer is suspect, however, because he claimed to have control over this universal force that he had discovered. At first he said that it was a magnetic force. Father Maximilian Hell (1720–1792), director of the Vienna Observatory, had recently returned from a trip to Lapland where he had observed the transit of Venus across the sun. Mesmer consulted him about magnets, but soon thereafter the two of them were publicly disputing each other's claim to have discovered the effects of magnets on people's health. The pragmatic showman Mesmer then took the view that it wasn't ordinary magnets anyhow; it was this inexplicable universal force over which only he had control through his own personal "magnetism." The Viennese physicians were naturally antagonistic, so he moved to Paris, where before long the French government appointed a commission, including Benjamin Franklin and Antoine Lavoisier, to investigate his claims to cure disease. He held very popular and lucrative sessions, carefully staged around a "magnetic battery" (a tub of liquid with wires attached).

As early as 1745 the German inventor Georg von Kleist made an electrical condenser for storing electric charge, and in the next year the Dutch physicist Pieter van Musschenbroek (1692–1761) independently invented the same device, usually referred to as the Leyden jar. Mesmer's *baquet* (little tub) was undoubtedly a variation on this. With some mass psychology and electrical discharges, he laid a very shaky foundation for hypnotism, suggestive therapy, and shock treatments.

Mesmerismus, as Mesmer so modestly named it, was soon modified in name and in fact by the British surgeon James Braid (1795–1860). The name was changed to "hypnotism," a lot of the misconceptions were removed, and the emphasis was shifted as indicated by the title of his 1846 book *The Power of the Mind over the Body.* It could have been the title of a book of the twentieth or twenty-first centuries, as Dr. Braid pioneered research on the unconscious mind. In France, Jean Martin Charcot (1825–1893), a French physician, carried on the work, investigating hysteria through hypnosis. He also had a distinguished student, Sigmund Freud (1856–1939). So many books have been written, and so many psychiatrists' couches worn out, that here we shall not add another word about Dr. Freud, except to note that, unfortunately, he too became an Authority.

Mesmer's primitive shock therapy has been expanded and redirected over the years. It is hoped that it will be made more specific; at present it overloads all of the circuits in the brain, sometimes with positive results not yet clearly understood and with unpredictable side effects.

An English physician, Thomas Beddoes (1760–1808) was fascinated with the idea of treating patients by giving them various gases to inhale. For this purpose, he founded the "Pneumatic Institution" at Bristol, generously invested in by James Watt (1736–1819), who had made some money with his steam engine and related inventions. Beddoes hired the young Humphrey Davy (1778–1829) to test various gases by breathing them himself. Davy soon discovered that pure "laughing gas" (N_2O) could be breathed without harm, putting a person in a state which he described as "absolute intoxication." None of the gases that were tried, however, seemed capable of curing any malady, and the Institution went out of business. As Sir Humphrey put it later, Dr. Beddoes was "a truly remarkable man, but more admirably fitted to promote inquiry than to conduct it." We referred to Davy's safety lamp in chapter 8, and we mention here his preparation of potassium, sodium, and calcium by electrolytic deposition. He also argued against the caloric model and, as noted in the next chapter, argued from the correct sides of a number of issues with the Swedish chemist Jöns Berzelius.

For years, laughing gas was used solely for amusement, as also was ether after Davy's assistant, Michael Faraday (1791–1867), discovered its properties in 1815. A few years later, the English physician Henry Hill Hickman (1800–1830) performed painless operations on animals by forcing them to breathe carbon dioxide, but the English establishment of physicians laughed at him; so also did the French when he moved to Paris in disgust. Hickman was not allowed to present his findings to the French Academy of Medicine either, and so he returned to England where he died within a year.

The American surgeon, Crawford Williamson Long (1815–1878) appears to be the first person to have used ether as an anaesthetic in an operation that he performed in 1842, followed in 1846 by the Scottish surgeon Robert Liston (1794–1847), and in England by John Snow (1813–1858). Also in 1846, the American dentist William Thomas Green Morton (1819–1868) etherized a patient, while Dr. John Warren removed a vascular tumor from the side of his neck. Morton had himself used ether in his dental practice, and in that same year another American dentist Horace Wells (1815–1848), former partner and teacher of Morton, published his claim to be the discoverer of anaesthesia. The claim was complicated even further by the chemist Charles Thomas Jackson (1805–1880) who had suggested to Morton to use ether instead of nitrous oxide and who, again in 1846, lodged his claim to have been the first person to develop surgical anaesthesia. The struggle to understand became a struggle to establish priority. In the next year another Scottish surgeon, Sir James Young Simpson (1811–1870), introduced chloroform as an anaesthetic. It was a "new" gas synthesized a few years earlier in several laboratories, and Simpson and his colleagues literally gambled their lives by inhaling it. Indeed, when the English specialist in diseases of the ear, Dr. Joseph Toynbee (1815–1866), tested chloroform on himself several years later, he tragically did not recover.

Dr. Simpson introduced a new element into anaesthesia—not just the new chemical $CHCl_3$, but the use of it by women in labor. What happened to the edict, "In sorrow thou shalt bring forth children," Genesis 3:16? The pulpits rang with condemnations of this impious action, so contrary to the Holy Scriptures. Fortunately, the clerics no longer had the power of two hundred and fifty years earlier when, as related in the last chapter, Agnes Sampson and Eufame Macalyane were burned to death in the same city of Edinburgh. Dr. Simpson reminded the good pastors of Genesis 2:21: "So the Lord God caused a deep sleep to fall upon Adam, and he slept; and he took one of his ribs and closed up the flesh instead thereof." That, it was argued, was before man had sinned and therefore before he had experienced any pain, so it didn't count. It was the last year of the life of the great Scottish theologian and preacher, Thomas Chalmers (1780–1847), who had led 470 ministers from the general assembly to form the Free Church of Scotland. Chalmers spoke eloquently in favor of this new painkiller, and the theological opposition subsided.

The nineteenth-century revolution in medicine and surgery continued with the discovery by Louis Pasteur that the long-neglected bacteria were in fact very important. As a chemist and microbiologist, Pasteur discovered that microorganisms were responsible for the spoiling of wine and beer, for a fatal disease of silkworms, and for the disease of anthrax in cattle—a fantastic boon for the wine, beer, silk, and meat industries. A great discovery for medicine also, as the devils who caused disease were coming into the focus of the microscope. Pasteur also saw how to control these germs by heating, and by developing less toxic varieties for use in inoculation. As we have noted, the German physician Robert Koch studied many bacteria scientifically and systematically. He isolated the anthrax bacillus, discovered the bacillus responsible for Asiatic cholera, and developed a technique of staining for identification of specific types of bacteria. For this and much more he received the 1905 Nobel Prize for physiology and medicine.

Robert Lister (1827–1912), British surgeon and biologist, brought these discoveries to the art and science of surgery. If bacilli caused disease, they may be floating around in the air and landing on open wounds, for he noted that simple fractures did not lead to blood poisoning. The belief in devils floating in the air was replaced by the more accurate idea of bacteria so transported, but covering up the wounds didn't help. Lister then supplemented the spraying of carbolic acid (phenol) into the air with the requirement that hands and instruments should be washed in it. The results were excellent, and thereafter it was possible to perform operations with much lower risk of infection. It must be remembered, however, that way back in 1848 the Hungarian obstetrician, Ignaz Semmelweiss (1818–1865), had saved the lives of many mothers by requiring that the physicians wash their hands in calcium chloride before leaving the dissecting room for the delivery room. When he reported that this procedure led to a spectacular decrease in the number of deaths from puerperal fever, the editor of the *Journal of the Medical Association of Vienna* emphatically rejected his manuscript. Sadly, Semmelweiss contracted the disease himself, and with no hope of letting the world know what he had discovered, he broke down and was committed to a mental hospital where he died. It was the year in which Lister performed his first experiment on a compound fracture by dousing it with phenol.

15

Establishments, Paradigms, and Progress

> If at some period in the course of civilization we seriously find that our science and our religion are antagonistic, then there must be something wrong either with our science or with our religion.
>
> —Havelock Ellis (1859–1939)

Three years older than Galileo and Shakespeare, Francis Bacon strongly and eloquently criticized the old scholastic methods of trying to understand nature, roundly condemning the muddled thinking and reliance on authority that had stifled progress for centuries. Knowledge, he insisted, is obtained from experience and from probing nature in a systematic way. Knowledge so gained should be integrated by induction into a theory from which apparently unrelated facts could be explained by deduction. Induction had consisted merely of drawing a general conclusion from a number of facts—the sun has risen every day in living memory, therefore it will rise tomorrow; the weather has been great for the past week, therefore it will be great tomorrow. Sometimes it works and sometimes it doesn't, and Aristotle had rightly pointed out its defects as a route to understanding. Bacon and his successors David Hume (1711–1776) and John Stuart Mill (1806–1873) expanded this notion, both by postulating that identical causes under identical conditions produce identical effects, and by advocating that data obtained from an observation could be used to suggest, by analogy, what data would emerge from a larger class of observations. This of course is subject to error, but with care it can be a powerful tool. Kepler's analysis of the orbit of Mars had ultimately led to the conclusion that it was an ellipse. The inductive step suggested that the orbits of all of the planets, including the earth, were also ellipses. The later step to analyze the orbits of electrons in atoms as ellipses did not work nearly as well, but it ultimately led to the quantum theory.

The postulate that identical causes lead to identical effects—the postulate of the uniformity of nature—implies that if you are unable to do an experiment yourself, you should publish your ideas and results so that others may take it from there. It doesn't matter who does the experiment, provided that he or she is trained and careful. Francis Bacon did in fact carry out a few experiments, mostly in determining the specific gravity of some materials and in showing that a definite amount of acid will dissolve only a certain weight of a particular metal. His main interest in science, however, was to turn people away from the time-honored methods of contemplation, argument and quotation towards involvement in an unbiased study and measurement of natural phenomena. Bacon called the old methods "idols"—idols of the tribe (commonly believed paradigms), idols of the cave (individual prejudices), idols of the marketplace (excessive dependence on words), and idols of the theater (dependence on the traditions handed down from past authorities). He didn't rest with this, however. He wrote many "apothegms"—from the Greek word meaning "to speak out." These were short, humorous comments, mostly about actions and sayings of people such as Queen Elizabeth, Walter Raleigh, Thomas More, and other notables. He published a hundred of these in the year 1624, revealing his own sense of humor and that of his times:

> Plato was wont to say of his master Socrates that he was like apothecaries' gallipots, that had on the outside apes, owls and satyrs, but within, precious drugs.
>
> Charles the Bald allowed one whose name was Scotus to sit at the table with him for his pleasure. Scotus sat on the other side of the table. One time the king, being merry with him, said to him, "What is there between Scot and sot?" Scotus answered, "The table only."
>
> (Translated by Stanley Lane-Poole. UA 8, 41)

Francis Bacon also gave us fifty-eight essays, along with works in history and law. His legal knowledge and persistent personality ultimately brought for him the highest legal office of lord-chancellor—and banishment from parliament and court for bribery. His comment to his secretary, William Rawley (d. 1667): "I was the justest judge that was in England these fifty years. But it was the justest censure in Parliament that was these two hundred years." He was pardoned by King James I and retired to his estate near Saint Albans (Roman Verulamium) in Hertfordshire, where he wrote his books on law. Francis Bacon, Baron Verulam, died in 1626, bequeathing to us his anti-Platonian philosophy to understand nature by means of contact with nature, by observations and measurements, and by inductive as well as deductive reasoning. Get rid of this two-thousand-year-old habit of starting with some assumptions, Bacon encouraged his peers, deducing their consequences and never looking for the correct answer in the book of nature. Besides, knowledge of nature brings power over nature.

While Bacon was giving this advice, his older contemporary, William Gilbert (1544–1603), was supplementing his activities as a physician with pioneering experimental studies in magnetism and electricity. He distinguished clearly between the two, noting that a number of materials—named "electrics" by him and "insulators" by us—were capable of attracting small objects when rubbed. Gilbert showed that a compass needle points in a definite direction when free to rotate around a horizontal axis as well as a vertical one. Since it points towards the earth, he recognized that the earth is a giant magnet and showed by experiment that a spherical magnet would produce that effect. "Look for knowledge not in books," he wrote, "but in things themselves."

For many years philosophers had believed that the motions of the planets were centered on the earth, and that the magnetism that oriented the compass needle was centered in the heavens. That it was the other way around was about to be established in the year 1600, when Johannes Kepler and Tycho Brahe met for the first time at Benatek Castle, near Prague, and Gilbert published his book on magnets. Kepler and Brahe had met on February 4 of that year; in Rome, Giordano Bruno was burned at the stake on February 17.

Aristotelianism and astrology were still barriers to understanding when Galileo's secretary and successor, Evangelista Torricelli (1608–1647), began some experiments with air. (He was secretary for only a few months, and was Galileo's successor not in persecution by the Church, but in the role of philosopher and mathematician to the Grand Duke of Florence.) The Aristotelian idea that a vacuum could not exist had become more firmly engrained in philosophers' minds than ever before. If a vacuum existed, it would allow for the possibility of atoms moving around in it. That concept of atoms was too mechanical and, besides, it had been supported by those evil atheists Epicuros and Lucretius. Furthermore, air's nature was seen by Aristotle to require it to move upwards, or perhaps sideways. It could not therefore have any weight. Torricelli demonstrated with the mercury barometer that there was a near vacuum at the top of the tube, and that it was the weight of the air that caused it to exert a pressure that would hold up a column of mercury. That was in 1643, a year after the death of his latest and greatest teacher.

Five years later Blaise Pascal (1623–1662) repeated and extended the experiment, comparing the heights of the mercury columns at different elevations. This led to the argument with Father Noel referred to in chapter 4. Pascal's father, Étienne, a mathematician and a provincial judge, decided to educate his precocious son himself. Noting Blaise's remarkable mathematical ability, he forced him to study other areas of knowledge instead, lest he become too one-sided in his thinking. That didn't stop Blaise from getting a piece of charcoal and drawing diagrams on the floor to help him visualize the propositions of Euclid. By the time his dad found out, young Blaise had mastered the first thirty-one of these and was working on the thirty-second. Thereafter, mathematics was included in his education, and within a few years he had invented a calculating machine. Later, he studied the remarkable properties of a hexagram inscribed in a conic, founded the theory of hydrodynamics and, with Pierre de Fermat (1601–1665), the theory of probability. At the age of thirty-one, following a long illness, Pascal experienced a conversion and renounced scientific and secular pursuits in favor of the study of theology. He was impressed by the writings and followers of the Dutch Roman Catholic theologian Cornelis Jansen (1585–1638).

Jansen's was a theology similar to Calvinism, based on the argument that as a sinner man is unable either to obey God's commandments or to resist divine grace. The Jansenists were bitterly opposed to the Jesuits (although, like everyone else, they agreed that the world was created around 4000 B.C.). In 1656, Pascal wrote a withering criticism of the morality of the Jesuits, but he did not publish it under his own name. Six years later Pascal died suddenly under circumstances that led some people to believe that he had been poisoned. Jansenism was declared a heresy, but its members continued to believe that they were returning to the teachings of St. Augustine, from which the Catholic Church, and the Jesuits in particular, had departed.

Pierre Gassendi (1592–1655) wrote a fine biography of Epicuros, and revived and extended the atomic hypothesis until it became a workable theory. The nineteenth century astronomer J. P. Nichol refers to him as "the most learned among the philosophers, and the ablest philosopher among the learned, of the seventeenth century." He adds, "Gassendi's personal character was of the highest order; gentle, serene and dignified; modest, not withstanding his wide repute; impartial and forebearing." He certainly impressed the people of Digne in the lower Alps where he was Provost of the Cathedral—nearly two hundred years after his death they erected a statue of him. As a philosopher Gassendi was a "Sensationalist"—the school dating back to the early Sophists and to Epicuros—asserting that all our knowledge originates in sense perceptions. As a physicist, he saw that all of the atoms of gas, colliding with the walls of a container, could produce a pressure. He also recognized that motions continue indefinitely if there is no interference; he measured the velocity of sound, and argued eloquently against Descartes and Aristotle. He didn't believe Harvey's work on blood circulation either.

Gassendi and René Descartes (1596–1650) disagreed about practically everything, including the circulation of the blood. In his essay on *Laughter*, Descartes noted "the surprise of wonder, which, being added to joy, is able to open the orifices of the heart so promptly that a great abundance of blood, entering all at once on its right side through the *vena cava* is rarefied, and passing thence through the arterial vein, inflates the lungs." That, at least, was his idea of one aspect of pulmonary circulation. While his conclusions were sometimes incorrect, he emphasized the importance of reason in understanding nature: "the chief ground of my satisfaction with this Method, was the assurance I had of thereby exercising my reason in all matters, if not with absolute perfection, at least with the greatest attainable by me." The Method to which Descartes referred was to accept no statement that he did not know clearly to be true, to divide each of the difficulties that he examined into as many parts as possible, to begin with the simplest objects and move up to the more complicated, and to be complete and meticulous in listing all of the details of a problem. He opposed Aristotelian physics, yet argued that a vacuum could not exist, bringing him into conflict with Gassendi's atomic theory and Galileo's discoveries. Well aware of the danger of expressing original thought—even though he was opposed to many ideas to which the Church was also opposed—he wrote: "Recalling my insignificance, I affirm nothing, but submit all these opinions to the authority of the Catholic Church." In mathematics, Descartes' development of algebraic geometry is a wonderful synthesis that is now taught in high schools, though not always appreciated. In physics and astronomy he had the universe populated with mysterious vortices, hoping thereby to account for gravity and magnetism. He studied medicine and meteorology, optics and motion, and based his philosophy on the belief that the easiest part of nature to know is one's own mind. (To experience, perhaps, but to understand? By his own rules, we should begin with simpler things.)

Scientific societies had not yet been established, so communication between scientists was by travel and by letters. Marin Mersenne (1588–1648), a friend of Descartes, was particularly helpful in disseminating information by meeting with other scientists in Paris and by sustaining a voluminous correspondence with scientists throughout Europe. He also extended the early studies of the notes from a stretched string that Pythagoras and his school had developed. Mersenne's laws of the violin string state how the frequency of the fundamental note (the note emitted when all parts of the string between its supports are moving together in the same direction as the string vibrates) depends on its length, its mass per unit length, and the tension it is under. That frequency, Mersenne discovered, is inversely proportional to the length and to the square root of the mass per unit length, and proportional to the square root of the tension. Double the length and the frequency is halved, and the pitch drops by an octave. Make the string four times as thick, or otherwise make its mass per unit length four times as much, and the pitch also drops by an octave. Make the tension four times as great and the pitch goes up by an octave. These facts seem to be unrelated, but nearly two hundred years later (in 1822) the French mathematician Jean Baptiste Fourier (1768–1830) developed the method of dimensional analysis by which these relations and many others may be demonstrated as simply derived consequences of theory. In this case, the theory shows that *if* the frequency depends only on the length, the tension, and the mass per unit length, then Mersenne's laws describe the *only* way in which it can possibly depend on them. This very powerful and elegant analysis had to wait for Newton to clarify the meaning of force—in this case tension—and mass.

There were, of course, famous universities to serve as sources and transmitters of scholarly knowledge, and one might hope that they would be leaders in the fight for freedom of thought. While visiting Oxford University in 1848 Ralph Waldo Emerson (1803–1882) wrote:

> On every side, Oxford is redolent of age and authority. Its gates shut of themselves against modern innovation.—Here, on August 27, 1660, John Milton's *Pro Populo Angelicano Defensio* and *Iconoclastes* were committed to the flames. I saw the school-court or quadrangle where, in 1683, the Convocation caused the *Leviathan* of Thomas Hobbes to be publicly burnt.
>
> (*English Traits,* ch. 12, HCl 5, 417)

Looking at the university nearly two centuries after these events, Emerson added, "I do not know whether this learned body have yet heard of the declaration of American Independence, or whether the Ptolemaic astronomy does not still hold its ground against the novelties of Copernicus,"—but he goes on to praise the high standards of the Oxford examinations in Greek and Latin.

Much earlier, Martin Luther had written, "The universities also require a good, sound reformation. I must say this, let it vex whom it may. The fact is that whatever the papacy has ordered or instituted is only designed for the propagation of sin and error." (HCl 34, 321) All would be well, it would seem, if universities were taken over by the Protestants. Unfortunately, for many years Oxford University, indeed the whole of England, continued to feel the heavy hand of William Laud (1573–1645), Prime

Minister (1628), Chancellor of the University (1630), and Archbishop of Canterbury (1633). The arrival of the Mayflower in America in 1620 has reminded generations of the earlier persecution in England of the unorthodox Protestants, and Archbishop Laud continued to be an enthusiastic and powerful supporter of much of it. The "Star Chamber" was the Anglican equivalent of the Inquisition—judges without a jury, accepting rumor as evidence and authorized to apply torture.

The puritan William Prynne (1600–1669) was one person who was brought before the Chamber, and the hatred of the established church leaders for Prynne and his colleagues resounds in the words of one of the judges, Sir Edward Sackville, fourth Earl of Dorset:

> Mr. Prynne, I declare you to be a schism-maker in the church, a sedition-sower in the commonwealth, a wolf in sheep's clothing; in a word *omnium malorum negussimus* (the most worthless of all evils). I shall fine him £10,000, which is more than he is worth, but less than he deserves. I will not set him at liberty, no more than a plagued man or a mad dog, who though he can't bite will foam; he is so far from being a social soul that he is not a rational soul. He is fit to live in dens with such beasts of prey as wolves and tigers like himself; therefore, I condemn him to perpetual imprisonment; and for corporal punishment I would have him branded in the forehead, slit in the nose, and have his ears chopped off.
>
> (CB, 738)

Taken bleeding from the scaffold to jail, Prynne continued to write against Laud, who by then (1637) had used the power of the Star Chamber to destroy the liberty of the press and make himself a virtual dictator. Prynne was condemned to have the remaining stumps of his ears hacked off, and he was to be branded on both cheeks with the letters S. L. (Seditious Libeler). Prynne said that it stood for *Stigmata Laudis.* He had never pulled any punches, declaring that all who went to the theater renounced their baptism—and that included the royal family. As Voltaire put it, Mr. Prynne was "a man of most furiously scrupulous principles who would have thought himself damned had he worn a cassock instead of a short cloak, and have been glad to see one half of mankind cut the other to pieces for the glory of God." (*Letters on the English* xxiii) Times and attitudes changed, sometimes very rapidly, especially at the beginning of England's civil war. Archbishop Laud was executed on January 10, 1645. The struggle to understand by means of edicts based on religious dogma has indeed followed a tortuous and torturous path.

After a brave but aborted attempt four centuries earlier, science in Christendom began to emerge from the closet during the second half of the seventeenth century. Suddenly in England there appeared the establishers of physics, chemistry, and astronomy—John Wallis (1616–1703), Robert Boyle (1627–1691), Christopher Wren (1632–1723), Robert Hooke (1635–1703), Isaac Newton (1642–1727), John Flamsteed (1646–1719), and Edmund Halley (1656–1742), to list them in the order of their birthdates. Others, some already mentioned, became part of the international community of scholars throughout Europe, learning about nature by experiment, observation, and related theory. In 1644, some of them in London began to meet each week, calling themselves "The Invisible College." In 1662, the "College" was officially blessed by Charles II and given a charter as "The Royal Society of London." Fortunately it did not restrict its membership to physical scientists—the botanist John Evelyn (1630–1706) was an early member, along with John Ray (1627–1705), the founder of the study of natural history in Britain. Later John Woodward (1665–1728), the geologist, was a Fellow of the Royal Society; others too joined, not primarily scientists—Samuel Pepys, John Dryden, and several bishops and lords of the realm. Henry Oldenburg (1618?–1677), coming from Bremen, became the first editor of the *Philosophical Transactions of the Royal Society,* still published today.

In 1668, in response to an invitation from the Society to investigate the laws governing collisions between moving bodies, independent studies were submitted by Wallis, Wren, and Christiaan Huygens (1629–1695) from Holland. The "impulse" of Philoponos, Buridan, and Benedetti was now clear to all three of them and recognized as the law of conservation of momentum. The analysis by Wallis was the most general because he considered collisions that were both elastic and inelastic. Huygens extended the analysis to include elastic collisions between many objects, giving precision to the remark that Descartes had made: "God preserves just the quantity of motion and rest in the material world that he put there in the beginning." The law of conservation of momentum was soon seen to be a direct consequence of Newton's First Law of Motion.

Another important pioneering scientist of this age was Robert Boyle, who was born to wealthy parents and was the second son of Richard, Earl of Cork. He is known to high school physics students for his discovery that the volume of a fixed amount of gas in a container is inversely proportional to the pressure exerted on it—a law that he credited to his student, Richard Towneley. Boyle had heard of the invention of the air-pump in Germany by Otto von Guericke (1602–1686) and of Guericke's spectacular demonstration in 1654 of the effect of atmospheric pressure. (Emperor Ferdinand III, Holy Roman Emperor and King of Hungary and Germany, and all the assembled crowd at Magdeburg were astonished when Guericke placed two hollow copper hemispheres together, evacuated the space between them with his pump, and showed that teams of horses could not pull them apart.) Boyle was also an enthusiastic supporter of the old and discredited idea that matter is made up of small corpuscles. He saw God as a Great Engineer who had built the universe and caused it to run mechanically, even in the bodies of men, where the beat of the heart was that of a pump. Francis Bacon had recommended the experimental method; Boyle used his talents and his money to make it a reality.

Boyle did other things with his money also. He financed a great deal of missionary work for spreading the gospel throughout the colonies and the Orient, paying for the printing of many Bibles, and he left money to sponsor an Oxford lectureship to defend Christianity against nonbelievers and to teach about "the being and attributes of God." He wrote religious essays as well as scientific treatises; indeed his *Occasional Reflections* later appealed so much to Lady Berkeley that she asked her chaplain Jonathan Swift (1667–1745), of *Gulliver's Travels* fame, to read one to her each time he visited. Less than impressed by Boyle's writings, Swift inserted in her book a copy of his own satire on Boyle, "A Meditation upon a Broomstick" and solemnly read it to her, pretending it was Boyle's. According to Thomas Sheridan, "she was, every now and then during the reading of it, expressing her admiration of this extraordinary man who could draw such fine moral reflections from so contemptible a subject." When held up to ridicule, not all of us can be as merry about it as she was when Swift's handwritten manuscript was discovered between the pages of her copy of Boyle's book. Swift always enjoyed teasing scientists and lampooning the Royal Society.

Robert Boyle had provided the chief inspiration for the founding of the Royal Society. Similar establishments appeared on the continent of Europe—the Académie des Sciences in Paris enjoyed more than a mere blessing from Louis XIV. Short-lived academies that had opened in Italy during the previous difficult hundred years were followed by more permanent institutions. By the early years of the eighteenth century, centers of scientific learning, independent of the universities, had been founded in Sweden, Germany, and Russia as well.

The scientific genie was really out of the bottle. Robert Hooke, by his experiments with elasticity, revealed the basic relation between the tension applied to elastic material and the extension produced by it. He also discovered Boyle's Law independently, and he was not the only one to do so. It was also discovered by the French priest Edme Mariotte (1620–1684), who added the important caveat that it applied only at constant temperature. Hooke, too, argued for the wave nature of light from studying thin films; Francesco Gremaldi (1618–1663) at Bologna discovered the diffraction of light; Huygens gave a detailed theory of the interference of waves. Newton demonstrated many properties of light but did not believe that it was a wave. Huygens also understood the acceleration of an object moving in a circle, a fundamental advance in the study of motion. Boyle again lifted chemistry out of the superstitions of alchemy. In his *Sceptical Chymist* he emphasized the need for experiments, rejected the old earth-air-fire-water theory, and the mystical analyses of alchemy.

A few years younger than these men, Isaac Newton (1642–1727) showed that a single theory could describe the motion of objects on the earth, and the motion of the earth and other planets around the sun. A fundamental law of the universe lay behind these phenomena. No need to suppose that angels were pushing the planets around; no need to pay attention to the physics or astronomy of Aristotle—his science had been erroneously wedded to Christian beliefs; no need to be afraid of comets, because they obeyed the same basic law. It was time to recognize that Galileo had been right, and his accusers dead wrong—the earth *did* move around the sun, and it was not the fixed center of the universe that everybody had supposed.

Earlier, as we have noted many times, those in Christendom who made any discovery about nature were afraid to publish it out of fear of persecution, unless, of course, it was a "discovery" based only on the Bible and on officially approved theological arguments, in which case it was worthless but brought recognition and fame. Now, however, there was theological criticism, to which we return in a moment, but it was more career-threatening than life-threatening. In England the scientists were more interested in knighthood than sainthood, and throughout Europe there was no more anonymous publishing of scientific discoveries. The struggle to understand became the struggle to be the first to understand, and it remains that way to this day. Understandably so; it is not enough to be the first, you have to convince your scientific colleagues about the validity of your discovery, and about your claim for priority.

Robert Hooke was very unhappy that Newton had received all of the credit for the discovery of the law of gravitation. In 1674 he had, in fact, stated the inverse-square law of force between gravitating bodies. A similar idea had been suggested by Kepler, and indeed Newton had planned, but decided otherwise, to include the following statement in his *Principia.* "Pythagoras understood by the harmony of the heavens that the weights of the Planets towards the sun were reciprocally as the squares of their distances from the sun." So the idea wasn't all that new. Hooke was also angry at Johannes Hevelius (1611–1687) for not giving him credit for the idea that comets move in (nearly) parabolic paths under the influence of this force from the sun. Since most discoveries were published in the *Philosophical Transactions of the Royal Society,* its editor, Henry Oldenburg, automatically became part of the controversy.

To show numerically that the acceleration of the moon in its near-circular orbit, divided by the acceleration of a falling body on earth, was equal to the square of the radius of the earth, divided by the square of the distance between earth and moon centers, had to await the accurate measurements of the French astronomer and Roman Catholic priest Jean Picard (1620–1682). To tie this in completely with the inverse-square law also required Newton to develop the differential and integral calculus, which was also needed to show that this same law could give rise to elliptical and parabolic as well as circular orbits. This raised another controversy, since Gottfried Leibniz had independently discovered the calculus at about the same time. Certainly the notation he used was superior to Newton's, and it is now common everywhere calculus is used.

Arguments between the two and their respective followers were not limited to claims for priority, however. Leibniz, greatly influenced by Aristotle, could not imagine that a force between sun and planets could be active across a vacuum. He believed that Newton had introduced "occult qualities and miracles into philosophy," and that the inverse-square law of force was "subversive of natural and inferentially of revealed religion" (quoted by Charles Darwin, *The Origin of Species,* ch. 15). Newton recognized the philosophical difficulty but not the religious one: "That one body can act upon another at a distance, through a vacuum, without the mediation of anything else . . . is to me so great an absurdity that I believe no man who has in philosophical matters a competent faculty of thinking can ever fall into it." Action at a distance was emphasized not by Newton, but by Roger Cotes (1682–1716), his assistant in the preparation of the second edition of *Principia.*

On the issue of priority in science, there was an argument between Leibniz and John Clarke (1682–1757), Dean of Salisbury, about the discovery of the law of conservation of kinetic energy, the dean espousing the cause of Newton. The Dutch physicist and histologist Niklaas Hartsoeker (1656–1725) argued with Huygens about whose method of calculating the distance to a star came first, and he also contested Leeuwenhoek's claim for priority in observing spermatozoa through his microscope. In Switzerland, the brothers Jakob (1654–1705) and Johann (1667–1748) Bernoulli were the first to develop some aspects of integral calculus, in particular the calculus of variations in terms of which it was later found that fundamental laws of physics may be elegantly expressed. Naturally they sided with Leibniz in the matter of priority, Johann defending the claim before London's Royal Society. Johann's son Daniel (1700–1782) would provide the theory behind the empirical Boyle's Law, based on the reinstated corpuscular theory.

Superstition, religion, and science were all changed in a very real way after the 1687 publication of Newton's *Principia Mathematica,* in which many cherished paradigms of both religion and science were revealed as superstitions. Of course Newton had no desire to attack the Christian religion, or at least the Anglican version of it. (It would not be until nearly a hundred years after Newton's death that the cardinals of the Roman Church allowed that "the printing and publication of works treating of the motion of the earth and the stability of the sun, in accordance with the general opinion of modern astronomers, is permitted at Rome," Edict of September 11, 1822.) Newton's professorship at Cambridge required him to be an Anglican, and he attended services regularly. His writings on religion and, as we have seen, on alchemy were more extensive than his works on physics and astronomy. He was very much impressed by the books published earlier in the seventeenth century by Jakob Boehme (1575–1624), the German mystic and theosophist, i.e., "wise in the ways of God." Newton encouraged the view that the unity of the universe was evidence of the power of God. Others weren't so sure.

The leader in the theological attacks on Newton's science was the Yorkshireman John Hutchinson (1674–1737), author of *Moses's Principia* (1724 and 1727). He was very upset with the model of the planets moving through a vacuum, arguing that it was mechanical and godless. Instead, he argued, there is a spiritual matter pervading the whole creation, and through it Jehovah is the master of the material universe. Writing about it in the middle of the nineteenth century, Elihu Rich noted that, "this philosophical doctrine which is supported by the recent discovery of an interplanetary ether, was in the work of Hutchinson a pure deduction from the Scriptures." Later in this chapter we return to this "paradigm of the ghost." Hutchinson had a number of followers, including the biblical scholar George Horne (1730–1792), who would have been wise to stay with his biblical scholarship. Instead, in 1751 he published the first of a series of ironical attacks on Newtonian science. Along with his *Commentary on the Book of Psalms,* they were enough to get him appointed as Vice Chancellor of the University of Oxford, Dean of Canterbury, and Bishop of Norwich.

Of course theological criticism of Newton was not limited to England. John Turretin (1671–1737), professor at Geneva and one of a family of Calvinist theologians, "proved" from the Bible that the earth is at rest while the objects in the heavens move around it. In some European universities, especially in Spain, professors were forbidden to tell their students about the facts revealed by the telescope. As late as 1771 the administrators of the University of Salamanca refused to allow the teaching of the physical sciences, and Ferdinand VII (1784–1833), king of Spain, fired the rest of the professors of science. Spain and its former colonies have been held back a lot in scientific studies, the official attitude in Mexico that had silenced Juana Inés de la Cruz* (1651–1695) persisting for many years.

As late as 1873, the ex-president of the Lutheran Teachers' Seminar bitterly attacked the astronomers for teaching that the earth moved. On the other hand, way back in 1721, while Newton was still alive, Cotton Mather, hearing from his colleagues in England, had decided to accept the discoveries of Newton into his theology.

According to Webster, the word "paradigm" means a model, or an example, or a pattern. Fundamental to the communication between people who are trying to understand anything at all is a common paradigm, a common belief system that is sometimes tentative but often accepted as indisputable. We all have this "noble curiosity, this restless and reckless passion to understand"—to requote Will Durant from page 7 of this book. It is a passion that invades all areas of knowledge, from the most trivial to the most profound. Arguments about the correct meaning of quantum mechanics, or the best way to discover the top quark, or the incorporation of scientific findings into theology are attempts to understand; so also are more common arguments about why a particular baseball club is so successful, or why the weather is so bad today. To have any discussion, however, the participants must at least start with their own personal paradigms; to have any useful discussion those paradigms must overlap, if not at the beginning, then at least at the end. Of course, if nothing will change your paradigm, discussion with someone equally convinced of another will lead only to frustration. That is why religion and politics are banned as subjects for discussion in some social circles, and why religious dogma is in conflict with scientific enquiry.

There is no way to describe here in any detail the enormous development of physical science and its changing interaction with religious dogmas during the nineteenth and twentieth centuries. Instead I will classify the relevant paradigms, whether valid, questionable, or dead wrong, that people have cherished up to the present. I call them:

1. The Paradigm of Taxonomy
2. The Paradigm of the Big Boss
3. The Paradigm of the Ghost
4. The Paradigm of the Little Fleas
5. The Paradigm of the Wedding

(1) *The Paradigm of Taxonomy.* In this approach it is generally recognized that it is useful to classify the elements under discussion, putting them into categories that describe their common properties. We are doing it right now by grouping together different areas of knowledge that share a common methodology. A prime example is the classification of animal and vegetable life initiated by Aristotle and Theophrastos that forms the

basis of our understanding of the natural world; it has now been refined to the classification of genetic codes. As we saw in chapter 5, a classification of materials according to the way objects formed from them move "naturally" was not profitable, standing in the way of progress for two millennia.

Another form of taxonomy is provided by the division of all people into twelve groups, according to the position of the earth in its orbit at the time of their birth or, possibly, conception. Useless as it is from a scientific viewpoint, it is popular because it purports to relate some personality traits to the "natural behavior" of an Aries, or a Pisces, or what have you. Superstitions like this flourish where education is lacking; the people of India believe in it even more than do those of the United States.

Other forms of the paradigm of taxonomy include stamp collecting and coin collecting. Merely collecting and classifying are not very conducive to understanding, and caused the great physicist Lord Rutherford to snort, "There is physics . . . and there is stamp collecting." He was referring to the accumulation of a pile of data in physics without any attempt to understand its significance. Similarly, stamp and coin collecting can be stimulating ways of learning about geography and history. Taxonomy is useful, but on its own it doesn't teach us much about understanding.

(2) *The Paradigm of the Big Boss.* We have encountered this approach to knowledge many times in this book. Here it is assumed that there is an Authority who knows it all, and all that you have to do is to tap in to what that Authority says. Aristotle, Thomas Aquinas, Isaac Newton, James Ussher, the Oracles, the Bible, the Koran, the *Upanishads*. . . . It is very comforting to many people to select a set of words written by others and to deify them, making little gods out of the very words themselves. It closes the mind, however, and it has led to much of the evil which we have chronicled in this book.

One way to become a Big Boss is to state that you have an inside track to God, and that you are merely the mouthpiece. It is very hard to argue against that! Moses condensed the Egyptian rules into the Ten Commandments and undoubtedly thought that he was divinely inspired. Whether he was or not, people even up to this time would not have paid so much attention to those commandments if it had not been claimed that they came from a Higher Authority. Earlier, it was believed that Hammurabi received from the god Shamash the laws that he gave to the Babylonians. Those laws had a big impact too. If you can convince people that you are speaking with the words of the gods or of God, you can exert an enormous influence for good—or for evil.

Even without invoking a god or any form of religion it can happen that the writings of a particular person are believed to be correct in all details. The theories of gravitation and of motion developed by Isaac Newton are spectacular in their predictive power, and before the beginning of the twentieth century they were thought to be one hundred percent correct. Newtonian mechanics became the paradigm of the Big Boss, and the physics establishment saw to it that there would be no departures from it—with very good reasons, it should be added, as success built on success, while mysteries of spectra, ether, and radioactivity slowly accumulated.

Another example of this paradigm is the domination of the chemists of the first half of the nineteenth century by the doctrines of the very fine Swedish chemist Jöns Berzelius (1779–1848), who distinguished himself by his scrupulously accurate chemical analyses. The English physician and chemist Robert D. Thomson (1810–1864) wrote this about him:

> It would be difficult to overestimate the value of the contributions made to the science by this indefatigable chemist, whose body and mind seem to have been an incessant action for the best part of half a century. . . . The ingenious generalizations which he sometimes made, *although generally ultimately found to be untenable,* were productive of vast benefit in encouraging and stimulating enquiry. Among those may be noticed his ideas about the compound nature of chlorine, his theory of electro-chemistry, of isomerism, of catalytism, etc. It is much to be regretted that the free enquiry and liberty of deduction which he claimed for himself he did not always allow to others. . . .
>
> (CB, p. 96)

Berzelius was an important member of the Establishment. He did not accept the electro-chemistry of Faraday, or the hypothesis proposed in 1811 by Amedeo Avogadro (1776–1856) and supported by Ampère that at a given temperature and pressure equal volumes of all gases contain the same number of molecules. Berzelius argued with Humprey Davy on the nature of chlorine, which Davy insisted was an element; Berzelius believed that it was a compound. They ended up disliking each other intensely. Fortunately, Avogadro's hypothesis was picked up later by Stanislao Cannizzaro (1826–1910), who used it to determine many atomic and molecular weights in terms of hydrogen.

(3) *The Paradigm of the Ghost.* In this belief system there exists some all-pervading intangible substance that exists throughout all or part of the universe. An early example of this was the indeterminate first principle of all matter postulated in the sixth century B.C. by Anaximandros. It wasn't hot like fire or cold like air, moist like water or dry like earth. It had all of the qualities and none of the qualities, because the opposites were

all balanced in it. It had no characteristic at all, yet it was the source of all being. Two centuries later Aristotle added aether to the earth-air-fire-water model in order to provide a source for objects in the sky. This aether pervaded everything from the moon up, and as we have seen, the five-element paradigm became entrenched in the collective consciousness for many centuries.

Other ghosts came into physics much later. In the seventeenth century Descartes's vortices pushed the planets around because there was believed to be some subtly concealed matter that pervaded all space, and the vortices were like tornadoes or hurricanes in this elusive medium. Newton and his critics were worried about the action-at-a-distance aspect of his theory of gravitation. How could the sun influence the planets except by pulling on a medium which pulled on them? Edmund Burke (1729–1797) put it this way:

> When Newton first discovered the property of attraction, and settled its laws, he found it served very well to explain several of the most remarkable phenomena of nature; but yet, with reference to the general system of things, he could consider attraction but as an effect, whose cause at that time he did not attempt to trace. But when he afterwards began to account for it by a subtle elastic aether, this great man . . . seemed to have quitted his cautious manner of philosophizing.
>
> (*On the Sublime and Beautiful,* Part IV, HCl 24, 103)

Newton's contemporary, Johann Becker (1635–1682) departed from the five-element theory to formulate a model based on three types of earth (mercurial, glasslike and greasy, or flammable). He was another one to contribute to a vast range of knowledge—medicine, chemistry, mineralogy, economics, finance, and education. He was even commanded by the Dutch authorities to use sand in order to turn silver into gold. He died four years after that and is remembered here for his postulate of the existence of the greasy flammable substance that he called "fat earth" (terra pingius), and that Georg Ernst Stahl (1660–1734) called "phlogiston." It was at once a material substance and a principle of combustion. A combustible material had phlogiston in it and during combustion the phlogiston left it. A metal would give off phlogiston to become what we call an oxide.

The paradigm of the elusive substance phlogiston worked very well because it consistently had things wrong. Where phlogiston was supposed to be emitted we know now that oxygen is absorbed; where phlogiston was absorbed, it is really oxygen that is being emitted. It was hard to tell the difference until the careful measurements of Joseph Priestley* and Antoine Lavoisier* at the end of the eighteenth century. It was found that if phlogiston existed, it had to have a negative weight because a substance gained weight by emitting phlogiston—it was really picking up oxygen. Priestley's measurements played a key part in revealing phlogiston as a substance that did not exist, but the paradigm was too strongly embedded, and Priestley believed in phlogiston until his death. Lavoisier, on the other hand, recognized the chemical nature of combustion and in particular the combination of oxygen with the burning material. There remained a problem, however, since objects tended to expand when heated. Perhaps there was some other ghostly substance that flowed into them when they were heated, some spirit of heat named "caloric." Although Francis Bacon had argued many years earlier that heat is a form of motion, his theory was not convincing.

As early as 1821 there had been an attempt by John Herapath (1790–1868) to show that heat is a form of motion by relating the temperature of a gas to the average speed of its molecules. Unfortunately, he assumed that these quantities were proportional to each other, and that led to disagreement with observed data. In 1846, however, a paper written by the Scottish physicist John James Waterston (1811-1883) shows that he had got the theory right—the temperature of a gas is proportional to the average kinetic energy of its molecules, $\frac{1}{2}mv^2$. For a particular gas it is therefore proportional to the *square* of the average speed of the molecules. This made all the difference, and a number of empirical laws obtained from experiments with various gases emerged directly as consequences of the theory—the laws of Boyle, Mariotte, and Avogadro mentioned earlier, and a law noted first by Jacques Alexandre César Charles (1746–1823) and studied in more detail by Joseph Louis Gay-Lussac (1778–1850), John Dalton, and others. This law states that at constant pressure the volume of a gas decreases by approximately one part in 480 of its value at 32° F for each decrease in temperature of one degree Fahrenheit. This suggests that a temperature of absolute zero would occur at 32 – 480 = –448° F, at which point the molecules would come to rest if they hadn't already liquefied.

John Waterston's fine paper was submitted to the Royal Society of London for publication—and it was scornfully rejected. When rescuing it from oblivion forty-six years later, Lord Rayleigh quoted some referees' comments: "the paper is nothing but nonsense, unfit even for reading before the Society." "The original principle involves an assumption which seems to me very difficult to admit." (WP 1, 634) It had to await the work of James Prescott Joule (1818–1889) before the relation between heat and motion was clearly established and the ghost of caloric laid to rest.

There were other strange fluids permeating matter. Stephen Gray (1696?–1736), experimenting with electricity, recognized the difference between conductors and insulators, and showed how to move electricity from one point to another. In France, Charles DuFay (1698–1739) recognized what we call positive and negative electricity, and the repulsion between two electric charges of the same sign. It would seem that there were two mysterious fluids that moved through conductors. The great Benjamin Franklin (1706–1790) saw through this, as he saw through so many things, and postulated the existence of a single electrical fluid. Too much of this fluid and the object has a positive charge; too little and it has a negative charge. This particular ghostly substance is no longer a mystery, although of course in metals it is the negatively charged electrons that constitute this remarkable fluid. Electrochemistry was born when the Scottish physician and chemist William Cruikshank (1745–1800) passed an electric current through water and observed the release of oxygen and hydrogen at the metallic poles.

Discoveries of the properties of light brought another all-pervading intangible ghost—the nineteenth century ether. Thomas Young (1773–1829) and Augustin Fresnel (1788–1827) established beyond doubt that light consists of a wave motion, Fresnel even showing that the oscillations were at right angles to the direction of motion. But what was waving? Young introduced the luminiferous, i.e., light-carrying, ether as the all-pervasive medium that was doing the oscillating. Further confirmation of the wave properties of light came from the amazing François Dominique Arago (1786–1853), who also demonstrated the laws of polarized light, made precise measurements of the speed of sound and of the densities of some gases, came close to discovering electromagnetic induction, studied lightning, and measured the diameters of the planets. It is not for this that we have called him "amazing," although these accomplishments were first class in quality and many in number. If we have the image of a quiet dedicated scientist, systematically studying these problems in his ivory tower, we are very wrong. He deserves much more than the few words we shall give him; he rates a long biography, preferably with movie rights. Here are some of the adventures that interfered unsuccessfully with his struggle to understand natural phenomena.

As a brilliant young astronomer, he was sent to Majorca to measure the length of an arc of the meridian, since it is on practically the same meridian as Paris. War broke out between Spain and France, and the local populace, thinking he was signalling to the French fleet, came after him. Disguised, he escaped to the coast where a French ship was in fact stationed, but the captain would not let him on board. He therefore gave himself up to the local prison warden, receiving some protection, though stabbed by one of the mob. Some intrigue brought his escape in a boat that managed to avoid the British fleet that was blockading the coast until it docked at Algiers. From there he took a ship to Marseilles, but, within sight of that city, ship and passengers were captured by a Spanish privateer. For a while, Arago was imprisoned in a foul Spanish dungeon and threatened with execution. Fortunately, on the ship coming from Algiers were two lions, a present from the Dey of Algiers to Napoleon. One of the lions had died, and Arago managed to get a letter to the Dey to tell him that the Spaniards had starved the animal. The Dey expressed his anger at the Spanish government and in the subsequent negotiations Arago was released. Headed for Marseilles, the incompetent pilot landed the ship at Bougie, ancient Saldae, in North Algeria. Arago dressed himself as an Arab and headed for Algiers, but the Dey had just died and his legitimate successor was killed, the usurper decreeing that all the French there would be sold as slaves for the galleys. Arago managed to ship out of Algiers to Marseilles on a corsair, a Muslim privateer, part of a convoy that was immediately seized by two English frigates. Fortunately, for Arago and for physics and astronomy, the two frigates could not hold all of the convoy, and his corsair finally got him to Marseilles and safety. In the revolution of 1830 he was an ardent and powerful supporter of democracy, and in 1848 he refused to take the oath of allegiance to Louis Napoleon (1808–1873). His prestige was so great that the new emperor left him undisturbed as head of the Paris Observatory.

Arago's contemporary, Baron Augustin Louis Cauchy (1789–1857) was forced into exile from France between 1830 and 1837 because of his Catholic faith and royalist sympathies. In addition to his magnificent contributions to mathematics, he provided a detailed theory of waves, which he applied to the motion of light through the postulated ether. In 1819 Hans Christian Oersted (1777–1851) had discovered that an electric current influences a magnetic needle, so that electricity and magnetism were closely related to each other. Very excited by this discovery, André Marie Ampère (1775–1836) showed that a solenoid carrying a current behaves like a magnet, and that two neighboring wires that carry currents exert a force on each other. Michael Faraday (1791–1867) discovered that a moving magnet induces a current in a neighboring coil and by very careful and imaginative experiments revealed other secrets of the newly discovered electromagnetism. Finally, James Clerk Maxwell (1831–1879) developed a theory that explained all of these effects and much more, including the amazing conclusion that electromagnetic waves are radiated from changing currents. Furthermore, the speed of these waves was the same for all frequencies if there was no matter present to distort their motion significantly, and this speed could be measured by experiments performed in a laboratory.

Maxwell showed that these waves were oscillations of the electric and magnetic fields, but to him these were too intangible to be the only elements that were oscillating. Pressure and density oscillate in a sound wave, but there has to be a medium of some kind through which the waves pass. As he put it:

> We have therefore some reason to believe, from the phenomena of light and heat, that there is an aethereal medium filling space and permeating bodies, capable of being set in motion and of transmitting that motion from one part to another, and of communicating that motion to gross matter so as to heat it and affect it in various ways.
>
> (WP I, 851)

In the same way that the speed of sound varies slightly when the wind blows, the speed of light and other electromagnetic waves should therefore vary slightly if the ether-wind changed during the course of a year. The key experiment of Albert Michelson (1852–1931), assisted by Edward Morley (1838–1923), failed to detect this effect, and it seemed that either Maxwell's equations or Newton's laws of motion had to be modified. The paradox was resolved by Albert Einstein (1879–1955) with the modification of Newton's laws for velocities close to that of light, and the postulate of the theory of relativity that, no matter what your velocity, the speed of light relative to you has the same value. The paradigm of the ghostly ether, strongly supported by the physics establishment of the nineteenth century was soon replaced by the paradigm of the non-ether. It is generally believed that the paradigm of this ether has been laid to rest, although there are a few dissenters. For research in the sciences the most likely path to understanding lies along the direction of the paradigms that are currently embraced, but nobody can be really certain about this. Crazy ideas like the theory of relativity and quantum mechanics have revolutionized our knowledge of physics *because* they challenged commonly held beliefs.

Another "crazy" idea and another ether appeared in 1928 when Paul Dirac (1902–1985) discovered an equation which was consistent with both relativity and quantum theory, and at the same time provided a magnificent description of the properties of the electron. Unfortunately, it seemed, the equation led to the conclusion that there were many states of negative energy—even an infinite number—into which all of the electrons in the world would disappear if not prevented from doing so. It was therefore necessary to assume that all of these states were already occupied, providing a ghostly background of unobservable infinite charge and infinite negative mass throughout all space, the only observable aspect of this "sea" occurring when an electron missing from it was interpreted as a positron. By reinterpretation of the equation itself, however, this paradigm of the infinite sea of negative energy electrons was avoided and replaced by the model of a vacuum as consisting of infinitely many electron-positron pairs being continuously created and destroyed. This paradigm now provides the most accurate theory of physical phenomena yet known, but a single electron in a vacuum is a most complicated object.

Science is not alone in postulating a belief in an all-pervading ghost that is difficult to observe and may in fact be unobservable or nonexistent. Such paradigms are always invented in order to help people understand some aspect of the world without and within. The ubiquitous devils, waiting to invade living bodies, were invented to account for the effects of unknown bacteria and viruses, or to provide an explanation for otherwise inexplicable mental disorders. The spirits of the dead were invented in order to console the bereaved, and to account for creaking noises in old buildings. The hierarchy of invisible angels was invented not just to take over the job of some earlier gods in controlling the weather, and not just to push the planets around. From Old Testament times they were good spirits, invented to assure man that he was not alone, that there were positive powers throughout the universe supporting his efforts to lead a moral life, and perhaps ready to punish him if he did not. The ultimate paradigm of the ghost is the belief in the Holy Ghost and the other aspects of the Trinity. It is a great help to many people, but it is a faith that is the end result of many intellectual battles, and even more killings. It is the official paradigm of Christianity, and can be compared to a paradigm of science because if you do not accept it you are not regarded as a true member of the community. Scientists, however, have learned to change their basic paradigms, albeit reluctantly. Some religious dogmas change only when scientists prove that they are wrong, and sometimes not even then. Other religious dogmas are not subject to scientific analysis.

(4) *The Paradigm of the Little Fleas.* To quote Jonathan Swift again:

> So, naturalists observe, a flea
> Hath smaller fleas that on him prey;
> And these have smaller still to bite 'em;
> And so proceed ad infinitum
>
> (*On Poetry, A Rhapsody*)

Augustus De Morgan (1806–1871) paraphrased it this way:

> Great fleas have little fleas upon their backs to bite 'em
> And little fleas have lesser fleas, and so *ad infinitum*
> And the great fleas themselves, in turn, have greater fleas to go on;
> While these in turn have greater still and greater still, and so on.
>
> (*A Budget of Paradoxes*)

In this paradigm, matter is analyzed into smaller parts that themselves contain components that in turn have smaller objects inside of them. How long does it continue? We now think that electrons, neutrinos, and quarks are the ultimate permanent components, but we can't be sure. We do know, however, that everything around us is made of molecular conglomerates that we can analyze as combinations of molecules. For many of them a million or more, if lined up, would reach a centimeter, and they in turn consist of atoms, that we know are made of nuclei surrounded by electrons. The nuclei consist of protons and neutrons, and inside these are quarks, not yet directly observed. Could the quarks be the twentieth century version of the "paradigm of the ghost"? There have been some reports of "quark sighting," but they remain unconfirmed. In any case there is an argument that indicates that they cannot be separated from each other, since the attractive force between them increases as they get farther apart. A stretched spring has that property until its elastic limit is reached. We may find out soon if a similar limit applies to quarks. The paradigm of the quark—the present ultimate example of the paradigm of the little fleas—leads to consequences that sometimes agree very well with experimental data, and at other times disagree strongly. It is not a near-perfect theory like the theory of the electron, but it is the best theory that we have.

The paradigm of the little fleas has proved very valuable in physics and chemistry. Early proponents of the atomic hypothesis were not believed, partly because they had very little evidence to back it up and partly because of the belief that a vacuum could not exist. John Dalton (1766–1844) put the theory of atoms and molecules on a quantitative basis, supposing that atoms are indestructible particles that retain their individuality during chemical reactions. His atoms were supposed to be surrounded by the mysterious caloric.

(5) *The Paradigm of the Wedding.* This paradigm relates to the many ways that science, religion, and philosophy base the latest beliefs on the unity of opposites, the yang and the yin, the faith and the reason, the being and the nothing, the proton and the electron. The yang-yin paradigm represents a sexual view of cosmology, the brilliant strong male principle of mercury combining with the dark passive female principle of sulphur to form the blood-red cinnabar. We see this theme of man, woman, and child, or thesis, antithesis, and synthesis in the model of the hydrogen atom, which has new properties that were not inherent in the individual proton and electron from which it is made. We see it in the unity of electricity and magnetism to form electromagnetism, and in the synthesis of space and time, or of energy and mass, according to the special theory of relativity. We see it in Einstein's preoccupation in later life with finding a unified theory of electromagnetism

and gravitation, i.e., incorporating electromagnetic phenomena into his general theory of relativity. Despite years of effort on his part, this marriage failed. There are many happy childless marriages in real life, but in physics two ideas that are really united into a new theory must lead to consequences that either idea alone could not conceive. To this day we do not know how the great ideas of electromagnetism and gravitation fit together, partly because of the electron-positron and other pairs that continually appear and disappear with consequent rapid fluctuation in the gravitational field. In this case Einstein's work was premature—we need to learn a lot more before a meaningful marriage of electromagnetism to gravitation can be expected. As Wolfgang Pauli (1900–1958) remarked about Einstein's efforts, "What God hath put asunder, let no man join together."

Other weddings, shotgun and otherwise, include the resolution of the faith-reason opposites developed by Thomas Aquinas, a synthesis of Christ and Aristotle that to many seemed so perfect that for years it led people to believe that there was no more to learn. The same tension between opposites appeared in the tumultuous debates among geologists early in the nineteenth century and described in the next chapter. There were the Vulcanists, who emphasized the violent nature of the formation of the earth's crust versus the Neptunists, who attributed the development of the crust solely to the action of water, in particular the Flood that Noah survived. The correct answer, of course, was a synthesis of the two which became possible only when geologists learned to distinguish between igneous and sedimentary rocks.

We see the unity of opposites in the writings of Georg Hegel (1770–1831), and of Karl Marx (1818–1883), and sometimes in Soviet textbooks, because in principle it was supposed to be a way of thinking for the Soviet scientist. With Hegel it was the opposites "being" and "nothing" that gave rise to the concept of "becoming." Both philosophers presented a hierarchy of ideas, or physical situations that grew like a family's geneological tree and merged into a "greater flea" or "big boss" paradigm, as the society of citizens became a state which all must obey.

A final example of the unity of opposites, or the wedding paradigm, is a dichotomy that stemmed from the time of Newton, but became intense with the discovery of quantum mechanics. Was matter composed of particles or was it composed of waves? In the seventeenth and eighteenth centuries only the properties of light were in question—was it a lot of particles, as Newton had said, or a wave, as Huygens had maintained? This was resolved in the nineteenth century in favor of the wave theory—only waves show the interference phenomena that had been demonstrated for light. The answer seemed clear—matter was made of particles, and light was made of waves.

There was still a fundamental problem, however. It was a straightforward calculation to determine the overall spectrum (but not the finer details) of a hot body such as the sun. The result was fine for long wavelengths but dead wrong for short wavelengths. Not only that, it was impossible *in principle* to overcome this problem without introducing an entirely new physical quantity into the theory. Max Planck (1858–1947) showed how to do this and thereby discovered a fundamental constant of the universe, now always denoted by the letter h. It is very small and has the units of a spin—the spin on a baseball could be something like 10^{30} h. It was soon discovered that light was emitted in packets that were more like particles, each having a momentum equal to h/λ where λ is its wavelength. Each packet had to be a wave to have a wavelength, but it moved as if in many ways it really was a particle.

Louis de Broglie (1892–1987) asked a very simple but outrageous question: could the momentum of an ordinary particle also be equal to h/λ? That would mean that the particle had a wavelength, and therefore could behave as a wave, exhibiting interference effects, for example. The idea has dominated much of twentieth century physics, but not without arguments about whether an electron is *really* a wave or a particle, and what is it precisely that is waving. The basic equation of Erwin Schrödinger (1887–1961), that describes so well these de Broglie waves, can be put in a form in which it is almost indistinguishable from the equation that had been developed by the Irish mathematician William Rowan Hamilton (1805–1865) to describe the motion of a particle according to Newton's laws. The "almost" consists of an extra term proportional to h that appears in the wave theory but not in the particle theory. It therefore ensures that the two theories agree when h is very small compared to similar quantities that appear in the problem, as in the motion of a baseball. Operationally, the wave theory works magnificently, but there remain today several interpretations of what it all means.

16

Hell, High Water, and Evolution

> It is very disgraceful and mischievous and of all things to be carefully avoided, that a Christian, speaking of such matters as being according to the Christian Scriptures, should be heard by an unbeliever talking such nonsense that the unbeliever, perceiving him to be as wide from the mark as east from west, can hardly restrain himself from laughing.
>
> —Attributed to Saint Augustine (HFO, 27)

More than 2500 years ago, Xenophanes of Colophon (570-478) anticipated modern scientific thinking by denying that absolute truth exists, arguing that all knowledge of the world is approximate. He also introduced a "paradigm of the wedding" into the philosophy of the time—that everything was made of earth and water. More to our present interest, as noted in chapter 6, he recognized fossils for what they are, seeing them as evidence of earlier inundations of the land. Followed over the years by Theophrastos and Strabo (58 B.C.–24 A.D.) in particular, he had initiated the first studies of the earth's crust, the very beginnings of what we now call geology and paleontology.

Christianity soon put a stop to these early struggles to understand—the earth was corrupt and not to be studied, the fossils were either divine jokes or divine discards from the creation. Everything was created "in the beginning," and to suggest that fossils were not made at the same time as man was a sacrilege. We shall not repeat here the distorted and tragic logic of Catholic and Protestant leaders that we glimpsed in chapters 9 and 10.

"Here a doubt arises," wrote Leonardo da Vinci (daV 2, 208), "and that is: whether the deluge, which happened at the time of Noah, was universal or not. And it would seem not, for the reasons now to be given. . . ."

He then argues that, if it were a flood ten cubits deep over the whole sphere of the earth, the water would have nowhere to go after it stopped raining; the flood would never subside. He continued: "Why do we find the bones of great fishes and oysters and corals and various other shells and sea-snails on the high summits of mountains by the sea, just as we find them in low seas?" and "the stratified stones of the mountains are all layers of clay, deposited one above the other by the various floods of the rivers. The different size of the strata is caused by the difference in the floods—that is to say greater or less floods." "I find that of old, the state of the earth was that its plains were all covered up and hidden by salt water."—and other comments about his geological observations. Leonardo's 5,000 pages of manuscript were valued for centuries as collectors' items, but they were not systematically studied and translated into English until the nineteenth century.

The artist Bernard Palissy* (1510?–1589) examined fossils with the care of a craftsman and saw how similar they were to living organisms. He argued against the belief of Martin Luther and others that fossils were left over from the Flood, but that idea took over the thinking of many Christians for several centuries. Palissy also argued that volcanoes were due to a big fire under the earth, which is not too surprising, and that this fire was also responsible for earthquakes. His Swiss contemporary, Conrad Gesner (1516–1565) published illustrations of fossils—without understanding that they were petrified ancient animals—and worked incessantly collecting and classifying botanical specimens.

While scientific biology thus moved slowly forward, superstitious biology found a stimulus from Wolfgang Frantz (1564–1628), and religious biology from the Spanish Jesuit theologian Francisco Suarez (1548–1617). Frantz wrote a sacred history of animals, complete with dragons with three rows of teeth in each jaw, dragon number one being the devil himself. Suarez was the third well-known theologian to write about creation, following, but disagreeing with, Augustine and Aquinas. As noted in chapter 3, he argued that a "day" in Genesis meant just that, twenty-four hours, and that the universe was in fact created in six ordinary days. Suarez should be declared the patron saint of any institution that preaches creationism today. His diatribe against the English Reformation was burned at St. Paul's Cathedral by order of King James I, but his special creationist views were echoed by prominent British clerics and lawyers.

Parallel to these developments were the writings of Paul Petau (1568–1614), a chronologist at Orleans, France, and those of his nephew Denis (1583–1652), known as Petavius, regarded as one of the leading chronologists of his age—he was three years younger than James Ussher (1580–1656), who also appeared in chapter 3. These men turned the belief that everything was created about six thousand years ago into a blind faith. They did not agree with Saint Augustine, but they might at least have heeded his remark given at the beginning of this chapter. It would indeed be hard "to restrain from laughing" except that it is not funny.

John Pearson (1613–1686) was regarded, by the British at least, as the greatest theologian of the seventeenth century, honored ultimately as the Bishop of Chester. One of the claims to fame that led him to that bishopric in 1673 was the publication of his book *Exposition of the Creed* (Without it, as Sir Humphrey Appleby put it in *Yes, Minister*, it might have been "Long time no See"). Pearson declared the Egyptian records forged, at least those that were in conflict with Hebrew chronology. The earth and man were 5,600 years old at that time, he insisted; to believe otherwise was to deny a creed of Christianity. Sir Matthew Hale (1609–1676), Chief Justice of the King's Bench and a cruel witch-hunter, backed up these assertions later with the publication of *The Primitive Origination of Mankind Considered.* He continued the teaching of the British to pay attention only to the book of Genesis, and to ignore all those stories about fossils. In Germany, Augustus Pfeiffer (1640–1698), professor of Eastern languages at Leipzig, gave Lutherans the message that these mysteries can be understood only by a literal interpretation of every word of Genesis.

In France, the great pulpit orator Jacques Bénigne Bossuet (1627–1704) was literally flourishing with his flowery funeral orations. His oratory led to an appointment as the Dauphin's tutor (1670-81), a step up on his previous appointment as Bishop of Condom (a commune in the department of Gers, southwest France). Bossuet had an evil temper, and his treatment of his fellow cleric François Fénelon, Archbishop of Cambrai, was abominable. Bossuet turned both the pope and Louis XIV against Fénelon, whose writings were then banned. Later, the courageous Madame de Staël (1766–1817), persecuted by Napoleon, placed Bossuet at the top of the list of those who "can never be surpassed." He wielded a tremendous influence, supporting the belief in demons, vilifying Copernicus, and preaching the chronology of Ussher. His consistent attitude was to oppose anything that was unorthodox in any way.

In England, John Ray (1628–1705) collected many thousands of plant and animal specimens throughout England and the continent, offering a new classification of plants according to whether their first leaf came singly (as in grasses, palms, etc.), doubly (roses, oaks, etc.), or in clusters. He also catalogued English birds, fishes, metals, minerals, and insects. He recognized fossils as the petrified remains of dead, even extinct, creatures. Albrecht von Haller, the great Swiss physician and botanist, called him "the greatest botanist in the memory of man." Ray held firmly to the belief that each species had been fixed and that it remained unchanged over the centuries. As an Anglican deacon and priest, he saw nature in the way indicated by the title of the book that he published in 1691: *The Wisdom of God Manifested in the Works of Creation.* His younger contemporary, Nehemiah Grew (1641–1712), used his microscope to distinguish between pollen grains, and he was the first to recognize explicitly that flowers are the sexual organs of plants. Known mostly for his studies of comparative anatomy, Grew was one of the early Fellows of the Royal Society of London, serving for a period as its secretary.

A contemporary anatomist from Copenhagen, Nicholaus Steno (1638–1686), discovered the duct that leads from the parotid gland and opens on the inner surface of the cheek opposite the upper second molar—"the duct of Steno." His interests extended to geology and paleontology. He recognized that fossils were the forms of earlier living creatures, and he studied the layers of sedimentary rocks in Tuscany, arguing that those that had been displaced were formerly horizontal. He attributed the tilting of these layers and the formation of mountain ranges to volcanic activity and earthquakes.

Some primitive ideas about evolution were offered by Benoit de Maillet (1656–1738), but religionists thought him to be too scientific, and scientists judged him as too superstitious. The science lay in his insistence that direct observation of nature is a necessary prelude to understanding it. Maillet was bothered by the story of the Flood that left only Noah and his group intact—but surely the fish and marine plants survived. After the waters subsided, these must have become land animals and plants. The flying fish must have turned into birds, the sea lions into real lions, and the rest of mankind must have descended from mermaids and (presumably) mermen. Coming out of the water, they were all transformed very quickly to fit their new environment.

But the age of science was dawning, and not just the physical sciences outlined in chapter 15. The great Gottfried Wilhelm, Baron von Leibniz, argued that fossils were not freaks of nature but rather they were evidence of the history of nature and testimony to the immense age of the earth. He also communicated with Bossuet on the old faith-reason controversy, taking the point of view that faith must not abandon reason. In 1712, Leibniz ran into opposition from the Jesuits in his attempt to found the Academy of Science at Vienna.

Carolus Linnaeus (1707–1778), the Swedish botanist and taxonomist, was the son of Nils Brodersonia, an impoverished country clergyman, but somehow he found the money to study medicine. Botany was his chief love, however, and he traveled through Lapland and later to other European countries, collecting and classifying according to genus and species. His work provided a more detailed systematic taxonomy, a very useful first step in understanding plants and animals, as noted in chapters 5 and 6. Linnaeus was a very devout man, throughly versed in biblical theology, well aware that all of these creatures were believed to have been formed at the Creation and not to have changed thereafter. In all his writings, except the last, he insisted that the species were fixed. In the last edition of *Systema Naturae* that caveat was omitted. He was in enough trouble anyhow, since he had supported the view that plants, like most other living systems, propagate sexually. This led to the banning of his works in several European countries until near the end of his life, when papal permission was granted to discuss his ideas in Rome.

In France, Georges Louis Leclerc, Comte de Buffon* (1707–1788) was the first to argue seriously that species can change in response to the changes of the environment. He wrote that, "The least perfect species, the most delicate, the least active, the least armed, etc., have already disappeared or will disappear." And: "One is surprised at the rapidity with which species vary and the facility with which they lose their primitive characteristics in assuming new forms." (HFO, 193) Buffon began as a believer in the special Creation, was forced by the evidence to change his position, but was then forced by his academic colleagues to recant, as noted in the *In Memoriam* section. In later life he toned down his remarks for fear of more persecution. He became convinced, for example, that early races of men had fashioned the stone weapons and tools that had been found in many places over the years and that had been assigned a supernatural origin. He did not emphasize this opinion, however; he'd had enough trouble from both the Church and the Sorbonne.

Charles Bonnet (1720–1793) was a good biological scientist from the age of twenty to the age of thirty-four, making original observations on plants and insects. Unfortunately, he then became blind and was forced to speculate rather than observe. He saw inanimate and living matter as forming a continuous chain, from the smallest speck of dust, all the way up through man, to the highest cherubim and seraphim—an extension of Aristotle's concept of this hierarchy noted in chapter 6. Higher forms appear naturally from germs existing in the lower forms—these "germs" were little models, reminiscent of the ideas of Anaximandros. He introduced the word "evolution" to denote the movement of living organisms up this predetermined scale, and argued that they were urged up this scale by a continuing series of catastrophes. The French philosopher Jean B. R. Rabinet (1735–1820) pushed this speculation further by emphasizing the continuity of all things. He had found some stones that by chance looked like parts of animal and human bodies, "proving" that they were moving toward life. These changes, however, were supposed to occur continuously without the need for catastrophes, whether by divine intervention or otherwise.

Erasmus Darwin (1731–1802) was a physician whose passions were botany and writing poetry. He cultivated a large botanical garden, classified plants according to the method of Linnaeus, and published *Botanic Garden* and *Loves of the Plants* in verse form. He rejected supernatural explanations and suggested that evolution could occur by a series of accidents. Man may have developed from a family of monkeys with, for example, the thumb muscle accidentally changing and developing to produce our opposing thumbs. Inspired by Lucretius, he expressed it this way:

> Untipt with claws the circling fingers close,
> With rival points the bending thumbs oppose. . . .

It is a form of "survival of the fittest"—the survival of this and other changes that have come by chance. He recognized the changes that occur in plants and animals due, as he put it, to their own exertions and assumed that these changes were transmitted to their progeny. Henry F. Osborn (1857–1935) pointed out that this is the first record that we have of the discredited hypothesis that acquired characteristics are inherited.

Independently, it seems, and a little later, the same hypothesis was advanced in France by Jean de Monet, Chevalier de Lamarck (1744–1829), a student of Buffon. Like Erasmus Darwin, he studied medicine but was intrigued with botany; eventually he was appointed as botanist to the ill-fated King Louis XVI, moving to a professorship of invertebrate zoology in the year (1793) of the king's execution. In this role he pioneered studies of living and fossilized invertebrates, and developed his theory of evolution. Parts of the body of a living creature grow or atrophy according to their use, so over the years the neck of the giraffe would grow as it stretched to reach high branches, each giraffe therefore giving to its offspring, by inheritance, an ever longer neck. It doesn't work that way, but, as he put it, "it rests with those who do not accept it to substitute another, with equally wide application . . . but this I believe is hardly possible." (HFO, 225)

He and Erasmus Darwin were in accord with the teachings of Aristotle noted in chapter 6—the performance determines the form—although sometimes it led them astray. Lamarck, for example, believed that snakes had lost their feet because they found them useless. He went beyond Aristotle's static conception of the hierarchy of living creatures, seeing changes that resulted from use or disuse that in turn resulted from need or lack of it. He was also aware of current developments in geology, recognizing the longer time scales that these discoveries revealed. That also left more time for these biological changes to occur. Sadly, Lamarck went blind during the last years of his life, finishing his last book by dictation to his daughter.

This period, the eighteenth century, heady from the discoveries of

Newton and not yet aware of new fundamental problems, was the time for consolidation, for working out exquisite details in physics and astronomy. The Swiss mathematician Leonhard Euler (1707–1783) applied Newton's dynamics to the rotation of a rigid body, analyzed in detail the motion of the moon, contributed to hydrodynamics, and developed with Joseph Louis, Comte de Lagrange (1736–1813), a fundamental formulation of the laws governing the motion of a system of particles. All this and much more, in addition to a series of very practical developments in mathematics that are in wide use today.

We pass over other eighteenth century research in physics and astronomy except to note one idea that is relevant to changes that occur over a very long period. The possibility of the evolution of an initially inanimate part of nature, the solar system—as opposed to the belief that everything was put there "in the beginning"—was raised in 1755 by the German philosopher Immanuel Kant (1724–1804), more details being provided in 1796 by Pierre Simon Laplace (1749–1827). It offered a mechanism for the formation of the planets by condensation of the enormous atmosphere supposed originally to surround the sun. If the solar system could evolve, it wasn't put up there like that "in the beginning," and perhaps living beings could evolve also.

For biology and geology, it was a time for new ideas and the recognition that the geological timetable was essential for understanding the origins of biological systems. In geology, there was a conflict between those who emphasized the role played by volcanoes and earthquakes in forming the earth's crust, and those who saw it all as a consequence of Noah's famous flood. It was a conflict between the Vulcanists (or Plutonists) and the Neptunists, the god of the fiery element versus the god of the sea. Was the crust of the earth formed by the fires of hell or by the biblical high water?

The Scottish geologist James Hutton (1726–1792) proved that some rocks that we now call "igneous" were in fact formed under intense heat. One particular proof of this that he cited was an example in Glen Tilt of granite that had clearly flowed into a sedimentary area. His discoveries made it apparent that the earth was very much older than had been thought. His work received strong support from his student John Playfair (1748–1819) whose book, *Illustrations of the Huttonian Theory of the Earth*, appeared in 1802. Playfair stated that he had heard Hutton lecture on geological formations until "the mind grew giddy by looking so far into the abyss of time."

Just before he died in 1859, the astronomer J. P. Nichol wrote about Hutton's work:

> Sustained by phenomena at once palpable, numerous and conclusive, Hutton's important views rapidly made way among men of science: and notwithstanding their novelty and the stupendousness of the vista they open into the past, *the popular belief has now accommodated itself to them,* and revolts no more at the notion of the unfathomed Antiquity of the Earth than at the august thought that the myriads of lustres in the Firmament, are worlds. This consummation came not without a struggle, but *thanks to the "press"* which could not aid Copernicus, *the struggle in this case was neither severe nor prolonged.* Hutton may be said to have revealed the second of the two dimensions of the Material Universe—the dimension Time.
>
> (CB, p. 410)

The italic type was not Dr. Nichols's—he saw no need for it. Two centuries after the death of James Hutton, with well-developed techniques for determining geologic times from the abundance of radioactive isotopes of potassium, uranium, and other elements, popular belief has been stirred up to reject all of this knowledge.

Hutton had a French ally in Nicholas Demarest (1725–1815). According to the French naturalist Baron Georges Cuvier (1769–1832), as paraphrased by his biographer Mrs. R. Lee, Demarest was "the antagonist of Werner and the champion of volcanoes; he in whose discoveries originated the famous disputes between the Plutonians and the Neptunians, and which disputes not only placed the whole world between fire and water, but occasioned more animosity than any question which has hitherto agitated the learned world."

Abraham Werner (1750–1817), the German geologist, was the chief opponent of Hutton and Demarest, arguing that all rocks had been formed by the action of water, primarily from the flood that he supposed engulfed the whole world in the time of Noah (like almost everyone else in Christendom, he had been brought up on the Hebrew legends, to the exclusion of those from Greece, Sumer, or other lands and cultures, noted in chapter 3). For Dr. Werner, volcanic activity was of comparatively recent origin, and definitely postdiluvian.

A mass of geological data was collected by the English geologist and civil engineer William Smith (1769–1839). Like Leonardo, Smith learned some geology while working with canals and irrigation systems. He published a large *Geological Map of England and Wales* in 1815, the basis for a more detailed geological atlas with 21 maps of individual English counties. He spent his life traveling around England, collecting samples and information, with "scarcely any home but the rocks." When Trinity College, Dublin, conferred on him the LL.D degree in 1835, it was discovered that he was penniless. His friends saw to it that the Crown would grant him a pension of £100 a year for the last five years of his life.

In Germany, Johann Wolfgang von Goethe (1749–1832) was and is better known for his poetry than for his studies of the life sciences. He saw plants and animals as evolving, even up to man. "The conviction that everything must be in existence in a finished state . . . had completely befogged the century." (HFO, p. 267) He compared human and animal skulls and jaws, proving to himself the relationship between man and animals, but not convincing his contemporaries. As H. F. Osborn put it, "The germ of the idea of Evolution and the proof of this idea through compartive anatomy and embryonic development were contained in Goethe's first scientific writings and discoveries between the years 1781 and 1790." (HFO, p. 271)

William C. Wells (1757–1817) went from South Carolina to Edinburgh to study medicine, then on to London to practice it. Apart from his medical accomplishments, with erysipelas and dropsy in particular, he distinguished himself in physics by being the first person to explain the occurrence of dew. He also received a note of appreciation in the Historical Sketch that begins *The Origin of Species* (1859) by Charles Robert Darwin (1809–1882):

> In 1813, Dr. W. C. Wells read before the Royal Society . . . [a paper in which] he distinctly recognizes the principle of natural selection, and this is the first recognition which has been indicated; but he applies it only to the races of man and to certain characters alone. . . . He observes, firstly that all animals tend to vary in some degree and, secondly, that agriculturists improve their domesticated animals by selection; and then, he adds, but what is done in this latter case "by art, seems to be done with equal efficiency, though more slowly, by nature, in the formation of varieties of mankind, fitted for the country which they inhabit."
>
> (HCl 11, 11)

It does not seem relevant that a contemporary of Dr. Wells, the political economist Thomas Robert Malthus (1766–1834), published in 1798 *An Essay on the Principle of Population as it Affects the Future Improvement of Society*. He pointed out the mismatch between the rate at which a population tends to grow in numbers, and the rate at which food and other essentials become available, bringing members of the population inexorably into a "struggle for existence." Malthus was thinking about people—his predictions are sadly confirmed in Ethiopia today—but when Charles Darwin read this book in 1838 he saw that it applied to the evolution of all living creatures.

Gottfried Treviranus (1776–1837) took a strong stand against simple taxonomy that is devoid of any attempts to understand what all these facts and classifications meant:

> The teachings of Natural Science have long been standing isolated like the pyramids in the deserts of Egypt, as if the value of Natural History were not rather the application than the mere possession of facts. What have Botany and Zoology been hitherto but a dry register of names, and what man who has not lost his sense for higher work can find time for these gymnastics of memory?
>
> (ADW 1, 234)

On the other hand, he warned against idle speculation. Treviranus saw that, to survive, a species must produce more progeny when more of its members are killed. He saw that problems of animal and vegetable life were related, and he coined the German word for "biology" to cover both cases. He recognized the effects of physical environments on evolutionary descent, but his speculations about the spontaneous generation of some living beings were of course way off target, falling into the trap about which he had warned others. However, they were not, perhaps, as wrong as the claim of his contemporary, Lorenzo Oken (1779–1851) that mankind originated on a distant seashore, where water, land, and air meet, emerging fully developed like Sandro Botticelli's "Venus."

For centuries, people had picked up strange stone objects shaped like the heads of arrows, spears, and axes and attributed to them supernatural powers and origins. The most obvious conclusion was that they were left over from the wars of the gods or, later in Christian countries, the war between Satan and God. Andrea Cesalpino (1519–1603) studied fossils, and his pupil Michele Mercati (1541–1593) catalogued the geological collection at the Vatican. He saw these shaped stone objects as weapons or other instruments of man rather than "thunder-stones" of divine origin. In the years that followed there were many more discoveries of handmade implements—John Frere (1740–1807) found some in Suffolk that were even mixed with the remains of prehistoric animals—but often the finders were ridiculed by geologists and clergy alike. There were "theories" of course —people hadn't given up trying to understand one way or another and they never will. Dr. White said of these theories that, "each mixes up more or less of science with more or less of Scripture and produces a result that is more or less absurd."

One strong and articulate opponent of those who thought that these finds were evidence of a very early human existence was Baron Cuvier. He was greatly respected for his work on comparative anatomy, especially the anatomy of fishes, and for grouping together the classifications of Linnaeus into large divisions that he called "phyla" (from the Greek word for a "tribe")—living organisms of common descent. Cuvier classified fossils and named the pterodactyls ("wing-fingered"—compare helico*pter*, "spiral-winged"). Cuvier also wrote very sympathetic biographies of other scientists. He describes eloquently the search for a spiritual home made by Joseph Priestley—from Calvinism to Arminianism (Methodism), to the old heresy of Arius, thence to the heresy of Faustus Socinus (1539–1604) that we call Unitarianism. "The subtle shade between two heresies for thirty years occupied that head which was required for the most important questions of science and, without comparison, caused Priestley to write more volumes than he ever produced on the different species of air." (Lee, p. 104)

Other subjects of Cuvier's éloges included Henry Cavendish (1731–1810): "All that science revealed to him seemed to be tinctured with the sublime and the marvellous"; René Hauy (1743–1842), the father of mineralogy; Louis Claude Marie Richard (1754–1821), the French botanist who brought 3000 plants back from the Antilles and Guiana; Antoine, Comte de Fourcroy (1755–1809); and many others. Fourcroy was a chemist and natural scientist, an eloquent and convincing lecturer, and an honored member of the French Academy. Cuvier defends him against the allegation that he could have saved the life of Antoine Lavoisier, but was glad to be rid of a powerful rival. In fact, Fourcroy did save others but was powerless to save Lavoisier. Cuvier wrote, "If, in the strict researches we have made, we had found the slightest proof of so horrible an atrocity, no human power could have forced us to sully our lips by his éloge." (Lee, *Memoirs,* p. 109)

Cuvier was very tough on muddled thinking, particularly on the thinking of others. His comments about Lamarck, who laid himself open to that charge, were devastating. Ami Boué (1794–1881) found human bones with fossilized mammals buried in deposits near the upper Rhine River. He showed them to Cuvier, who later pronounced the human bones as "recent." A few years later (1830) Geoffrey St. Hilaire (1772–1844) read a report to the Academy of Sciences at Paris on the similarity between different animals, in this case between vertebrates and some cephalopod molluscs, like cuttlefish. Geoffrey felt that the similarity between different species represented a universal unity of nature. Cuvier pointed out that his facts were wrong, and strongly recommended to all biologists that they quit speculating. He had a point, of course. Yet thirty years later Charles Darwin was kinder: "Geoffrey St. Hilaire . . . suspected, as early as 1795, that what we call species are various degenerations of the same type. It was not until 1828 that he published his conviction that the same forms have not been perpetuated since the origin of all things."

The Baron Cuvier had become an Authority. He agreed with Bonnet about catastrophic changes and special creations, and was strongly opposed to any ideas that departed from this orthodox view. "If there be anything positive in geology," he wrote, "it is that the surface of our globe has been the victim of a great and sudden revolution, the date of which cannot be carried back further than from five to six thousand years." Here he encountered strong opposition from Charles Lyell.

The main thesis of the British geologist Sir Charles Lyell (1797–1875) was that geological changes occur uniformly according to immutable laws: "Never was there a dogma more calculated to foster indolence and to blunt the keen edge of curiosity than this assumption of the discordance between the ancient and existing causes of change." (HCl 38, 416) That meant that there was no place for divine miraculous intervention, a point of view that naturally led to a great deal of criticism in certain quarters. There were those besides Cuvier who argued that the shape of the earth was determined by a series of catastrophes; Lyell took the view that it was one continuous change. As with the Vulcanist-Neptunist controversy, there was some truth on both sides, although the truth about the age of the earth lay much closer to Lyell than to Cuvier. In 1825 Cuvier published his pseudo-theological views in his *Theory of the Earth.* A few years later (1830–1833) Lyell's *Principles of Geology* became the standard text on the subject, and the Baron was discredited.

Lyell later wrote about the geology of North America after extensive travels there. He was not one to jump to conclusions about the antiquity of the human species. In 1847, the French archaeologist Jacques Boucher de Crèvecoeur de Perthes (1788–1868) discovered some prehistoric flint hatchets near Abbeville, at the mouth of the River Somme in northern France. The find did in fact show that man existed far back in the Ice Age, thousands of years before 4004 B.C. Unfortunately, Boucher had shown a tendency to jump to unjustified conclusions and the geology establishment would not believe him. Lyell and others visited Boucher and examined his finds and the circumstances under which they were made. Lyell was convinced, and because of his reputation for careful analysis, other geologists were convinced too. Later (1863) his *Antiquity of Man* established beyond reasonable doubt that the chronology of the earth and its inhabitants that was published by Suarez, Ussher, Petavius, Hale, and Pfeiffer, and was to be publicized by a long list of others throughout the twentieth century, was dead wrong.

The general reaction among the prominent clergy, however, was either to ignore these developments and hope that they would go away, or to use the pulpit for verbal bullying of these geologists and other scientists who were trying to lead people from the theological "truth." More and more evidence of man's antiquity was literally unearthed—human skeletons mixed with bones of extinct animals, and products of human workmanship found in deposits many thousands of years old. Vilification grew worse for British geologists like Mary Somerville (1780–1872), and American geologists like Edward Hitchcock (1797–1864), himself a pastor of a Congregational Church—he had found fossil footsteps in a red sandstone deposit in Connecticut. Pressure was so great that the British Egyptologist Sir John G. Wilkinson (1797–1875, identical with the birth and death years of Sir Charles Lyell) modified his data to fit the chronology of the Bible. Karl von Raumer (1783–1865) announced from his professorship of geology at Breslau that fossils were the remains of imperfect plant embryos.

In 1836 William Buckland (1784–1856) published his *Geology and Mineralogy Considered with Reference to Natural Theology*. In that and other writings this Oxford professor and later Dean of Westminster tried to reconcile recent discoveries with the text of the Bible and with Cuvier's anti-evolutionary thesis. Nicholas Wiseman (1802–1865), Cardinal and Archbishop of Westminster, saw the inevitable and, perhaps remembering St. Augustine's warning, began to embrace these new findings. Mindful of the way that the Roman Catholic Church had made a grave mistake in its treatment of Galileo, although not admitting it, the Cardinal propagated the idea that these geologic truths were what the Church had been saying all along. Nevertheless, Dr. White quotes some epithets hurled at the geologists from various pulpits and recorded in Lyell's *Principles of Geology*: "infidel," "impugner of the sacred record," "assailant of the volume of God." *Geology and Scripture*, by Pye Smith, lists some ecclesiastical descriptions of geology: "a dark art," "not a subject of lawful enquiry," "dangerous and disreputable," "a forbidden province," "infernal artillery," "an awful evasion of the testimony of revelation." Geologists were "attacking the truth of God." No longer allowed to kill the scientists, some churchmen resorted to these obscenities.

Others, like the German theologian David Friedrich Strauss (1808–1874), were condemned by their colleagues for departing from orthodoxy. In a series of books, Strauss described the conflict of those times between science and Christian dogma, and advanced the idea that much of the biblical history is mythical. He was forced to leave his professorship at Zurich because of the not-surprising opposition to this thesis. In France, the great biblical scholar Joseph Ernest Renan (1823–1890) was removed from his professorship of Hebrew at the Collége de France after the publication of his *Life of Jesus* (1863) (over 300,000 copies sold). He excluded the miraculous in this work and throughout his *Origins of Christianity*—he even excluded any form of divine intervention. The conflict with science could be resolved only by recognizing these views of the Bible, or by denying well-documented scientific discoveries.

The Scottish naturalist Robert Chambers (1802–1871) must have feared reprisals because in 1844 he published his *Vestiges of the Natural History of Creation*—anonymously. It was not until 40 years later, when this book went into its twelfth printing, that the editor released the author's name, since by then Chambers had died. He had rejected the miracles of fundamentalism; instead he saw evolution as unfolding according to the laws of nature. Some of his ideas were weird, and some were crazy enough to be a century before their time, such as his belief that there are many other worlds similar to our own circling distant suns. That made him the first exobiologist. His *intro*biology wasn't as good—he thought that man was descended directly from the apes rather than having common ancestors with them. He had the imagination, however, to see that mankind will evolve further. (Some evidence for that may exist now with the occasional birth of an incredible genius who far outstrips the rest of us.) Chambers certainly paved the way for a new look at creation, and Charles Darwin (1809–1882), grandson of Erasmus, gave him credit for that.

He was not the only one to whom Darwin gave credit. The "Historical Sketch" that appears at the beginning of *The Origin of Species* lists many of them. Darwin was meticulous in analyzing the work of others. Although isolated on the Beagle for five years he moved from belief in the fixity of species to the formulation of mechanisms for evolution, with a minimum of external stimulus except for the direct observation of nature.

Darwin mentions Charles Victor Naudin (1815–1899) who published *Hereditary Traits in Plants* in 1852. Naudin recognized that many species have been produced by nature from, he believed, a small number of types that existed initially. He recognized the similarity between artificial breeding and the way that nature works, always controlling the changes like some external godlike breeder.

Darwin also refers to Herbert Spencer (1820–1903) whose latest book at the time was *Principles of Psychology*. Earlier, Spencer had argued that the concept of "creation by evolution" was much superior to that of "creation by manufacture," as he labeled the special creation hypothesis. "Any existing species—animal or vegetable—when placed under conditions different from the previous ones, immediately begins to undergo certain changes fitting it for the new conditions. There is at work a modifying influence of the kind they assign as the cause of these specific differences." Spencer tied evolution to sociology, and with the evidence for evolution that came from Darwin's work he applied with more conviction the idea of "survival of the fittest" to human society. He tried to do too much—he took this theme and applied it to *everything*, psychology, political science, and ethics.

The story of Darwin has been told so often that there is no need to repeat it here, except to emphasize that *The Origin of Species*, published in 1859, has been called the greatest scientific achievement of the nineteenth century. I do not know how to compare it with medical advances, or the discovery of entropy, or with J. C. Maxwell's *Treatise on Electricity and Magnetism* (1873) or the understanding of the mechanical equivalent of heat by J. R. Mayer (1842), and in more detail by J. P. Joule (1847). (Unfortunately, Mayer was given the cruel treatment for the insane, outlined in chapter 13, while Joule was given the "treatment" by the Establishment noted in chapter 15—he had to publish his first results in a *newspaper*.)

Like Mayer and Joule, the evolutionists Darwin and Alfred Russel Wallace (1823–1913) had similar ideas, in their case arising from their travels. In 1831 Darwin began his famous voyage on the Beagle to return in 1836; Wallace traveled to the Amazon (1848–1852), and to the Malay Archipelago (1854–1862). Both read Lyell and Malthus, both were trained and very observant naturalists, both were aware of some of the current ideas about evolution, and both saw natural selection as the mechanism behind it. Darwin was older and got to it first; like Newton he did not want to publish until he had worked out a convincing set of details; like Newton also he learned that another scientist was working on the same idea. In 1858 they sent two papers together to the Linnaean Society, announcing their independent but identical conclusion that evolution occurs through natural selection of changes that arise by chance. The wording in parts of their two papers was almost identical. Darwin: "There is in Nature a struggle for existence. Any minute variation in structure, habits or instincts, adapting that individual better to the new conditions would tell upon its vigor and health. In the struggle it would have a better chance of surviving." Wallace: "The life of wild animals is a struggle for existence. If any species should produce a variety having slightly increased powers of preserving existence, that variety must inevitably in time acquire a superiority in numbers."

Darwin had a wealth of information available and he was able to publish *The Origin of Species* during the following year. Wallace's book, *Natural Selection*, appeared in 1870, followed by *Darwinism* in 1889. The titles of some of his other books give a hint about his thinking: *Miracles and Modern Spiritualism* (1874), *Vaccination, A Delusion* (1898), *Is Mars Inhabitable?* (1907). Darwin, however, published treatises on orchids and other cultivated plants, but then shocked the world in 1871 with his *Descent of Man*. It wasn't just evolution of plants and animals, he showed, but we too are part of the evolutionary process.

Moses and Aristotle had taught that plants and animals were there to be of service to man; Christian patriarchs had taught that man was lord of the earth, and that the earth was the center of the universe. All of this had been built in 144 hours, nearly 6,000 years ago. Galileo and others had shown that the earth is not the center of the universe, Lyell and others that the earth was much more than 6,000 years old, and that geological processes take many thousands of years, not a few hours. Now

Darwin argued that man was not specially created—over an enormous number of years he had descended from animal ancestors. People were upset because it went right in the face of what they had been hearing; the clergy were upset because it went right in the face of what they had been telling. It is often an aspect of human nature that, the closer you are to being 100 percent wrong, the more stridently you claim to be 100 percent right.

Some biologists were not convinced—indeed they argued strongly against the theory of evolution. Jean L. R. Agassiz (1807–1873), the Swiss zoologist and geologist, came to teach at Harvard, and later he became a U.S. citizen. He wrote extensively on fish from various parts of the world, on fossils, and on glaciers, establishing the former existence of an Ice Age. His work justly brought him renown; he had been perhaps the most brilliant of Cuvier's students. Like his master, he was completely opposed to evolution, and believed that new species could arise only through divine intervention. Like Priestley, who never gave up his belief in the phlogiston that his own research helped to disprove, Agassiz never gave up his disbelief in evolution, which his own research helped to establish. The British ornithologist Philip Henry Gosse (1810–1888) wrote about his observations in Canada and Jamaica, and added his belief that geologic strata were formed all at once, indeed that the whole earth came into being in its present (at that time) state. He tried to reconcile scientific findings with his fundamentalist creationist conviction by inventing two types of time, the real time of Genesis, and the not-so-real time of the appearances by which geologists were being misled.

Unfortunately, any attempt to reconcile a literal interpretation of Genesis with the findings of geologists and biologists is doomed to make matters worse by its sophistry. William Ewart Gladstone (1809–1898), for many years the Prime Minister of Britain, may have found this out when he entered into public debate in 1885 with Thomas Henry Huxley (1825–1895) on the subject of evolution. By that time Huxley was a great supporter of Darwin, although he had naturally taken some convincing. Gladstone tried to show that at least there were hidden truths in Genesis, "affirmed in our own time by natural science." Genesis told him that the populations inhabited the waters, the air, and the land, in that order; Huxley pointed out that it was not that way at all, and presented the evidence. Gladstone also objected to the way that the theory of evolution implies that the world works according to law, with no need for divine intervention. Herbert Spencer reminded him that this objection could also be raised to the established theories of Newton—did he want to throw them out too?

Many of the clergy, both Protestant and Catholic, tried specious logic and verbal abuse to convince the public that evolution is an evil theory, and a few scientists argued that there was no conflict between science and Genesis. Men are very clever at twisting words to suit their ends. The English biologist St. George Jackson Mivart (1827–1900) tried to reconcile evolution with Roman Catholicism—and got into serious trouble with the leaders of both sides. He revived a scholastic argument of "individualism" —the development of the individual from the universal—a vague metaphysical concept that led the evolutionists to reject him, and the Catholics to excommunicate him. In Canada, Sir John William Dawson (1820–1899), professor of geology at McGill University, was an expert on fossils of plants and early reptiles, particularly those found in Canada. He opposed the idea that man had descended from early animal ancestors, and made up a table in which he tried to show that the geological and theological orders of creation coincided. The English biblical scholar Samuel Rolles Driver (1846–1914) pointed out many cases in which "what was created before what" was different in the two versions. But the subject of the discussion was not whether the account in Genesis is literally true, but instead whether the account gives any valid forecast of later geological findings.

At that point, as we have seen, scientists felt that they had won the battle and that was the end of it. Little did they dream that a century later many millions of Americans, Australians, South Africans, South Americans, and others would have been persuaded to go the authoritarian route and believe in the literal meaning of every word in the Bible, despite overwhelming evidence to the contrary. One of many reasons for this is that some evolutionists have lost credibility by jumping too quickly to erroneous conclusions, as in the case of the Piltdown Man cited in chapter 1. There have been other examples—wild extrapolations from the slimmest evidence, as in the case of a tooth found in Nebraska, or another found in Colorado, which was diagnosed as being from an ancient human, but was later found to be of animal origin. As noted in chapter 1, there have been times when experts have been too speculative, and others when they have been too cautious. You have only to look at a map of the S-shaped Atlantic Ocean to suspect that its two curved sides once fitted together. In 1912 the German physicist Peter Paul Wegener published the hypothesis that the continents had slowly drifted apart. "Nonsense," cried the Establishment. "Utter nonsense!"

Of course, it is hard to believe that the magnificent and complicated structure of a human, or even of a sparrow or a gnat, could evolve by chance. Throwing molecules at each other, and hoping to form one of these creatures that way seems like a hopeless operation. So it would be if *simple chance* decided the outcome. Instead, it is *rewarded chance*, and that makes all the difference in the world. If you were to play roulette with the condition that one particular numbered slot became just the smallest amount larger each time the ball landed in it, uniformly compressing the others, you would eventually find that random chance was favoring that number. Ultimately that number would come up practically every time. In real life, a characteristic that gives an animal the slightest edge over its fellows may help that animal to prosper, and perhaps make it more likely that his offspring will inherit that characteristic. Some recent ideas on the mechanisms of evolution are discussed in the next chapter.

It is not easy to be close to 100 percent correct in scientific studies. The exciting thing about it is that it is a continuous search. Some incorrect paradigms have been handed down from one generation to the next, but they are eventually discarded when a better and more accurate model is found. But for some people it is much easier to accept the assurance that the Bible is 100 percent correct—it saves a lot of intellectual effort and makes it no longer a struggle to believe that you understand.

1. VOLTAIRE

2. BENJAMIN FRANKLIN

3. ISAAC NEWTON

After centuries of superstition, witch burnings, persecution of heretics, and suppression of non-traditional ideas, a few brave and brilliant thinkers finally ushered in the Age of Enlightenment. Voltaire (1), "the apostle of unbelief," seriously challenged many long-established beliefs of Christendom and defended victims of religious intolerance. His new ideas and attitudes helped bring about the social and intellectual paradigm shifts that culminated in the French Revolution. His contemporary in America, Benjamin Franklin (2), had an equally active mind and independent spirit, and like Voltaire he served as a catalyst for a revolution—on this side of the Atlantic. His scientific interests led to important discoveries about the nature of lightning and electricity, and many useful inventions like the lightning rod.

But the man whose ideas probably had the most profound effect was Isaac Newton (3). His *Principia Mathematica* irrevocably altered our understanding of the natural world and laid the groundwork for modern physics and mathematics. The ideas of Ptolemy and Aristotle, which for millennia had been the basis for man's conception of the universe, were finally abandoned.

1. CAROLUS LINNAEUS
2. COMTE DE BUFFON
3. BARON CUVIER
4. GOETHE

Other important scientists of the age were Carolus Linnaeus (1), whose systematic taxonomy of plant life established the science of botany, and Georges Louis Leclerc, Comte de Buffon (2), whose ideas about evolution anticipated Darwin and whose suggestions about stone-age man would later be proved by anthropologists. In the next generation, Baron Cuvier (3) wrote ground-breaking works on comparative anatomy and the classification of fossils. His contemporary, Johann Wolfgang von Goethe (4), who is known mainly for his literary genius, also contributed to science by comparing human and animal skulls and thus anticipating the theory of evolution.

17

Past, Present, and Future

Thought is the care with which men consider things past, and those present, and those which are to be.

—Alfonso X (1221–1284), King of Castile and Leon

We care nothing for the present. We anticipate the future as too slow in coming, as if we could make it move faster; or we call back the past, to stop its rapid flight.

—Blaise Pascal (1623–1662)

I believe the future is only the past again, entered through another gate.

—Sir Arthur Pinero (1855–1934), *The Second Mrs. Tangueray,* Act IV

"Why worry about all these dead people? We should look to the future. It's a waste of time to wonder about the issues of the past. As we enter the twenty-first century, it makes no sense to look backwards." Many people in our pragmatic society have this attitude, but there are those of us who find it fascinating to trace ideas held by our ancestors, and to obtain thereby a more complete view of the ideas and problems of today.

There are two reasons why I believe it is important to be aware of the conflicts recorded in this book. First, we are often reminded by churchmen, particularly those of the Roman Catholic persuasion, that Christian martyrs were persecuted, tortured, and murdered for their faith. It is right that we should remember them, because their bravery and commitment to their belief are a permanent indictment of intolerance. The Jewish people also remind themselves and others of their terrible trail of persecution from their captivity in Egypt to the Nazi horrors and beyond. They do not want the world to forget, nor should they. To remember is to be better prepared to prevent other atrocities from happening. Black people too have their history of persecution, shorter, but still dehumanizing. They also are becoming more aware of their past; many will make it a springboard for their future.

Religious and racial oppression is compounded by political oppression—even implemented by it. We honor those throughout the world who have fought for political freedom, and we mourn those who died in the fight. The fight for intellectual freedom, however, has not involved as many people, and the scientists, philosophers, and other scholars who have been persecuted for their beliefs tend to be forgotten. They don't number in the millions, or even the thousands; still, the section *In Memoriam,* indeed this whole book, is a tribute to them.

The second reason for retelling these conflicts is that they have not disappeared. Of course, the laws of Western Civilization have changed; it is no longer permissible to torture scholars or anyone else for disagreeing with the beliefs of a particular church. The Catholic Church lost a great amount of respect because of its treatment of Galileo and others, and it has wisely backed away from its former infallible position. It no longer makes pronouncements about the physical sciences—except for matters like the origin of the universe—and indeed it is careful to avoid statements (for example about the age of the Shroud of Turin) that can be proved or disproved unambiguously by scientific measurements. Although the televised exorcism on April 5, 1991, causes one to have doubts, the Catholic Church is genuinely trying to move into the twentieth century.

The same cannot be said about the growing fundamentalist movement, now some 25 million strong in the United States. The very fact of its growth shows that it serves a human need. For one thing, it is more exciting to many people than moderate Protestantism, which is not gaining members, and liberal Protestantism, which is losing them. One cannot but applaud the way that these conservative Protestants bring joy to their lives—but they should keep their hands off of science and other scholarship and the teaching thereof. Their attitude towards science is akin to that of the seventeenth century Catholic church. They don't worry about falling objects and the motion of the earth like people did then, but they are blind to anything that isn't in the Bible. The subject matter of this book's chapter 16 is seen by them as a threat to their faith, just as surely as Galileo's discoveries were seen as a threat by the Catholic Church of the time.

Education is in enough trouble in America without having these people purging our school textbooks, and demanding that the myths of the Old Testament be taught to our young people as science. Their intellectual fascism is a genuine threat to this country. It is easy to dismiss it as unimportant, noting that nearly all of these problems were settled during the nineteenth century conflict between religion and science. And so they were, but there is a growing revival of these primitive ideas. Passions are very strong, and it would be a disaster if these people gained more political power. We do not want to see the *In Memoriam* list extended.

To illustrate how the problems and ideas of the past are often significant today, I have assembled some clippings from recently published newspapers and magazines, ranging from news items to the opinions of various individuals. These quotes are arranged according to the chapter of this book to which they are most relevant, although some apply to several chapters. I have also added some comments of my own.

Chapter 1: Gods, Goddesses, and Time

> Last summer, the Supreme Court ruled that Allegheny County couldn't display a lone creche inside its courthouse in Pittsburgh, because it gave the appearance of an unconstitutional government endorsement of a religion. At the same time, though, the court said a menorah celebrating Hanukkah in front of Pittsburgh's City-County Building was all right because the display included a Christmas tree and a sign by the mayor about freedom. Some people, like the writer across this page [see below] say these rulings are anti-Christian and an attack upon a Christian nation. They say creches should be allowed everywhere without any restriction. That zealous view of our society offends millions of people—Jews, Buddhists, Moslems, believers and nonbelievers. It would rob us of our heritage of religious freedom by having government tell us what we should worship. . . .
>
> (Editorial, *USA Today,* December 22, 1989)

> The fanatical, scorched-earth policy, insisted on by some, which seeks to expunge from every aspect of U.S. "public" life any mention or symbol of Christianity, is nothing new. . . . Leaders of the Third Reich were particularly vigorous in their efforts to de-Christianize various public celebrations. . . . Reich propaganda minister Joseph Goebbels and Reich leader of the dreaded SS Heinrich Himmler, agreed that Christ had to be removed from Christmas. . . . Sound familiar? You bet it does. This is precisely what has been happening for decades in the U.S.A.—and with a vengeance that would undoubtedly have Goebbels and Himmler smiling if they were still alive, instead of burning in hell eternally. Make no mistake about it, friends. What this annual fight over creches on public property is about is just one more skirmish in the ongoing war against Jesus Christ in this country.
>
> (John Lofton, *USA Today,* December 22, 1989)

You may care to judge on which sides the editor and Mr. Lofton would have been in the conflicts between St. Cyril and Hypatia, Saint Bernard and Abelard, Jerome of Ascoli and Roger Bacon, Bishop Tunstall and William Tyndale, John Calvin and Michael Servetus, Theodorus Beza and Sebastianus Castellio, The Inquisition and Giordano Bruno, Galileo, and Tommaso Campanella, the Sorbonne faculty and Georges de Buffon, and the court and John Thomas Scopes (see *In Memoriam*). Mr. Lofton would certainly have been on the side of Saint Bernard of Clairvaux, because Bernard also used guilt by association by comparing Abelard to the "heretics" Arius Pelagius and Nestorius, rather than to Goebbels and Himmler.

Short biographies of approximately 5,000 saints of the Catholic Church are given in John L. Delaney's *Dictionary of Saints,* and there are almost as many minor saints not listed. The practice of revering saints began with the commemoration of the martyrdom of St. Polycarp in the year 155 or a few years later; it spread to other martyrs, was given theological significance by Origen, and then was extended to those "who now are in heavenly glory. . . . [They] may be publicly invoked everywhere, and their virtues during life or martyr's death are a witness and example in the Christian faithful" (quote from the *Modern Catholic Dictionary*). There is at least one saint each for actors, astronomers, boy scouts, cab drivers, comedians, dentists, gardeners, grocers, journalists, pawn brokers, poets, scientists, even for eye trouble, or television, and of course much more. Many of them are the patron saint of a country, many are identified in paintings by their characteristic symbols, from tree or candle to dragon or lion. This is the modern equivalent of the hierarchies of the pagan gods. Like some gods, these saints were real people once; now they are endowed with individual specialties and heavenly glory.

> In no other country has the "holy war" against social revolutionaries and Catholics won over more souls than in Guatemala, the most populous country in Central America. About 400 sects have overrun Guatemala with a network of tens of thousands of evangelical temples, and it is expected to become the first predominantly evangelical republic in Latin America by the end of the 1990's. . . . "Salvation lies in a reformation," says Francisco Bianchi, general secretary of the Verbo (the Word) sect. "Prosperity begins with Luther."
>
> (*World Press Review,* March 1991, p. 30, translated from *Der Spiegel,* Hamburg)

Chapter 2: Birth, Death, and Resurrection

> It is Christianity's most irreducible tenet: On the third day, Jesus arose from the dead. From the very beginning, Christians have proclaimed the bodily resurrection as a validation of all that Jesus taught and all that they believe. . . . Yet despite its centrality, the resurrection has been, for believers and nonbelievers alike, one of the most problematic of Christian doctrines. . . . For modern readers of the Gospels, as for the disciples on the first Easter, the resurrection is largely a deduction drawn from two pieces of data: The discovery of an empty tomb and reports of postcrucifixion appearances by Jesus in Jerusalem and Galilee. . . .
>
> There are discrepancies in the accounts. Were there three women visitors (to the tomb) (Mark), two (Matthew), or one (John)? Did they arrive before dawn (Matthew, John) or after (Mark)? Was the stone rolled away after (Matthew) or before they arrived (Mark, Luke, John)? Were angels present (Matthew, Mark and Luke, yes; John, no)? . . . Ultimately the events of the "third day" in Jerusalem must remain, as they have for 2,000 years, in the realm of things unprovable—matters, like the very existence of God, to be grasped only by faith.
>
> (*U.S. News and World Report,* April 16, 1990, p. 53)

On August 7, 1989, a PTL-TV audience was whipped into frenzies as a narrator described to them the battle between Satan and Jesus. It was told as one might report a wrestling or boxing match between two professionals. Satan makes some evil remarks and Jesus replies, "Go ahead, make my day." But a blow of death sends Jesus to the ground. It looks to be all over, but Jesus rises, demolishes Satan. He has won. He has won. (A thunderous cheer from the crowd.) Satan is defeated and Jesus is the Champion!

> Within minutes of Neil Armstrong's first steps on the moon, I—working in a remote corner of Turkey—discovered the most famous Temple of Aphrodite.—We found a concert hall and a senate house; a monumental altar on which mystery rites had been celebrated and a sanctuary of those rites. . . . A woman named Love had discovered the lost temple of the goddess of love. A man had walked on the moon at the same time. And Aphrodite's worship had always been connected with the moon. . . .
>
> (Iris Love, *Parade Magazine,* July 9, 1989, page 16)

> Dr. Michael Harrington, acting chair and associate professor of philosophy and religion at Ole Miss, said prior to the formation of the first-century church the Europeans observed a winter solstice festival—Mithrism, a rival religion of Christianity, also observed a December 25 celebration of "the

undying sun," he said. ". . . In order to keep members, the church gradually accepted the festivals but changed them to celebrations of the undying Son, Jesus of Nazareth."

(*Mississippi Press,* December 14, 1990)

Chapter 3: Creation, Fall, and Flood
Chapter 16: Hell, High Water, and Evolution

A few years ago, the state of Louisiana passed a law requiring that "creation science" be taught in public schools in any class in which the theory of evolution was part of the curriculum. The Supreme Court of the United States struck down this law, rightly pointing out that creationism is a religious belief, and that to require this belief to be taught in the schools is unconstitutional.

The Supreme Court refused to hear the final appeal of the seven evangelical Christian families in Tennessee who maintained that their children's religious freedom was violated when they were required to read textbooks that expressed the "secular humanist" outlook. Among the reading selections objected to were *The Wizard of Oz,* and *Rumplestiltskin.* By refusing to hear the appeal, the Supreme Court left unchanged the August 1987 decision of the Court of Appeals. That decision had overturned Federal Judge Thomas G. Hull's ruling that the children could be excused from classes and learn to read at home!

In 1982, a suit was brought in federal court in Mobile, Alabama, asking that the state's school-prayer statutes be declared unconstitutional. Chief Judge W. Brevard Hand granted an injunction, thereby abolishing school-prayer activities, but with a twist of logic that would have done credit to Machiavelli, he wondered out loud why other religions should not get the same treatment. After all, he noted, "It is common knowledge that miscellaneous doctrines such as evolution, socialism, communism, secularism, humanism, and other concepts are advanced in the public schools. These are, to the believers, religions; they are ardently adhered to and quantitatively advanced in the teachings and literature that is presented to the fertile minds of students." The six hundred parents who had been admitted as defendants by Judge Hand got the hint and immediately sued to have the "religions" evolution, materialism, atheism, humanism, etc. removed from school teaching.

The learned judge then lifted his injunction against school prayer, arguing, it is hard to believe, that this application of the First Amendment did not apply to the states. He then arranged for the defendants to become plaintiffs, and the school executives to be defendants in a suit demanding that the "religion" of "secular humanism" not be taught in schools. He eventually found that forty-four school books established a religion of secular humanism, and that, indeed, was unconstitutional. Those forty-four textbooks in social studies, history, and home economics were subject to an order that prohibited their use by the state. Fortunately, Judge Frank Johnson of the Eleventh Circuit reversed this decision almost immediately on August 26, 1987, two days after the U.S. Court of Appeals reversed the district and decision in the Tennessee textbook case. It is clear that these well-meaning but thoroughly misguided Christians are regrouping, and will continue to push America into the straitjacket of their narrow views. Fortunately, they don't have the power of earlier religious leaders. They are forced to prosecute rather than persecute.

Perhaps we can learn tolerance from other countries. For example:

> Leading church groups in Australia supported Bill Hayden's decision not to swear on a Bible when he was made Governor-General of Australia in March. Hayden, a self-described atheist, affirmed his allegiance to the Queen and promised to carry out faithfully the duties of the vice-regal office at his swearing-in ceremony. He is believed to be the first Governor-General to take advantage of this option, provided under the Australian Constitution.
>
> (*Free Inquiry,* 9, no. 3, 1989)

(Let it be known, however, that Australia is second only to America in its growth of fundamentalism.)

A pastoral letter warning the Roman Catholic population in Alabama and Mississippi against "fundamentalism" was officially released yesterday by the four ruling bishops in the two states. In the letter, they contend that the "Fundamentalist stance of literal interpretation of Scripture by each believer violates the history and tradition of Scripture itself." They say it offers "an unreasonable certainty about the meaning of Scripture texts regardless of their content. Fundamentalists, because of their literalist mindset, have led others by using brief Scripture quotations taken out of context to world views and judgments very much opposed to our Catholic understanding." (Or, for that matter, to our scientific understanding.) (Press Release, July 16, 1989).

> A resemblance to the Bible's recounting of the descendants of Adam appears in the clay prism known as the Sumerian king list, compiled in the second millennium B.C. . . . This Sumerian list records legendary kings of fantastic

> longevity: in Eridu, A-lulim is described as ruling 28,800 years. Two kings ruled for 64,800 years. Then, we are told, "the flood swept over the earth."
>
> (Nelson Glueck, *Horizon,* vol. 2, no. 2, November 1959, p. 4)

After much of this book had been written, I sent $19.95 plus to Barnes and Noble for a copy of their *Wall Chart of World History.* The books they sell are usually very interesting and reliable. I had gone through considerable effort to check each date referred to in this book, but I thought that a chart like that would help me to double-check.

I was appalled. This chart of "world history" gives the history as seen by the fundamentalists. It begins with 4004 B.C. From then until 1426 B.C. it displays a horizontal family tree from Adam to Moses. Long thin sausage-like markings denote the life-spans of the record holders (Methuselah 969 years, Adam 930—they talked with each other for 243 years, it tells us—Cainan, Enos, Noah, and Seth, all over 900 years), as well as those less fortunate, e.g., Abel. Apparently not much else was going on until the descendants of Noah began to multiply, living only about 200 years each, on the average. About the same time, the rest of civilization sprang up, apparently originating in the Tower of Babel. The rest of the chart is quite reasonable, although mythical and real people are not clearly distinguished, and we are warned that "exaggerated chronologies are common to a large number of nations; but critical examination has [at any rate *in all cases but one*] demonstrated their fallacy." It emphasizes "Difficulties in Egyptian Chronology" (because it sometimes conflicts with the sacrosanct Hebrew chronology), completely ignoring twentieth-century studies that date events in Egypt back to 2500 B.C. with a probable error of only twenty-five years. "The chronological system of ARCHBISHOP USSHER, followed here, and upon which a large majority of our histories are founded, enables us to obtain a *correct* outline of the sequence of events and of the relation of one period to another!" The chart does not, however, quote a younger contemporary of Ussher, Sir John Marsham (1602–1685), the Egyptologist who wrote in 1672: "The most interesting antiquities of Egypt have been involved in the deepest obscurity by the very interpreters of her chronology who have jumbled everything up so as to make them match with their own reckonings of Hebrew Chronology. Truly a very bad example and quite unworthy of religious writers." (ADW 1, 255) This was more than three hundred years ago, and they are still doing it.

Fortunately, there is another World History Chart published by International Timeline which is much more accurate. It traces the history of fifteen parts of the world, also beginning at 4,000 B.C., but with reference to humans living thousands of years earlier and with clear distinction between periods of "educated speculation" and those of "historical or relative historical accuracy." This chart mentions the Hebrew patriarchs and notes the approximate times when they lived but makes no claim for their alleged longevity. The other chart does a great disservice to anyone who is trying to learn from it; in the hands of children it could become downright dangerous.

Chapter 4: Matter, Atoms, and Vacuum

It is pleasing to note that the study of matter, atoms, and the vacuum no longer is the cause of theological disputes. Thanks to the understanding that is provided by quantum mechanics, individual atoms and conglomerations of them in the form of matter are studied intensely and profitably, and the vacuum has proved to be incredibly complex. Much attention is centered on the objects that are inside the nuclei that are inside the atoms. By constructing ever larger and more expensive accelerators, we can come closer to connecting chapter 4 with the modern successor to chapter 5—cosmology. The incredible possibility exists of reproducing, over an extremely small region of space and interval of time, conditions akin to those that existed at the beginning of the universe. Theology will clearly offer an input in that area. Compared with some weapons of war, the superconducting supercollider will be cheap, but the understanding that it offers will be immense. However, we quote a word of caution: "The U.S. Secretary of Energy, James Watkins, has convened a 'blue ribbon' panel of 15 'eminent' physicists to advise him on the Superconducting Super Collider (SSC).—He should have convened a panel of 15 eminent poultry farmers, as they would have had the common sense to avoid two classic errors: (1) 'Don't put all your eggs in one basket,' and (2) 'Don't count your chickens before they hatch'. . . ." (Robert J. Yaes, *Nature* 344, no. 6263, 15 March 1990, p. 188.)

When one small part of a hierarchical structure of belief that a person holds very dearly has been shown to be factually incorrect, the response can include one or more of the following:

(1) Ignore the fact.
(2) State that the fact is not a fact.
(3) Muzzle, or even kill the messenger.
(4) Recognize the error and cut it out of the belief system.

These leave the nagging doubt, even an unconscious concern, that other parts of the belief system may be wrong too, and where will it all end?

Thus, as we saw in chapter 4, Pascal's demonstration of the existence of a near-vacuum was seen to pave the way for the evils of atheism. The Jesuit who believed that was very sincere; he too was struggling to understand, within the restricted framework of his Christian beliefs and the theology of Aquinas. Trying to understand within that framework is like trying to square the circle, using only ruler and compass. Many bright minds have worked on that one too.

Chapter 5: Sun, Planets, and Motion

> No other astronomy project has taken so long to develop, proven so technically challenging, or cost so much as the Hubble Space Telescope. At a development cost of about $1.5 billion, the HST is big science by the standards of modern physics. . . . Building the HST was quite unlike building most big-science accelerators. Essentially every element of the HST required new developments, with corresponding cost and schedule uncertainties. . . . The earth's atmosphere only transmits radiation in the radio and visible regions; the rest of the astronomical signal is absorbed. Also, the atmosphere distorts the wavefront arriving from a distant star or planet. . . . At 2.4 meters diameter, the HST is not a giant telescope by today's standards. . . . The designers wanted to exploit the mirror's performance in the ultraviolet. The primary mirror was finished to better than 1/60 of the neon-laser wavelength (i.e., to better than a millionth of a centimeter). . . . The HST should be able to detect stars at least 30 times fainter than can be 'seen' with the largest operational telescope, the 5 meter Hale telescope on Palomar Mountain. . . . The HST should remain unequaled in this regard compared with foreseeable ground-based telescopes. . . . Five specific scientific instruments have been developed solely to make use of the telescope's image to do science: Wide-Field and Planetary Camera, Faint Object Camera, Faint Object Spectrograph, Goddard High Resolution Spectrograph, High Speed Photometer. . . ."
>
> (Excerpts from C. R. O'Dell, *Physics Today* 43, no. 4, April 1990, p. 32)

> What one hopes for is that worn tracks, or even partially worn tracks, will eventually be abandoned for what is new and unexpected, which by its nature is at present impossible to discuss. This hope cannot be better expressed than it was in 1946 by Lyman Spitzer: "It should be emphasized, however, that the chief contribution of such a radically new and more powerful instrument would be, not to supplement our present ideas of the universe we live in, but rather to uncover new phenomena not yet imagined, and perhaps modify profoundly our basic concepts of space and time."
>
> (Sir Fred Hoyle, *Nature* 344, no. 6269, 26 April 1990, p. 810)

> *Hubble probe finds "astonishing" error.* Washington (AP)—An error of about one millimeter—called 'astonishing' by one expert for its large size—has been found in a measuring device used to guide the manufacture of a flawed mirror in the Hubble Space Telescope. A NASA committee investigating the defect that has crippled the $1.5 billion telescope announced Thursday that the millimeter mistake was found while testing the measuring device, called a null corrector. The space telescope, touted as an orbiting observatory that would be able to see objects up to 14 billion light years away, was launched in April.
>
> (News Item, August 10, 1990)

A millimeter is a hundred thousand times the planned tolerance mentioned in the *Physics Today* article.

Between the SSC and the HST there is no way to know yet what understanding lies ahead. We've come a long way from the egocentric model of the universe taught by Christian theologians. The universe is not a paltry 6,000 years old, the reality is much grander—more than 15,000,000,000 years. The earth is not fixed at the center of the universe—that conceit has gone the way of the theological impossibility of men living in the southern hemisphere. There are so many other galaxies besides our Milky Way, and so many stars in each of them, that it is very likely that life has evolved in some other parts of the universe. If Christ is the son of God, did he visit them also? Perhaps, someday, we will find out—but first we have to find one of these civilizations, and learn about their superstitions, their religions, and their culture.

> The U.S. National Aeronautics and Space Administration (NASA) is developing a powerful radio-wave receptor under a program called SETI (Search for Extraterrestial Intelligence). The receptor is designed to observe 1000 stars similar to our sun and to sweep the sky in search of artificial signals that could have been produced by a far-away civilization. If everything goes as planned, the receptor will be connected to the great radiotelescope in Arecibo, Puerto Rico, in October 1992. . . . Even if we were unable to decipher a message from another world, the message's detection would still overwhelm us.
>
> (Françoise Harrois-Monin, *L'Express,* Paris, reprinted in *World Press Review,* January 1991)

How much experimental work did Pythagoras and his followers do? It would be hard to learn about notes that sounded well together without playing them on a monochord. For an instrument with several strings, the tensions may be unequal and that would mask the regularities that the Pythagorean school noted.

> Boethius [see *In Memoriam*] recounts a story of unknown origin in which Pythagoras, after going into a metal working shop, conducted impromptu experiments to learn how different hammers produce specific tones. . . . According to the story, Pythagoras returned home from the shop and conducted experiments into the relationship between objects' physical proportions and the tones they produce. His experiments included plucking strings of different lengths and widths and hitting vessels filled with varying amounts of liquid.
>
> (*Science News,* vol. 137, May 12, 1990, p. 295)

> There is no way to keep Christopher Columbus from getting credit for today's holiday, but some Norsemen just can't help pointing out that Leif Ericson beat him to the New World by 507 years. Ivan Christensen, president of the Leif Ericson Society, said the society is devoted to advancing the argument that the son of Eric the Red landed with his family in what is now known as Greenland in the year 985.
>
> (Staff Wire Reports, *Mississippi Press,* Oct. 9, 1989)

> She sent Christopher Columbus on his voyage and opened up the New World for the spread of Catholicism, but did Queen Isabella I of Spain have the stuff of saints? Jews and Muslims are outraged by the idea. The Vatican is studying a petition that supporters hope will put the queen on the road to possible sainthood in time for 1992, when the church plans to celebrate 500 years of Christianity in the Americas. . . . Isabella is also remembered as the queen who expelled the Jews and Muslims from Spain and for the dreaded Inquisition that began during her reign with her husband, Ferdinand II.
>
> (Victor L. Simpson, Associated Press, *Sun-Herald,* January 13, 1991)

Chapter 6: Animals, Vegetables, and Minerals

"*Eyebright:* A small scrophulariaceous plant of Europe, formerly much used as a remedy for diseases of the eye. *Celandine:* A papaveraceous herb with yellow flowers. It was formerly used as a cure for warts, jaundice, etc." *Webster's Dictionary.* "Birds use the juice of the herb Greater Celandine to heal a wounded eye. Celandine has been used since the middle ages to 'sharpen the sight.' " I know that taking large *internal* doses can be dangerous!

Many years ago we prepared an extensive report on another herbal cure for eye problems. This was the eyebright herbal eyewash. I know this is safe when used properly. Dozens and dozens of people have gotten miraculous results with it. . . . "I heard about the eyebright formula. I used it for four months. I used it three times a day. One day I felt like rubbing my eyes and a cataract came off in each finger." No one should use the formula without reading the comprehensive 98-page report *The Regeneration of Human Vision*. Send $48 plus $4P and H to. . . .

(Mailed advertisement, 1989)

The spirit of the rhizotomists (see p. 84) lives on with the authors of this and similar material. Generally these are cures not accepted by the medical profession, and it is difficult to judge which claims are valid and which are false. Herbs that were used earlier as cures are generally replaced by a whole pharmacopoeia of official medicines that have been judged to be superior. Perhaps there are times however when the old fashioned remedies work better for some people.

Research at two American colleges has discovered that the old salts were right. They munched on ginger to avoid seasickness. Researchers at Brigham Young University, and Mount Union College gave one group of subjects "highly prone to seasickness" a placebo; another got powdered ginger root, and a third Dramamine, a common drug for motion sickness. The subjects were, ugh, placed in whirling tilted chairs. "None of the volunteers who took Dramamine or a placebo lasted in the whirling chairs for six minutes without becoming sick." The ginger volunteers did much better. *Health Plus* reports more good news from animal tests: ginger lowers blood cholesterol, it works as an antibiotic, is a blood thinner, and may have anticancer properties.

(*Washington Spectator* 15, no. 15, Aug. 15, 1989)

Abraham Lincoln's mother, who died in Little Pigeon Creek, Illinois, when the future president was only seven, may have been the most famous person ever to die of the "milk sick" or milk sickness, so-called because it seemed to be caused by drinking milk. . . . Settlers early recognized that milk sickness occurred in areas where cattle suffered from a disease called trembles. Trembles seemed to be related somehow to grazing in rich woodlands. . . . In many areas, cattle with trembles died by the thousands. . . . The first frontier experimenter appears to have been a woman named Anna Pierce who, at a time when women were not allowed to attend medical school, had taken midwife and nursing courses and settled in Illinois as a doctor. . . . Pierce made a number of critical observations . . . (and) finally learned the identity of the poisonous plant when she befriended a Shawnee woman . . . (who) took the doctor into the woods and showed her white snakeroot, saying that it caused both milk sickness and trembles and that it was used by the

Shawnee for the treatment of snakebite. Other tribes are also known to have used snakeroot for ague, diarrhea, fever and urinary infections.

(David Cameron Duffy, *Natural History,* July 1990, p. 4)

A Kuna Indian argues against cutting down the forest in Panama:

If I go to Panama City and stand in front of a pharmacy and, because I need medicine, pick up a rock and break the window, you put me in jail. For me the forest is my pharmacy. If I have sores on my legs, I go to the forest and get the medicine I need . . . without having to destroy everything, as your people do.

(*Amicus Journal* as quoted in *The Washington Spectator,* vol. 7, no. 1, January 1, 1991)

Chapter 7: Miracles, Surgery, and Medicine

You are being healed! God is coming into your life telling you what to do. Plant seeds of faith (with good works and money) and expect a miracle. If you don't look for it you won't recognize it. Expect a miracle all the time. . . . Two days ago a miracle! A seed faith gift was sent in and the person received a government check for $883. "I earned credit income," he said, "and harvested my seed."

(Oral Roberts, TV Program, July 30, 1989)

Literary criticism, long thought to be peripheral or even irrelevant to biblical study, has emerged during the mid-1970s as a new major focus of academic biblical scholarship, but there is a major split among theologians—between the serious and fundamentalist scholars—that can be illustrated with regard to the questions of miracles in the New Testament. The serious position has been well put by Ernst Käsemann: "Over few subjects has there been such a bitter battle among the New Testament scholars of the last two centuries as over the miracle stories of the Gospels. . . . We may say today that the battle is over, not perhaps as yet in the arena of church life but certainly in the field of theological science. It has ended in the defeat of the concept of miracle which has been traditional in the church. . . . The great majority of the Gospel miracle stories must be regarded as legends."

On the other side, we may put a book that pretends to seriousness but is in fact fundamentalist: *Gospel Perspectives: The Miracles of Jesus* edited by David Wenham and Craig Blomberg. . . . The pretense of the book's intellectual seriousness begins to become clear in its preface which asserts that "serious historical and literary scholarship allows us to approach the gospels with the belief that they present an essentially historical account of the words and deeds of Jesus." But this is, of course, precisely what serious historical and literary scholarship does *not* do; the preface is intellectually dishonest. . . . Fundamentalist biblical study is in the literal sense of the word "unprincipled," its real goal, dogmatic argument, being disguised as objective information. . . . The New Testament's intertextual relationship with the Old Testament prompts (Northrop) Frye to ask "How do we know that the Gospel story is true? Because it confirms the prophecies of the Old Testament. But how do we know that the Old Testament prophecies are true? Because they are confirmed by the Gospel story." . . . Early Christians turned the Old Testament into a book about Jesus, finding in it stories that they read as "prophecies" about him. The miracle stories about Elijah in First Kings provided the basis for a number of miracle stories about Jesus. Elijah performs several striking miracles, among them the creation of much food from little and the resurrection of a dead son. (Compare Luke 7:11–16 with First Kings 17:8–10, 17, 19–23.) This remarkable similarity has been known for some time. See J. W. Draper, *History of the Conflict between Religion and Science* (1897).

References:

1. *The Literary Guide to the Bible,* Robert Alter and Frank Kermode, eds. (Harvard University Press, 1987), p. 6
2. *Essays on New Testament Themes,* Ernst Käsemann, trans. by W. J. Montague, (SCM Press, London 1964), p. 48

(extended excerpt from *Free Inquiry 9,* no. 2, 1989, p. 49)

As an introduction to the Catholic Book Club, *The Sun Danced at Fatima* (1989) is offered to prospective members. It tells the story how, seventy-two years earlier, three children were minding sheep near their homes in Fatima, Portugal, when they were startled by an apparition of a beautiful woman who told them that she was the Virgin Mary. "I am from heaven, and I have come to ask you to say the beads to obtain peace for the world and the end of war and sin." The sun moved and sent out brilliantly hued rays, as the Virgin Mary stood at the top of a carrasquiera tree!

The Book of Miracles by Zsolt Aradi (Monarch Books, 1961) bears the *nihil obstat* and *imprimatur,* the official declarations of the Catholic Church that a book or pamphlet is free of doctrinal or moral error. Miracles of the Old and New Testament, other miracles from the past, and modern miracles are described in detail. As the author states at the beginning of chapter 1:

> Assent to miracles has never constituted a problem for me. . . . The Church grants the greatest possible freedom to search for meanings in things and

> events. . . . But we know that in the Church we have an authority upon whom we can and should rely.

Which is it to be, "greatest possible freedom" or "authority"? On p. 249 Aradi quotes canon law:

> If anyone should say that no miracles can be performed . . . or that they can never be known with certainty, or that by them the divine origin of the Christian religion cannot be rightly proved, let him be anathema (Vatican Council session III, Canon 3, 4)

He also reports (p. 218) that "the *Osservatore Romano,* the Vatican daily newspaper, published a study in which the author explained the presence of Christ in the Eucharist, i.e., in the bread and wine, by applying the new facts about the structure of matter." He offers a long and specious argument, with a few twentieth century discoveries about physics added to a sixteenth century logic:

> It is possible to analyze the species [meaning atomic structure] of the bread and the wine. There is, however, beneath this species a *substance* that is hidden from any kind of experimental probing. . . . Physics being an experimental science knows only the palpable appearances . . . about the *substance* per se, physics claims to know nothing. Modern physics would be contrary to the dogma of the Eucharist if physics stated that *material substance* does not exist. . . .

You can impress a lot of people by using words like "substance" that you do not define. The claim to "explain the presence of Christ in the Eucharist by applying the new facts about the structure of matter" is a dishonest attention-getter. Some theologians look for prestige-drainage from science by claiming that their arguments are in accordance with one of its disciplines. In matters of sheer faith—like the existence of God, or the body and blood in the Eucharist—it seems that we have had too much rationalization. Faith and reason will never meet on these issues. If people want to believe in something without any evidence, let them do so if it doesn't harm anyone. And let them not try to justify their belief by appealing to science. The faith-reason controversy is the conflict between the right and left halves of our brains.

> In large cities and small towns, the emergency room is the abused child of American medicine. Overburdened, understaffed and underfinanced, emergency departments across the country are reeling from multiple blows. Start with 37 million patients who have no health insurance. Add a graying

> population with a growing need for expensive equipment. Subtract government reimbursements, which often cover only half the cost of treating the poor. Factor in the effects of the AIDS epidemic and drug violence. Under such pressures, the miracle is that the system shows any vital signs at all.
>
> (*Time,* March 28, 1990, p. 59)

> U.S. health care "is by any measure—the most expensive in the world. And the least cost-effective" (*Public Citizens Health Letter*). *The Economist* states that Americans "spend nearly twice as much per head (on health care) as the French or Germans, three times as much as the British. . . . Yet Americans are less healthy, and less well-looked-after than other rich Westerners. Their average life expectancy is shorter, and their infant mortality rate is higher. For the inner-city poor, things are much worse. Such a child born in Washington, D.C., has less chance of reaching his or her first birthday than a child born in Jamaica. . . . For all the expense, America's high-cost medical industry produces lousy results," the *Economist* states bluntly—some 37 million Americans have no health care insurance. "Poor people who stay off welfare find it hard to qualify" (*Economist*). Canada spends less than nine percent of its GNP on health care and delivers it to "virtually all who seek it." The U.S. spends twelve percent of its GNP on health care and serves a smaller percentage of the population every year.
>
> (*The Washington Spectator* 16, no. 6, March 15, 1990)

The death is reported of a patient who went to a plastic surgeon for chemical treatment of her face. The assistant was untrained and had to monitor her breathing under an anaesthetic. He was questioned, "Do you know what the normal breathing rate is?" *Answer:* "120/80." *Questioner:* "That is the blood pressure!" *Answer:* "Oh." Examples like this have been brought up in a congressional investigation of plastic surgeons. (news item, Sept. 20, 1989.)

Chapter 8: Astrology, Alchemy, and Numerology

> An estimated 72% of all American adults believe to some extent in astrology and/or psychic phenomena. Ninety-two percent of all newspapers carry horoscope columns—compared to 78% a decade ago. Yet with all this interest, there is no investment-related forecasting. How does one create a responsible astrological/psychic investment periodical?
>
> Beginning with Nostradamus (and going back even further), psychics and astrologers have had a pretty good record. With the really big events (World War One and Two, the great depression, etc.), they've been chillingly

accurate. Non-Psychic/Astrological predictions for 1983–1988: 97% wrong — only 3% correct.

(Advertisement for the *Psychic Forecaster*)

We in India join all Third World people by entering the new century with more fear than hope, more frustration than aspiration. . . . Appalling poverty is still accepted as inevitable fate; the educated middle class still run away from realities to take shelter under an ever-widening umbrella of superstitions; false beliefs and religious ritual are manifestations of rapidly spreading feelings of helplessness and frustration. The last decade and a half have witnessed an astonishing rise of religious fundamentalism, casteism, and interregional strife all over my country."

(Indumati Pariteh, *Free Inquiry* 8, no. 3, 1988, p. 25)

The French have long thought of themselves as rationalists deeply imbued with the spirit of the seventeenth-century philosopher René Descartes, who exalted the mind and vowed never to accept any idea unless he had first subjected it to rigorous critical analysis. . . . It might come as no surprise to some of the whackier regions of the U.S. West Coast or the zanier elements in the Soviet Union which has been in the grip of hysteria about visiting spacemen, but the reality is that France appears to be taking its cue from Nostradamus rather than Descartes these days. Large and established companies turn to graphologists, birthdate interpreters, and astrologers before hiring job candidates. A leading computer company only hires people after a tarot-card reading. A big insurance group uses a swinging pendulum to judge whether a candidate is honest.

(Barry James, *Skeptical Inquirer,* 14, no. 3, 1990, p. 233)

A new problem confronts (Argentina's President Carlos) Menem. Legions of superstitious Argentines have become convinced that he is cursed with a hex. The alleged evidence: several of his Cabinet ministers have died in office and after Menem shook the hand of a driver before a powerboat race, the driver crashed and lost an arm. The clincher came when Menem showed up on the field with the Argentine soccer team the day before its World Cup match with Cameroon. The heavily favored South American team was then humiliated 1-0. Menem has heard the talk that he brings bad luck and is said to be worried, since he is superstitious and regularly consults a fortune-teller.

(Paul Gray, *Time,* July 2, 1990, p. 13)

Throughout history, the moon has been associated with a great many human activities. . . . The moon is especially hypothesized to be associated with notable crimes and disasters, and Lieber (1978, p. 12) contends that the most violent crimes have occurred at the time of the full moon. . . . We have conducted a small-scale exploratory examination of this possibility. . . . A statistical analysis revealed no evidence of a relationship between frequency of type of disaster (U.S. railway accidents, U.S. mine disasters, marine disasters worldwide, fires worldwide, aircraft disasters worldwide, and explosions, assassinations, and kidnapping worldwide) during five days centered on each side of the four phases of the moon. . . . These findings are consistent with other reviews that have failed to find a relationship among psychological, physical and sociological behaviors and the synodic lunar cycle.

(I. W. Kelly, D. H. Saklofsho, and R. Culver, *Skeptical Inquirer* 14, no. 3, p. 298, 1990. Reference: Lieber, A.L., *The Lunar Effect,* New York: Anchor Press/Doubleday, 1978)

Joan Quigley, Nancy Reagan's astrologer for seven years, claims to have advised the former first lady, "Gorbachev's Aquarian planet is in such harmony with Ronnie's you'll see, . . . They'll share a vision." The astrologer added, "I was responsible for timing all press conferences, most speeches, the State of the Union Addresses, the takeoffs and landings of Air Force One. I timed congressional armtwisting, the second inaugural oath of office, the announcement of Anthony Kennedy's Supreme Court nomination. I delayed President Reagan's operation for cancer from July 10, 1985, to July 13 and chose the time for Nancy's mastectomy."

(Associated Press, Steve Wilstein, March 16, 1990)

Any attempt by scientists to tell why a popularly held idea or theory is not valid inevitably leads to complaints from the wounded of authoritarianism and scientific elitism. All this has made the scientific community reluctant to enter the fray. And this in turn has allowed occultist ideas to go largely unchallenged and unevaluated in the publicly visible arenas in which they flourish. But now there is something new on the scene. . . . The formation of the Committee to Scientifically Investigate Claims of Paranormal and Other Phenomena was announced on April 30. . . . Co-chairman Paul Kurtz . . . has long been concerned about what he calls the "enormous increase in public interest in psychic phenomena, the occult and pseudo science. Often," he states, "the least shred of evidence for these claims is blown out of proportion and presented as scientific proof."

(Kendrick Frazier, *Science News* 109, May 29, 1976, p. 346)

The quasireligious character of the UFO movement has been underscored recently by the publication of two popular UFO books: *Communion,* by Whitley Strieber, and *Intruders,* by Budd Hopkins. Both books report stories of visitations and abductions by semi-divine, intelligent "space-aliens." Hop-

> kins, who conducted "hypnotic regressions" upon numerous alleged abductees, paints an alarming picture of extraterrestrials performing genetic experiments on the human race. Strieber, whose book (subtitled *A True Story*) tops the *nonfiction* list, describes his odyssey as "an intense spiritual search."
>
> The committee for the Scientific Investigation of Claims of the Paranormal has found these two books to be wholly unreliable. . . . Both books fail to provide or document convincing corroboration or physical evidence for their claims. But it is apparent that the transcendental religious temptation is not easily discarded by rational considerations.
>
> (Paul Kurtz, *Free Inquiry* 7, no. 3, 1987, p. 35)

A Collection from Mysteries of the Unknown is a 95-page booklet published in 1989 by Time-Life Books. It provides information on many forms of superstition and mysticism under the chapter headings Omens and Auguries, Portents in the Palm, Penmanship and Personality, Charting the Four Basic Numbers, Casting Your Fate, Dominoes and Destiny, Other Ways of Seeing, The World of the Psychic, and Psychics at Center Stage. The book is illustrated with photographs, from a Gurung (Nepal) shaman examining the shape and color of the lungs of a sacrificed chicken, to a full-length portrait of the "mentalist" Washington Irving Bishop (1856–1889). It gives a list of forty-four ways that have been used for trying to predict the future, from aeromancy (atmospheric phenomena), to zoomancy (imaginary animals). There are many anecdotes recorded, seemingly miraculous, but mostly recognized as due to natural causes. Detailed illustrations reveal the "secrets" of palmistry, and the analysis provided by a "reputable palmist" is described. The earth-, air-, fire-, and water-type hands are illustrated, but no information is given on how one is to read the palm of a monkey. Handwriting analysis used to detect forgeries is a legitimate science, it is noted, but "more questionable . . . is the contention of many graphologists that they can deduce character traits from a sample of script." The nonsense of numerology is presented as if it were a legitimate endeavour, and coupled with it are dice, dominoes, and destiny.

Stories of clairvoyance and psychic powers, of seances, of miraculous appearances and disappearances follow one another; one psychic offers a triple route to mystic understanding—palmistry, numerology, and the tarot. "Today several hundred thousand psychic advisers are operating in the United States alone" and "the critics' responses range from mild to vitriolic." There is dishonesty and trickery mixed with a very sensitive awareness that allows the psychic to detect and interpret the body language and even small muscular movements of the client. Astrology is barely mentioned, probably because it is too big a business to criticize. But the stories of Ingo Swann, Eileen Garrett, Edgar Cayce, Patricia McLaine,

Irene Hughes, Kuda Bux, Wolf Messing, Joseph Dunniger, and of course Uri Geller and James Randi make very interesting reading. As Jacob Needleman put it, however, "You should be open-minded, but not so open-minded that your brains fall out."

> We may now see alchemy as 2,000 years of scientific failure. Yet the strange fact is that Isaac Newton, the greatest of all scientists, laboured for more than 30 years on alchemy. He built furnaces for his endeavours at Trinity College, Cambridge, collected a huge private library, copied manuscripts in his own hand and compiled dictionaries of myth and occult symbols, using all the powers of his unique genius to understand matter and matter's relation with mind. Newton wrote a million words on alchemy, yet did not publish a single one.
>
> (Richard L. Gregory, *Nature* 342, 30 November 1989, p. 471)

> Bottom of your foot tells your fortune.
> How long you'll live, how rich you'll
> be and what Cupid has in store for you.
> —Free, easy to use chart inside.
>
> (*Weekly World News,* February 12, 1991)

Once you believe that marks on your hand can tell your future, it is no great leap of faith to suppose that your feet can also reveal secrets. Besides, it can offer an interesting social diversion as we sit around examining each other's feet. Palmists will go to their graduate schools to study "pedistry" in order to double their knowledge and their income. Some genius will tie it to astrology through Rigel, the left foot of Orion.

Chapter 9: Emperors, Saints, and Scholastics

> Cleopatra's Beauty Secret: Cleopatra massaged aloe gel into her skin to make it shine. But if you are after beautiful skin, do what the legendary Egyptian beauty did—use the fresh leaf gel, not the 'stabilized' gel used in commercial shampoos and skin products.
>
> (Advertisement for *The Healing Herbs* by Michael Castleman, 1991)

> The pope and his clerical followers continue to perpetuate the legend of the Petrine foundation of the papacy and the church of Rome. They do this with full knowledge of the fallacious bases for these claims. . . . Their purpose can only be to protect their status and their claims to some sort of special authority.
>
> (Gerald Larue, *Free Inquiry* 10, no. 2, 1990, p. 26)

For the second time in two years, a Little Rock, Arkansas, engineer and Bible teacher claims to have pinpointed the time of Christ's coming. An event Christians call the Rapture—the snatching away of God's people from earth, leaving others to seven years of misery and the end of the world—will be on September 1, 1989, Edgar Whisenant says. Last year he proclaimed the Rapture would be sometime between September 11 and 13, 1988. About three million copies of the book explaining his theory were sold or distributed. . . . Either the Rapture will be September 1, 1989 or on September 30 when Jews will celebrate Rosh-Hashanah.

(Regina Hines, *Mississippi Press,* July 7, 1989)

Maybe the Bible was right and there really was a Tower of Babel. Or, at least, maybe there really was once a single human language, before we were all cursed with a confusion of tongues. Now, with the aid of computers, linguists believe that they are reconstructing the mother tongue—which they call "proto-World." . . . More than 150 words of the very first human language which was spoken in East Africa 100,000 years ago are now known, wrote linguists Vitaly Shevoroshkin and John Woodford in a recent issue of the U. S. magazine *The Sciences.*

(Gwynne Dyer, *Toronto Star,* quoted in *World Press,* December 1990, p. 67)

German researchers have unlocked the mysteries of Noah's ark. This Week in Germany reports that their research indicates that the ark "had the capacity to accommodate 33,396 animals the size of unshorn sheep. Basing their work on the description in the first book of Moses, the scholars estimate that the ark had three stories and approximately 41,000 cubic meters of space. They calculate that it was 135 meters long, 22.5 meters wide, and 13.5 meters high. They theorize that large specimens, such as elephants, were taken aboard as youngsters, and feeding problems were partially overcome through the animals' winter and summer hibernation."

(*The Washington Spectator,* March 1, 1991)

These numbers are practically the same as those given in chapter 9, using one cubit equal to 45 centimeters or 17.7″, instead of 18". Not what you would call a new research result—it was the smallness of these numbers (in cubits) that bothered Origen about 1750 years ago. Since 135 × 22.5 × 13.5 = 41,006 these scholars must have decided that the ark was shaped like a big box.

Chapter 10: Caliphs, Scholars, and Linguists

About 860 million people, almost 20 percent of the world's population, are Moslem. They include Arabs, Pakistanis, Bangladeshis, Indonesians, and Africans. And there are at least 40 million Moslems in the Soviet Union. The ones we see on TV burning American flags and carrying signs reading "Death to America," and calling the United States "the great Satan . . . ," the ones who are mistreating American hostages in Beirut—these terrorist fanatics are no more representative of the Moslem religion than the Ku Klux Klan is representative of Christianity.

(Paul Harvey paraphrasing the Jewish historian Douglas Steusand, October 31, 1989, *Mississippi Press.*

Fortunately, the Ku Klux Klan percentage is much less at present.

The religion of Islam, as set forth in the Koran, encourages people to explore the secrets of the Universe in order to understand better the wonderful creations of God and to strengthen faith. One of the most important channels leading to religious feeling is the study of God's creations. In Islam, there are no contradictions between religion and science.

(Dr. Hosny Gaber, director, Islamic Center of New York, *Science Digest,* November 1981, p. 43)

Colonel Moammar Qaddafi is getting a taste of his own fanatical medicine these days. It seems like he's facing armed resistance from Arab groups inside Libya whose commitment to Allah is even less compromising than his own. Which led to the amazing spectacle of his denouncing Islamic fundamentalism before the General People's Congress in Tripoli during October. Qaddafi likened Islamic fundamentalism to cancer or AIDS and delivered this ironic statement: "Once a ruler becomes religious it (becomes) impossible for you to debate with him. Once someone rules in the name of religion, your lives become hell."

(*Secular Humanist Bulletin,* no. 1, February 1990)

Britain, with a total population of some fifty million, now has more than a million Muslims, mostly from refugee immigration within the past quarter of a century. Thousands of those in their early twenties were raised in Britain as moderates but now espouse fanatical Islamic fundamentalism—probably because they feel marginalized by society at large and psychologically alienated from it. Therefore they were ready to respond fervently to the call of the late Ayatollah Khomeini for the assassination of the secularist author Salman Rushdie—a British citizen of Muslim origin—for daring to write a novel,

The Satanic Verses, critical of Islam. The book has been publicly burned in several parts of the country, the author and his family have had to go into hiding, his publishers have also received death threats, and booksellers and libraries have been forced to withdraw the book. . . .

(Barbara Smoker, *Free Inquiry* 9, no. 4, 1989, p. 38)

Understanding the Arab Mind

Americans are fighting a war in a land where the currents of Arab culture and history are stronger than U.S. leaders realize and could pull against America's efforts to forge a lasting peace in the Middle East. . . . We're Americans first, Christians second and Presbyterians third. They are Moslems first, Arabs second and Jordanians or Iraqis third. . . . The crusaders were in the Middle East hundreds of years ago, but to the average Arab and the average Moslem they were just there yesterday. . . . "We understand the Holocaust as the never-closing wound that continues to have an impact on Jews, we understand slavery as the never-closing wound that still defines blacks, but we fail to understand how the Palestinian issue is the never-closing wound that defines the Arab experience," said (James) Zogby, director of the Arab-American Institute in Washington.

(Jim Nesbitt, National Press Service, *Mississippi Press,* January 25, 1991, p. 2D)

Chapter 11: Faith, Reason, and Doubt

Over the centuries, the battles between science and theology have consistently been won by scientific skeptics, yet traditional religions seem to be winning the larger war. For religionists have the sustaining power to survive their defeats; their cathedrals outlive their detractors. The Vatican doctrine on human sexuality, birth control, abortion, surrogate motherhood, artificial insemination, and freedom of inquiry may defy all reason but the church has outlasted its dissenters in every age.

(*Free Inquiry* 1, no. 3, 1987, p. 7)

Next week in Miami, Pope John Paul II begins a ten-day visit to the U.S., the 36th major journey by this most peripatetic of Roman Catholic Pontiffs. . . . Once regarded by Rome as among the most dutiful sons and daughters of the church, many American Catholics now believe they have a right to pick and choose the elements of their faith, ignoring teachings of the church they disagree with. Nonetheless, more than in most Western nations where dissent is widespread, American Catholics continue to be committed to the church, though increasingly on their own terms. . . .

The inevitable excitement about the first papal tour of the U.S. (in 1979)

> overshadowed the stern admonitions that John Paul delivered on church teachings and discipline. Since then, the Pontiff and Vatican officials have taken a number of widely noted actions to apply those admonitions. Some of the most controversial: temporarily limiting the authority of Seattle's liberal Archbishop Raymond Hunthausen, firing the Rev. Charles Curran from his professorship of moral theology at the Catholic University of America, threatening nuns with expulsion for declaring that pro-choice opinions on abortion are legitimate, and directing bishops to cut church ties to gay Catholic groups.
>
> (*Time,* September 7, 1987, p. 48)

Who created the universe—the Father, the Son, or the Holy Ghost? There were big arguments about that, but the Apostle's Creed expressed belief in "God the Father, Maker of Heaven and Earth," and that settled it. As we have seen, Lorenzo Valla (1406–1457) was the first to prove that this Creed postdated the apostles by several centuries, but it continues to be mumbled by many Christian congregations even today. "I believe in the resurrection of the body." It would be difficult to obtain accurate data on how many who repeat this almost every week really believe it. The Creed has inhibited the study of human anatomy, because you wouldn't want a person to be resurrected in the body after he had been dissected. The "resurrection of the body" is an early church myth and it has no place in the Credo of any church today. Remember I Corinthians (6:13)? "The body is not meant for immortality."

> Despite the Vatican hard line on feminism in the church, Roman Catholic women and their male allies are turning up the pressure to wash away age-old restrictions against women. . . . All, or almost all, though devoted to their faith, strongly favor an end to mandatory celibacy for priests and nuns and to strictures against birth control, both tenets of the Holy See. . . . "However rigid the pope is, he really isn't a major player anymore," said an important Catholic lay woman. "He is analogous to Reagan. He has set up a structure that carries out Church policy. The problem is now the structure". . . . One long-time fighter for the rights of Catholic women is not optimistic about official change. "The pope has put in the people who will elect his successor. The people he's left behind will pick a clone."
>
> (Jack Anderson and Dale Van Atta,
> *Washington Merry-Go Round,* February 2, 1990.)

> A prominent Roman Catholic priest Sunday celebrated the kind of ethnic Mass he thinks meets the spiritual and cultural needs of blacks. But Rev. George Stallings Jr. risked suspension and possible excommunication as he led the first Imani Temple service for several thousand followers at the Howard

> University Law School chapel. Sunday's service typified the kind of ethnic Mass Stallings wants expanded: a mixture of jazz, evangelical melodies and traditional African music, dancing and swaying by the congregation.
>
> (*USA Today,* July 3, 1989)

In the year 1989, a church was expelled from the Southern Baptist Denomination for accepting a woman as a pastor, and the installation of an Anglican woman bishop has, according to Pope John Paul, "effectively blocked the path to the mutual recognition of ministries."

> When the Washington, D.C.-based Catholic University of America had to choose between American traditions of academic pluralism and its formalities to the Vatican, there was no contest. Catholic University—the only American University directly chartered by the Vatican—ousted dissenting Father Charles E. Curran. He sued, maintaining that his tenure prohibited the school from withdrawing his teaching rights. D.C. Superior Court Judge Frederick Weisberg has ruled that the university was within its rights to terminate him. . . . Among those agreeing with Curran was Georgetown University theologian Monika Hellwig, a past president of the Catholic Theological Society of America. "I think it spells the end of academic respectability for Catholic University," Hellwig told the press. "I simply hope it doesn't affect other (Catholic) universities not chartered by the Holy See."
>
> (*Secular Humanist Bulletin* 5, no. 1, May 1989)

We have seen how, several hundred years ago, a philosopher/scientist who held an unorthodox or even heretical viewpoint on some issue could argue that he believed it *as a philosopher,* but *as a Christian* he didn't. Today some Catholic politicians are facing a similar schizophrenic situation in which they espouse a cause *as Catholics* but vote against it *as politicians.* This has surfaced in the recent (June 1990) public reminder by Cardinal O'Connor of New York that he has the right to excommunicate anyone who advocates abortion. Pro-choice Catholic United States politicians are therefore being threatened by a foreign power, because the Cardinal takes his orders from the Pope.

> "A united Europe is no longer only a dream," (John Paul) said in Prague. "It is an actual process that cannot be purely political or economic. It has a profound cultural, spiritual, and moral dimension. Christianity is at the very roots of European culture." Historically, that is simply untrue. Pagan Greece and Rome surely possess far stronger claims. Nevertheless, the central role of the church in the past 2,000 years of European history is indisputable. Looked at from the vantage point of a secular society such as Britain, the

Pope's vision of a political movement based on a belief in God—simultaneously rejecting impiety and materialism—seems an unworldly way of looking at the future. Judged from the Protestant perspective of many parts of Northern Europe and the Orthodox perspective of much of the East, it looks more sectarian. Seen from the point of view of the Muslem minorities within Europe and their coreligionist states beyond, the claim even threatens antagonistic religious fundamentalism. A successful outcome to a new crusade looks both improbable and downright dangerous.

(Martin Kettle, *World Press Review,* July 1990, p. 32,
reprinted from the *Guardian*)

Loachapoka, Alabama: Auburn University senior John Clark—pastor of Loachapoka Baptist Church—has been ousted by the deacons' board because he invited blacks to a revival at the all-white church.

(*USA Today,* July 3, 1989)

A panel of 12 churchgoers was selected to try Jim Bakker while the PTL founder's wife pleaded on the couple's television show for money for the faithful. "Our faith is in God," Bakker proclaimed. Bakker is charged with eight counts of mail fraud, 15 counts of wire fraud, and conspiracy to commit mail and wire fraud."

(Dennis Patterson, Associated Press, August 22, 1989)

Charlotte, North Carolina: David Taggart, a former Bakker aide who has been convicted for tax evasion, testified in U.S. District Court that the founder of the PTL evangelical empire was enchanted with real estate. "He told me he wanted to have ten homes," Taggart said Tuesday. . . . Bakker complained during a visit to Oral Roberts University in Tulsa, Oklahoma, that he didn't live as well as other evangelists, Taggart said. "Mr. Bakker said that he lived shabbily compared to Oral Roberts, the (Rex) Humbards and other ministers."

(*Mississippi Press,* August 30, 1989)

"It is easier for a camel [or "rope" if you follow Dr. Lamsa] to go through the eye of a needle. . . ." (Matthew 19:24)

Southern evangelists Mario "Tony" Leyva, Rias Edward Morris, and Freddie Herring pleaded guilty to charges that they had lured dozens of young boys into homosexual prostitution through a Columbus, Georgia-based ministry called the Tony Leyva Evangelistic Association. . . . Morris is Leyva's church organist, Herring heads a congregation in Douglasville, GA.

(*Secular Humanist Bulletin* 4, no. 3, December 1988)

A federal court panel has struck down a North Carolina law that exempts Bibles from the state sales tax. — In the unanimous ruling, a three-member panel of the court said that exempting the Bible, but not other works of literature, forces the state to discriminate on the basis of the contents of a book, text or other published work, which is intolerable under the First Amendment.

(Among the plaintiffs were) John S. Friedman of Durham, who is Jewish and was charged sales tax on the Tanakh, the Jewish Bible; Vasudha Gupta of Raleigh, a member of the Hindu faith, who had to pay sales tax on a copy of the Bhagavad Gita, a sacred Hindu text; Bruce Jacobs of Route 6 Hillsborough, a member of the Hare Krishna faith, who paid the sales tax for a copy of The Upanishads, the Hindu Spiritual treatises.

(Erin Kelly, *Mississippi Press,* May 12, 1990)

Former Arizona Governor Evan Mecham is well poised to stage his hoped-for comeback: He and his so-called Evanistas seem to have taken over the state's Republican Party. And one of their goals is to intertwine Christianity with government. The Evanistas joined with Pat Robertson's supporters to obtain passage of resolutions declaring that the U.S. is a "Christian nation" and that the Constitution created a "republic based on the absolute laws of the Bible, not a democracy." Although many in Arizona share Barry Goldwater's opinion of the group ("a bunch of kooks"), passage of their resolution was facilitated by no less a figure than Supreme Court Justice Sandra Day O'Connor.

(*Secular Humanist Bulletin* 5, no. 1, May 1989)

Approximately 35,000 persons attended the 1990 Southern Baptist Convention in the New Orleans' Superdome, where a civil war threatened to divide the second-largest denomination. . . . A steering committee of Baptist moderates met in Atlanta to hash over methods of dealing with their opposition and to develop a long-range proposal. . . .

When Catholic church officials discovered that Archbishop Eugene H. Marino, 56, had carried on a two-year relationship with a 27-year-old Atlanta woman (Vicki R. Long), Marino ended up in the psychiatric unit of a hospital. . . . The Long controversy fueled a celibacy debate within the Catholic Church.

(*Sun-Herald,* Sec. B, December 29, 1990)

Archbishop Rembert Weakland said he would ask the pope's permission to ordain a married man before allowing parishes to decline because of a growing priest shortage. . . . Pope John Paul II repeatedly has rejected suggestions the church ordain married men to counter a sharp decline in the number of priests. . . . Weakland . . . recently was barred by the Vatican from accepting an honorary doctorate from a Swiss university because of his remarks on abortion. . . . The archbishop did not address the possibility of ordaining women.

(Jodie de Jonge, Associated Press, *Sun-Herald,* January 10, 1991)

VH-1 takes in-depth look at censorship

. . . Viewers also get a close-up look at last year's goings on in Church Hill, Tennessee, where parents sued the school board to remove "certain books" from the school library. Sharing the hit list with the beleaguered *Catcher in the Rye* were *The Adventures of Huckleberry Finn, The Hobbit,* and *To Kill a Mockingbird,* to name just a few titles. . . . "How can we lose?" asks one Church Hill resident, "We have Jesus Christ on our side." . . . . At a book-and-records burning event in Damascus, MD, (*not* Syria) the interviewer stopped a parent to ask which titles were being incinerated. His reply, "I don't really know the names of any of them."

(Jean Prescott, *The Sun-Herald,* January 24, 1991)

Chapter 12: Physics, Astronomy, and Persecution

For meddling with the prayer book there isn't even the scholarly excuse. . . . Is it entirely an accident that the defacing of Cranmer's prayer book has coincided with a calamitous decline in literacy and the quality of English? . .- . Looking at the way English is used in Britain's popular newspapers, radio and television programs, even in schools and theatres, (a great many people) wonder what it is about our country and our society that our language has become so impoverished, so sloppy and so limited. . . . It leads me to wonder how Hamlet would deliver his grand "To be or not to be" soliloquy in the language of today. . . . What about this? "Well, frankly, the problem as I see it at this moment in time is whether I should just lie down under all this hassle and let them walk all over me, or whether I should just say: Okay, I get the message, and do myself in. I mean, let's face it. I'm in a no-win situation and quite honestly, I'm so stuffed up to here with the whole stupid mess that, I can tell you, I've just got a good mind to take the quick way out. That's the bottom line. The only problem is: What happens if I find that when I've bumped myself off, there's some kind of a, you know, all that mystical stuff about when you die, you might find you're still—know what I mean?"

(Prince Charles, *World Press Review,* March 1990, p. 44.)

Montgomery, Alabama: The mayor prayed on the 50-yard line as 10 protesters blew horns and jeered in one of the first gridiron confrontations over a court-imposed ban on invocations at high school athletic events. . . . Montgomery Mayor Emory Folmar's prayer ended with six people being escorted out

> of the football stadium for sounding airhorns and police arresting one protester accused of struggling as he was led out of the stands. . . . In Childersburg Thursday night a group of about 100 people led fans in a spontaneous recitation of the Lord's Prayer during a game between B. B. Corner High School and Childersburg High. All across Alabama, radio stations are broadcasting prayers at 6:55 p.m. C.D.T.—five minutes before most prep games start—and are asking fans to take portable radios to the stadiums.
>
> Earlier this summer, the U.S. Supreme Court let stand a decision by the 11th U.S. Circuit Court of Appeals that organized prayer before football games in Alabama, Georgia and Florida violates the separation of church and state required by the constitution.
>
> (Jay Reeves, Associated Press, and Richard Russell, *Press Register* Reporter, *Mobile Press Register,* August 26, 1989)

The ultimate in smugness occurred in 1989 when the Catholic Church made it official that it had forgiven Galileo. *Forgiven* him! Why don't they have the courage and honesty to admit their mistakes, instead of clinging to that myth of infallibility. You have an enormous influence in this world, Pope John Paul II, and much of it is for good. Why can't you bring a new message to the world and ask all Catholics to go on their knees before Christ and ask forgiveness for the sins of the Church? Why can't you make it clear that you condemn the murders and tortures inflicted by church officials in the name of Christ and the Church? Or are you afraid, like the Communists or the fundamentalists, that if you condone a little freedom, more will be demanded? Forgive Galileo! It is as if the Greeks were to forgive Socrates, the French to forgive Lavoisier, the Germans to forgive the followers of Einstein.

> (Cardinal O'Connor) is the kind of man who, if the church still had the power to burn people at the stake, would be right out there lighting a fire—and saying at the same time, "What can we do? These people are doing the wrong thing and the rules say we light a fire, so here I am," says Frances Kissling, president of Catholics for a Free Choice.
>
> (*Vanity Fair,* August 1990, page 161)

Chapter 13: Witches, Devils, and Lunatics

King James I of England was prominent in this chapter because of his witch-hunting, his book *Demonologie,* and for the Bible published during his reign. He was also a writer of prose and verse, being less unworthy of literary fame than of royal renown. His *Counterblaste to Tobacco,* published in 1604, gives us a nearly four-hundred-year-old message that is relevant and important today. Here are some extracts:

> This *Tobacco* . . . hath a certain venomous faculty joined with the heat thereof, which makes it have an Antipathy against nature as by the hateful smell thereof doth well appear. This stinking smoke being sucked up by the Nose and imprisoned in the cold and moist Brains, is by their cold and wet faculty turned and cast forth again in watery distillations. . . . If a man smoke himself to death with it (and many have done) O, then some other disease must bear the blame. So do old harlots thank their harlotry for their many years, but never have mind of the Pocks in the flower of their youth. And so do old drunkards think they prolong their days . . . but never remember how many die drowned in drink before they be half old. . . . And for the vanities committed in this filthy custom, is it not both great vanity and uncleanness, that at the table, a place of respect, of cleanliness, modesty, men should not be ashamed, to sit tossing of *Tobacco pipes,* and puffing of the smoke of *Tobacco* one to another, making the filthy smoke and stink thereof to exhale athwart the dishes and infect the air, when, very often, men that abhor it are at their repast? . . . Herein is not only a great vanity, but a great contempt of God's good gifts, that the sweetness of man's breath, being a good gift of God, should be willfully corrupted by this stinking smoke. . . . What sins towards God, and foolish vanities before the world you commit, in the detestable use of it. . . . It is a thought by you . . . that the brains of all men, being naturally cold and wet, all dry and hot things should be good for them, of which nature this stinking suffumigation is. . . . These Cures ought not to be used but where there is need of them, the contrary whereof is daily practiced in this general use of Tobacco by all sorts and complexions of people.
>
> (UA 13, 58)

> People who have discovered in their genealogical research that accused witches are their forebears have now formed at least two genealogical societies. The Associated Daughters of Early American Witches was organized in April 1987. . . . The group has about 75 members in 22 states, descended from about 25 of the victims of the witchcraft delusion in Connecticut and Massachusetts. Another society organized last year targets specifically the people involved in the Salem trials. The group is called the Sons and Daughters of Victims of Colonial Witch Trials. There are about 45 members. The organization, which recalls the trials as a sad time in American history, wants to plan an observance for the 300th Anniversary of the trials in 1992. They are also hoping to erect a monument to the accused witches in the Old Burying Ground at Salem.
>
> (Regina Hines, Certified Genealogical Record Searcher, *Mississippi Press,* October 25, 1989)

> There is a simple, if surprising, reason wild ideas tend to sweep across an area. University of Maryland historian Mary Kilbourne Matosian became interested in a terror that seized parts of France in 1798. Peasants, weeping and shouting, took to the woods with pitchforks and muskets searching for mythical creatures. Matosian discovered, lo and behold, that rye, a major food of the peasants, had been ingested with ergot, and that ergot produces a hallucinogen. This, in turn, causes delusions and hysteria.
>
> (*The Washington Spectator,* February 1, 1990)

Demons are still being exorcised, but now with the publicity offered by TV:

> Make your vow to this ministry. A vow is when you enter into a covenant with God. — Are you sick and tired often? God sent me to you today. I believe in miracles. I see a woman today—right now. You want to make a vow of $1,000. I know you cannot afford that, but you will pay your vows. Don't be afraid. You'll be able to pay this . . .
>
> (Robert Tilton, WPMI, January 17, 1991)

On April 8, 1991, superstition and religion teemed up to present on national television an exorcism, blessed by the Catholic Church, and performed by priests. The young female patient was shown to be much calmer later, but it was admitted, though not emphasized, that she had been given tranquilizing medication. One of the priests expected her to levitate, but it didn't work out that way.

> Los Angeles. The devil-worshiping "Night Stalker" who murdered 13 people during a summer-long rampage that terrorized California told the judge who sentenced him to death that Satan will avenge his execution. "You don't understand me," the 29-year-old Richard Ramirez told the crowded courtroom Tuesday. "You are not expected to. . . . I am beyond your experience. I am beyond good and evil. . . . I will be avenged. Lucifer dwells within us all."
>
> (Associated Press, September 30, 1989)

> Monrovia, Liberia. Liberia's defense minister, his wife and seven other people have been accused of slaying a young policeman and using his heart and other organs in black magic rituals . . . the statement alleged, "one of the suspects . . . made available the living body of Patrolman Melvin Pyne to be murdered and parts extracted therefrom, in satisfaction of the general's and his wife's request."
>
> (Associated Press, April 1990)

Chapter 14: Scriptures, Weather, and Diseases

By the 18th century, skeptics and believers alike were examining the Scriptures and other records in a quest for "the historical Jesus." They found disappointingly little to corroborate the Gospel texts, a compendium of oral traditions written 20 to 60 years after the events and sometimes differing on important details of the story.

That has changed dramatically in recent years. Since shortly after the discovery of the Dead Sea Scrolls some 40 years ago, theologians, Bible scholars, archaeologists and cultural anthropologists have refocused their search. . . . Some scholars have come to reject the Passion as pure fiction. For others, the account of Jesus' death, burial and resurrection some 2,000 years ago remains a story worthy of faith. . . . Still others have put forward intriguing evidence that the truth lies somewhere in between: That the Gospel narratives are a mix of legend and fact that attempt to describe a historical and mystical human encounter with one who called himself the Son of Man.

(*U.S. News and World Report,* April 16, 1990, p. 46)

Southern Baptists, believing a perfect infallible Bible should be the standard of the nation's largest Protestant denomination, Tuesday elected a fundamentalist president for the 12th consecutive year. The Rev. Morris Chapman of Wichita Falls, Texas, defeated the Rev. Daniel Vestal of suburban Atlanta, the candidate supported by the moderates by a vote of 21,471 to 15,753.

(Tom Bailey, Jr., Scripps Howard News Service, June 13, 1990)

Avarice, arrogance, sleaze, fraud, carnal sin. Various American televangelists have already been accused of almost every imaginable transgression. What more could media-star ministers possibly be charged with? Answer: sloppy theology. That is precisely the theme of a new anthology, *The Agony of Deceit,* published by Chicago's fundamentalistic Moody Bible Institute. The book's twelve contributors . . . have scoured books and sermon tapes and found the TV preachers guilty of egregious doctrinal heresy.

(Richard N. Ostling, *Time,* March 5, 1990, p. 62)

A Prophecy Seminar. Designed to meet your needs in a world of crises. Featuring the Book of Daniel, Revelation's Foundation. Understand: World Crises, Nuclear Build-up, Terrorism, Economic crisis, Bank failures, Runaway debt, Social crisis, Aids, Abortion, Drugs, Religious Crisis, Holy Wars, Scandals—See Ancient Prophecies Unfold—Revealing God's Plan for the Future. God told the Prophet, "seal the book, even to the time of the end." Now it's open!

(Advertisement for Prophecy Seminar, January 9, 1990)

Abingdon's modern version of Strong's *Exhaustive Concordance of the Bible* (1890) has probably established a world record for pedantry and trivial scholarship. It is indeed useful to have a reference to help you find out where significant words are found in the Bible—words like *God, love, witch, spirit,* and thousands more. But words like *a, an, and, he, her, him, his*? Did you know that the word *a* or *an* occurs some 13,000 times in the King James Version, and *and* more than 20,000 times; *his, her, him,* or *his* a mere 17,500. In case you are interested, the concordance will also tell you chapter and verse where each of these words appears.

> Virginia Beach, Virginia. When Pat Robertson ran for president, it was reported that he prayed away a hurricane, and he's at it again. Before it was ascertained where hurricane Hugo would hit the United States mainland, the televangelist swung into action. On his "700 Club" broadcast, Robertson claimed that the Virginia Beach area was "right in the path" of Hugo, and asked his television audience to pray.
>
> "We'll have to do some rather strong praying," he said. "We haven't had a hurricane hit here since the early 1960s. We've found that determined prayer has a great effect on them. We've found that they just don't come in here if people pray!" Apparently God likes Virginia Beach better than Charleston.
>
> (Freedom Writer Faxsheet, quoted in *Free Inquiry* 10, no. 1, 1989/90, p. 62)

Chapter 15: Establishments, Paradigms, and Progress

> Non-scientist colleagues . . . are not as comfortable with ignorance as scientists are. I was struck by the thought that scientists are different, even exceptional, because they are comfortable with the idea of not knowing the answer to a question, and with acknowledging that they know something with a degree of *uncertainty*.
>
> (Professor Adrienne Clarke, *Melbourne University Gazette,* 1990)

> Science is much more than a body of knowledge. It is a way of thinking. This is central to its success. Science invites us to let the facts in, even when they don't conform to our preconceptions. It counsels us to carry alternative hypotheses in our heads and see which ones best match the facts. It urges on us a fine balance between no-holds-barred openness to new ideas, however heretical, and the most rigorous skeptical scrutiny of everything—new ideas and established wisdom. We need wide appreciation of this kind of thinking. It works. It's an essential tool for democracy in an age of change.
>
> (Carl Sagan, *Skeptical Inquirer* 14, no. 3, 1990, p. 263)

Being a responsible adult means accepting the fact that almost all knowledge is tentative, and accepting it cheerfully. You may be required to change your belief tomorrow, if the evidence warrants, and you should be willing and able to do so. That, in essence, is what skepticism means: to believe if and only if the evidence warrants.

(James Lett, *Skeptical Inquirer* 14, no. 2, 1990, p. 160)

Look at how people actually behave when they have formally decided that there is no God. What often happens is that they become even more puritanical. Far from encouraging licentiousness, these societies permit much less. . . . Joseph McCabe checked out some prisons and something like ninety-nine percent of the inmates were religious. . . . During the years in which Americans are formally educated, . . . they spend more time watching television than they do reading or writing. That's scary, it's wrong, it's a pity and it's one of the reasons Americans are getting dumber.

(Steve Allen, quoted in *Free Inquiry* 9, vol. 1, 1988/9)

We have to infect the education system with enthusiasm for thinking skills. This includes creativity, problem-solving, and critical thinking. It also means automatically considering all sides of an issue and the long-term consequences of your actions. It means having a healthy dose of skepticism. Having a breadth of thinking skills is more important than critical thinking alone. . . . Not one person in fifty in this country understands the process of science or the scientific method. Also, kids must understand the processes of evolution; not necessarily the details, but the overall picture. They should understand the biological diversity of our planet and how limited the earth is. They should know how the human mind works. If we understand that we all have mental blinders, we are less likely to be tripped up by them. If kids were exposed to these topics in school, I think that would make a great difference.

(Paul MacCready, *Free Inquiry* 10, no. 1, 1989/90, p. 7)

The life of a physicist is not without disappointments, frustrations, intractable problems and the usual human pains and sorrows. . . . But when things are going reasonably well, and even sometimes when they aren't, there's a tremendous sense of joy in the games physicists play. We impinge here on a theological principle. . . . It is the idea of *deus ludens,* God at play. The notion is that God created the world for the fun of it; it's all a great game in the mind of God, and it's for us to find out the rules if we can. That, I submit, is what physicists are about. Of course, it's not necessary to believe that the game has an author in order to play, but you have to believe in the game at least. The religious can say it's our sacred duty to study the rules: the nonreligious may say it is one of the highest evidences of our humanity.

(Dietrich E. Thomsen, *Science News* 133, March 5, 1988, p. 156)

Lest anyone think that religious establishments have the sole monopoly on propagating erroneous ideas, it must be recognized that scientists are not blameless. Here I am not referring only to the paradigms clung to by leaders in the field—these paradigms are mostly correct, although a few important incorrect ones do hold up progress. And there is one incorrect belief that is still held by many high school physics teachers.

I had the good fortune to teach freshman physics at a college that had very high standards, attracting first-class high school graduates from many parts of the United States. I was appalled to discover that a large majority of the students believed that the period of oscillation of a pendulum is independent of its amplitude of swing. Galileo had been still in his teens when he noticed that the period of swing of a pendulum depends only on its length—at least that is how it is usually reported. It is only approximately true, but of course he did not have the resources and equipment to demonstrate that the wider the amplitude of swing the longer the period. A pendulum that keeps accurate time for very small oscillations will lose over 40 seconds in a day if it oscillates up to 5° on either side of the vertical. If it oscillates through 30° on each side it will be nearly 25 minutes slow at the end of the day. I don't think that the students believed me until they made measurements themselves—which is the way it should be—but somehow they had been uniformly misinformed by a wide variety of teachers, mostly in our better schools.

Edward Jenner and Ignaz Semmelweiss, Marcelino de Sautuola, Alfred Wegener, and others made important discoveries in preventative medicine, archaeology, and geology respectively, but, as we have seen, their works were rejected by their professional colleagues. One way to be rejected is to have the editor of a professional journal refuse to publish a report on a discovery. To help him decide whether a paper is to be published, the editor usually relies on confidential reviews from experts in the field. Without this peer reviewing, much more nonsense would be published than is now the case. This means that sometimes the struggle to understand is frustrated, on the grounds that the research violates not religious but scientific paradigms. Without this anonymous censorship there would be utter chaos in the scientific literature, yet genuine breakthroughs in our understanding of nature can often go against established beliefs. One is reminded of the Impressionists of the nineteenth century who were denied places in the Louvre—and also of the thousands of inferior paintings that also didn't make it to the Louvre.

Some recent published comments on the procedure of anonymous reviewing include:

"Is it possible that traditional sympathy for democratic traditions has led the scientific community spectacularly to overlook the inappropriateness of anonymous peer review?"

—Alexander A. Berezen

". . . as none of the reviewers was able to produce experimental evidence for the rejection of the results, it was frustrating that all the reports were anonymous."

—G. C. Fletcher

"Few editors even read unsigned letters. None publishes them. It is therefore inexplicable to me why scientists accept anonymous reviews of their work."

—S. Roth

(Reference for all three, *Nature* 337, 1989, p. 202)

"We scientists owe it to ourselves and to our financial patrons to be more open and equitable in the critical assessments of nonconsensual views and results."

—A. Thyagaraja, (*Nature* 335, 1988, p. 391)

"Nobody has yet presented me with a convincing argument in favour of anonymous refereeing."

—J. B. Wright, *Nature* 336, 1988, p. 10)

And so it goes on. The open mind on which we scientists pride ourselves has a secret agenda. The problem now is that there is so much scientific material published that it is difficult to know what to read and what to skip. Like other people, we turn to an Authority to guide us, and one such Authority is the carefully refereed journal. We can be sure that everything we read in it will be first-class professional work, thoroughly entrenched in the paradigms of the time. A brilliant student in his twenties will be encouraged to build on this—and if he does his success is assured. These guidelines are essential, but we must not forget that while they were in *their* twenties, Einstein, Heisenberg, and Dirac discovered the theories of relativity, quantum mechanics, and relativistic quantum mechanics, respectively.

More open discussion of heretical ideas in science could be healthy, although most scientists would regard it as a waste of time, and most of the time they would be right. At least lifting the veil of secrecy from reviewing would help and would prevent the vicious and sarcastic reviews that are sometimes issued *ex cathedra* by reviewers who hide behind their editorial secrecy.

Chapter 16: Hell, High Water, and Evolution

> A revised theory of evolution has been proposed by two British researchers. The struggles for survival and natural selection were not . . . the dominant determinants of man's evolution according to Professor Michael Crawford . . . and David Marsh (who) argue that nutrition was at least as important and sometimes more so. They say that man . . . found his "ecological niche" in coastal areas which were rich in lipids and linoleic acid. (Their) research shows more similarity . . . in terms of the brain's chemical composition and brain-to-body ratios . . . between man and dolphins and perhaps others, than between man and primates.
>
> (*World Press Review,* June 1990, p. 68)

In his book *Evolution of the Brain: Creation of the Self* (Routledge 1989), John C. Eccles writes, "each soul is a new Divine creation which is implanted into the growing foetus sometime between conception and birth." His critic Ralph Estling notes that "Parliament, now exercising itself over a similar conundrum would doubtless be interested to know exactly when this miracle occurs. At 14 weeks? 18 weeks? Can Eccles be a little more specific? At what stage of human development did hominids receive their souls? The matter is in fact covered by the 1950 papal encyclical *Humani Generis* which explains that the soul was acquired during the early Pleistocene, 800,000 years ago, apparently in Kenya." (*Nature* 344, 1990, p. 582)

> Washington (AP). Religious extremists and members of right-wing organizations are gaining in their battle to ban or censor library books and to restrict sex education in schools an anti-censorship group says in a survey released today. "Most would-be censors are not content with restricting their own children's freedom to learn. . . . Instead, the censors insist on the blanket banning of these materials for both their own and other parents' children," *People for the American Way* said in its report. The main targets of such challenges are literary classics such as John Steinbeck's *Of Mice and Men,* and J. D. Salinger's *The Catcher in the Rye* as well as plays by Arthur Miller and Aristophanes. Nearly half the challenges to instruction resulted either in removal of the material or in restrictions on its use, such as a requirement of prior parental consent. . . . The report found that religious extremists have intensified their campaign to force schools to teach creationism in science classes.
>
> (*Mississippi Press,* August 30, 1989)

> A renewed fight over the teaching of evolution and creationism in California will have a crucial test this week in a battle that has national implications. The focal point is the upcoming adoption of a new science curriculum

framework, a detailed statement of what should be taught in California and how it should be taught. . . . The new proposal would strengthen the teaching of evolution. "Material should be included in the textbooks by various scientists who say that evolution . . . cannot be substantiated," said the Rev. Louis Sheldon, chairman of the Traditional Values Coalition, a coalition of 6,000 California churches. But Bill Honig, the state superintendent of public instruction . . . contends no credible scientific evidence refutes evolution. Honig said evolution belongs in science classes because there is hard scientific evidence to back it, and creation theory belongs in philosophy or history classes because it is based on faith and is part of Western cultural heritage.

(Wire Service, September 27, 1989)

Thirty percent of *high-school biology teachers* polled by Dana Dunn believe that "creationism" should be taught in biology. Dunn, a University of Texas sociologist, surveyed 400 teachers selected at random from a list of 20,000 provided by the National Science Teachers Association. Some of her findings:

1. Nineteen percent believe that humans and dinosaurs lived at the same time (the Flintstone effect). Fourteen percent were not sure.
2. Twenty-seven percent thought the dead can communicate with the living.
3. Twenty-eight percent said people can predict the future.
4. Thirty-four percent believe in mind reading.

(*Secular Humanist Bulletin* 5, no. 1, May 1989)

Geologists and astronomers attempt to deduce the origin and evolution of the Earth and the Universe from their present states. Difficulties arise because astronomical and geological time-scales are so long that observations provide only a single snapshot of complex time-dependent processes. To help interpret this snapshot, both disciplines rely on the paradigm of uniformitarianism, which asserts that natural laws are constant in space and time, and that past evolution should be explained through presently observable processes. A weakness in this paradigm is that some important processes could occur rarely or last for brief periods, so that evolution proceeds mostly through a series of rare "catastrophes" rather than at a steady rate. Two examples of proposed geological catastrophes, of very different degrees of plausibility, are the Biblical deluge and the impact of a large extraterrestrial body at the Cretaceous/Tertiary boundary.

(Scott Tremaine, *Nature* 345, no. 6270, 3 May 1990, p. 31)

Two years ago, Boston scientists (led by John Cairns) committed an evolutionary heresy of sorts by suggesting that certain bacteria can mutate *on demand* to suit themselves. . . . Now, evolutionary biologist Barry Hall says . . . (that he) started with special strains of *Escherichia coli* that require external supplies of the amino acid tryptophan. . . . He grew the bacteria in a culture medium provisioned with a three-day ration of tryptophan, then "starved" them of tryptophan for the next 9 to 11 days. During that period, the numbers of bacteria mutating to produce their own tryptophan jumped 3 to 30-fold. "The phenomenon that Cairns describes is real," he asserts. "Mutations *occur more when they are useful* than when they are not. That I can document any day, every day, in the laboratory."

(*Science News,* vol. 137, June 23, 1990, p. 301)

. . . As a college freshman in 1925, I was sure that the Scopes trial, in which Clarence Darrow in effect made a monkey of William Jennings Bryan, had put an end to any serious debate. Even earlier, President Woodrow Wilson had confidently declared as much, and no important politician contradicted him until, in 1980, presidential candidate Ronald Reagan won cheers from the religious right by announcing, "Evolution is only a theory,"— meaning, of course, a mere hypothesis.

Soon the argument attained a higher judicial level. Supreme Court Justice Antonin Scalia with Chief Justice William Rehnquist . . . cited testimony that "creation science" merits equal class time with Darwinian evolution as a competing theory of the origin of life. . . . There is no theory of life's origin in Darwin's work. . . . Darwin confined himself to describing the process by which new species, including our own, have evolved from the old . . . he dismissed all theological pretensions on his part with the words "A dog might as well speculate on the mind of Newton."

(Philip Dunne, *Time,* January 15, 1990, p. 84)

A common ploy in dismissing the works of Darwin and his successors is to state that evolution is "only a theory." In scientific research, there are hunches, educated guesses, ideas, hypotheses, models, laws, . . . and at the top of the list are theories. In experimental as well as theoretical work, people come up with hunches, etc., to try out and to see if they work. The word *theory* is reserved for ideas that have been verified over and over again—the theory of gravitation and motion, for instance, or the theory of electromagnetism, the theory of continental drift, the theory of relativity.

To avoid confusion, however, the *theory* of evolution should be called the *fact* of evolution. Darwin established it as a fact with his monumental studies on the distribution of plants and animals, on the fossil records, on comparative anatomy, and on the evidence of adaptation of animals and plants. Since then the evidence has multiplied, and in recent years has been strengthened by the discovery of DNA and by comparisons of the DNA inheritance codes of plants, humans, and other animals. Humans and chimpanzees, for example, differ by only one percent or so in their genetic codes—a very important one percent, we would all hasten to add.

As to the theoretical part of evolution, there are various models—I would not call them "theories" yet. What is the mechanism by which evolutionary changes occur? The model of Lamarck, that acquired characteristics are inherited by the offspring, has been dead for a century. Darwin had no way to know about DNA, or cosmic rays and radioactivity, and thereby to see a mechanism by which spontaneous and unpredictable changes could continue to occur in the DNA makeup of living organisms. Changes occur by collision with a penetrating particle (J. B. S. Haldane argued that the incidence of hemophilia in the royal houses of Europe was due to a single cosmic ray.) Changes can also occur by copying errors, perhaps even by errors due to the fundamental uncertainties of quantum mechanics. Even without changes, some living creatures find themselves in an environment that deteriorates rapidly, even cataclysmically, or one that, for the lucky ones, provides ever more food and other amenities. Perhaps evolution is not the long, slow continuous process that Darwin imagined. Perhaps, as the model developed by Stephen Jay Gould suggests, the process is even slower than was thought, as long as a species finds itself in a very suitable environment, but a species is capable of much more rapid change when conditions are less favorable. It is a dynamic system. If scientists disagree sometimes about the relative importance of various mechanisms, that is par for the course of science. They are not denying the fact of evolution. Those who do so are like those who refused to look through Galileo's telescope to see the moons of Jupiter.

Chapter 17: Past, Present, and Future

We have cited many manifestations of the growth of superstition and fundamentalism in modern times; there are very many more. Two reasons for this are that scientific details are too complicated for all but the experts to understand, and although scientific thinking may offer a solid anchor, for some it is lodged in shifting sands. The conflict is one of method. One could hope that religion and science would not overlap each others' territories, but those territories overlap naturally, and there is no way to avoid disagreement.

To most people it is very depressing to think that we can live and die, and that is all there is to it, like putting your hand in a bucket of water, taking it out, and when the few ripples die down nothing has been changed. One way to avoid this is to have children—nobody can say that we haven't had an effect if we leave children behind us, children and their children who would never have existed if we hadn't. What else do we do to ensure that our own lives will not sink into oblivion? We make friends, work hard, make money, impress people, and excell in scholarship, sports, politics, entertainment, or preaching. It would seem, though, that the certain way to achieve immortality is to believe that life continues indefinitely after death, despite the fact that there is no solid evidence that such is the case. To really establish that as a fact would be the scientific breakthrough of all time, but it is such an attractive idea that many millions believe in it anyway. It is believed because it can be very comforting, and because it is handed down from one generation to the next and reinforced by preachers, charismatic and otherwise.

A scientist cannot avoid asking, however, "Where is the evidence for life after death?" The old faith-reason controversy surfaces, and science is seen as opposing religion and therefore to be evil. In fact it is a conflict between scientific methods and the belief, without evidence, in Authority. It is a conflict between an open mind and one that is closed; a conflict about whether judgment is to be made on evidence or by decree. Law courts have been compared to churches, their Supreme Court to the Council of Bishops, their flag to the cross. Analogies like this are often misleading; in this case there is at least one clear distinction—upholding the law is based on evidence. No church dogma could be sustained by a properly run court of justice.

Science is more the child of superstition than the offspring of religion; indeed, some early scientific beliefs are now seen as forms of superstition, while superstitious old wives' tales have often offered pathways to scientific discoveries. Superstitious people are trying to understand some aspects of nature; the only problem is that they have lost their critical faculties. That is a great asset to them—they don't have to be careful in their thinking, or waste time checking on information they have heard about. You can get so much more done if you work that way. Facts can clip your imagination.

Say rather that facts can stimulate the imagination, that facts revealed by scientists grow ever more spectacular, that the theories behind them grow ever more subtle and beautiful, and that there is excitement in observation and experiment, and wonder in reasoning and theory, and exaltation in discovery.

> We are now threatened by self-inflicted, swiftly moving environmental alterations about whose long-term biological and ecological consequences we are still painfully ignorant—depletion of the protective ozone layer, a global warming unprecedented in the last 150 millennia, the obliteration of an acre of forest every second; the rapid fire extinction of species and the prospect of a global nuclear war that would put at risk most of the population of the Earth. . . . We are close to committing—many would argue that we are

> already committing—what in religious language is sometimes called Crimes against Creation. Problems of such magnitude and solutions demanding so broad a perspective must be recognized from the outset as having a religious as well as a scientific dimension. Mindful of our common responsibility we scientists . . . urgently appeal to the world religious community to commit, in word and deed and as boldly as required, to preserve the environment of the earth.
>
> (Carl Sagan, *American Journal of Physics* 88, no. 7, July 1990, p. 615)

In time past, as we have seen, the official attitude of the Christian church was that all things above the moon were perfect and incorruptible, whereas the lower region, especially the earth, was full of sin, death, and decay. The emphasis was entirely on the heavenly life hereafter, the earth not being worthy of study and care, and especially not of worship. The pagan worship of Mother Earth had been transformed into the worship of the Mother of God, and the wonders and beauty of nature were ignored.

Isn't it time, as Professor Sagan suggests, that Christian pulpits—indeed all pulpits—condemn the Crimes against the Creation that we are committing throughout the world? The fundamentalists feel very strongly about their belief in the way that Creation occurred. If they really believe that, what a powerful force they could be in speaking up against the ravaging of God's Creation. Isn't it more important to preserve this magnificent creation than to advance a particular dogma? Love this creation and learn to understand it; speak up against those who are destroying it.

At present the church leaders are lagging far behind the scientists in addressing this problem. They are even lagging behind many superstitious people who feel strongly about environmental pollution, and behind many lovers of nature who are concerned only marginally with superstition, religion, or science, people who see and feel the devastation and back up their feelings with action. Preachers, remind your congregations of the beginning of Psalm 24:

> "The earth is the Lord's, and the fulness thereof
> The world, and those who dwell therein."

It would make a fine text for a sermon.

In Memoriam

We mourn the victims of the tortures and murders that have been committed throughout history by warring nations, and by warring religious and political factions. We mourn the members of racial and religious minorities who have been persecuted and systematically slaughtered, the innocent women accused of witchcraft, the innocent men taken as hostages, the modern victims revealed by Amnesty International—the list is endless.

In the following pages we mourn in particular some of the men, and two women, who have been persecuted, at least partly, because of their dedication to freedom of thought and intellectual honesty. The motives of their enemies were mixed, as all motives are, and some of these scholars made enemies by both their scholarly and by their less-unbiased writings and speech. These thinkers have in common that over a period of more than two thousand years they have been humiliated, harassed, exiled, imprisoned, persecuted, tortured, and even burnt at the stake. Some of them were scientists, some others theologians, some philosophers, and their studies ranged from astronomy to biblical scholarship. There were many more, most of them now unnamed, and others who were careful to avoid persecution but who lived in constant fear of it: Johannes Tholde, inventing a fifteenth-century alchemist as the author of his own work; Michael Maestlin, rationalizing his observation that the comet of 1577 lay beyond the moon; Nicholaus Copernicus, afraid of punishment for suggesting that the earth moves around the sun; René Descartes, very conscious of the fate of Giordano Bruno; and Friedrich Spee, publishing anonymously in another town. These are just five examples of the fear that filled the hearts of so many who were trying to understand nature.

If science were superstition, some of the people listed in the following pages would be its gods; if it were a religion, they would be its saints. This book is dedicated to them and to what they and their oppressors have taught us about true science and true religion.

PYTHAGORAS
(582?–497? B.C.)

Greek philosopher and mathematician, who probably studied in Egypt and Babylon. He founded a school of science and religion at Croton in southern Italy. He and his successors established the theory of numbers and of plane and solid geometry, speculated that the earth is spherical, discovered relations between the lengths of strings that produce harmonious notes, and pioneered many other studies. After being expelled from Croton with some of his disciples in 510 B.C., he settled in Metapontion, a hundred miles up the coast on the Gulf of Tarantum. His followers in both towns were persecuted, and some massacred, over the next sixty years.

ANAXAGORAS OF CLAZOMENAE
(500–428 B.C.)

Greek philosopher. First teacher of natural philosophy in Athens. Dissected the brain, contributed to medicine and mathematics, and maintained that each object is composed of innumerable seeds that contain all of the properties of the object. The chaos of seeds was put into order by Mind. Did not believe that the sun was a god. Charged with impiety and indicted for rationalism (432 B.C.). Saved from imprisonment by Pericles, but exiled.

PROTAGORAS OF ABDERA
(485–411 B.C.)

First Sophist Greek philosopher. Systematized the study of grammar and practical logic. Accumulated a fortune by lecturing and teaching. Did not believe in absolute truth, and questioned the existence of gods. After being accused in Athens of impiety, his books were publicly burned, and he was exiled (or escaped from prison). He died when the ship taking him to freedom was wrecked.

SOCRATES
(470?–399 B.C.)

Greek philosopher who developed an original method of study and inquiry consisting of a series of carefully chosen questions. One of the greatest of thinkers, Socrates was contemptuous of conventional wisdom and ways of life. Attacked by Aristophanes and others, he was accused of impiety and the corruption of youth. His defiant defense brought condemnation and he was forced to drink hemlock while in prison, surrounded by his disciples.

ARISTARCHOS OF SAMOS
(Third Century B.C.)

Greek astronomer. Recognized that the sun is at the center of the solar system and that the other planets, including the earth, circled it, the moon moving around the earth, and the earth spinning on its axis. The charge of impiety was brought against him by the leading Stoic philosopher, Cleanthes of Assos, for "moving the hearth of the universe." His magnificent idea was rejected for nearly two thousand years.

HYPATIA
(?–415 A.D.)

Neoplatonist scholar. Wrote commentaries on the algebra, geometry, and astronomy of earlier Greek scholars and developed a philosophy in the spirit of Plato and Plotinus. She was appointed to the chair of philosophy at the Museum of Alexandria, but was later accused by Archbishop (later Saint!) Cyril of conspiring with Orestes, the pagan prefect of the city. Members of Cyril's staff dragged her into a church, stripped her, battered her to death, then dismembered and burned her remains. Immortalized in Charles Kingsley's "Hypatia" (1853).

ANICIUS MANLIUS SEVERINUS BOETHIUS
(480?–524)

Roman philosopher who summarized in Latin the works of Euclid, Archimedes, and Ptolemy. He also translated and wrote commentaries for Aristotle's *Organon* and Porphry's book on Aristotle. As consul under Theodoric, Ostrogoth ruler of Rome, he tried to build a bridge between paganism and Christianity. But he was resented by Gothic officials, falsely accused of plotting to depose Theodoric, and was sentenced to death. In prison he wrote his greatest work, "On the Consolation of Philosophy," using arguments of Greek (i.e., pagan) philosophy in discussion of morals and free will. He was executed in an abominable way 23 October 524. His writings had an enormous impact on medieval scholars.

DAMASCIOS OF DAMASCUS
(458?–?)

Greek philosopher and mathematician. and last Director of the Academy at Athens founded by Plato nine centuries earlier. He wrote a commentary on Hippocrates. The academy was seen as a school of pagan and perverse learning, and a threat to Christianity, so it was closed in 529 by Justinian I, ruler of the Eastern Roman Empire. Damascios fled with other scholars to the court of King Khosrau I of Persia, taking with them some of the main documents of Greek science.

SIMPLICIOS OF CILICIA
(Sixth Century)

Greek natural philosopher and neoplatonist who wrote commentaries on the works of Aristotle, Euclid, and Hippocrates. He used arguments about impetus and gravity to try to understand the stability of celestial bodies, and was working at Athens at the time of the purge by Justinian. He escaped with Damascios and others to the court of King Khosrau I and returned to Athens in 549.

AVICENNA (IBN-SINA)
(980–1037)

Persian physician and philosopher. His summary of medical knowledge, *The Canon of Medicine*, was a standard text in Europe until the seventeenth century. He recognized the retina as the essential organ of vision, suggested that the speed of light is finite, and wrote on alchemy although he rejected the transmutation of metals. He was critical of Aristotle's ideas about motion. He was appointed Vizier at Hamadan (present Iraq), but then was banished, restored, imprisoned, escaped, and fled to Isfahan.

PETER ABELARD
(1079–1142)

French philosopher and theologian who established a school, called "The Paraclete," at Nogent-sur-Seinet. As a "nominalist," he was opposed to Plato's theory of Ideas and to some of the writings of Aristotle. He argued from this point of view that the Trinity was three Gods, not one, and appealed to logic and reason to solve problems. He was persecuted and condemned for heresy by the Council of Sens (Yonne, France) in 1140. He died on the way to Rome to present his defense. His body was given to Héloise, prioress of the Paraclete, to whom he had been secretly married; it was because of this marriage that he had been mutilated by the command of her uncle.

AVERROËS (IBN RUSHD)
(1126–1198)

Spanish-Arabian philosopher and physician, and one of the most important medieval Muslim thinkers. He maintained that science must be concerned with examining and reflecting on the material things of the world, and that religion was not a branch of knowledge to be reduced to propositions and systems of dogma. He was accused of heretical opinions by al-Mansur, Almohade Caliph in Spain and was banished in 1195. He died at Marrakesh. His writings had a major influence on European scholars.

ALBERT THE GREAT (ALBERTUS MAGNUS)
(1206?–1280)

Great pioneer of the physical and life sciences—chemistry, physical geography, and botany in particular. He was the first scholar to criticize the methods of scholasticism. He believed that there could be human life on the other side of the earth. Because of his views, he was subjected to suspicion and indignity by the Dominican Order and was accused of sorcery. In his later years he escaped persecution by using the scholastic methods of deducing the truth from Scriptural texts.

ROGER BACON
(1214–1294)

English philosopher and scientist. He studied the ebb and flow of the tides, the nature of starlight, the anatomy of the eye, the reflection and refraction of light. He argued for the combined use of experiments and mathematical reasoning in studying nature, and rejected the principle of truth by authority. Bacon was also deeply involved in alchemy and astrology, and was imprisoned by the Franciscan Order in Paris under suspicion of heresy and black magic (1278–1292) by order of Jerome of Ascoli, head of the Franciscans and later Pope.

RAIMUNDUS LULLIUS (RAYMOND LULLY)
(1235–1315)

Physician, alchemist, and philosopher from Mallorca who was a forerunner of modern logistic thought. He was opposed to the teachings of Averroës and became an expert on Arabian science. He also founded a Franciscan school for missionaries. Believing that he could prove that faith and reason are compatible, he wrote nearly 300 works on medicine, alchemy, mathematics, and philosophy. He was imprisoned twice, banished in Tunis for preaching to Muslims, and was finally stoned to death in the third attempt.

CECCO D'ASCOLI (FRANCESCO DEGLI STABILI) (1269–1327)

Italian poet, philosopher, astronomer, and professor of astrology at the University of Bologna. He believed that there could be people living on the other side of the earth. For this and other heresies, and under suspicion of sorcery, he was driven from his professorship and burned alive at Florence. The Campo Santo at Pisa has a fresco by Andrea di Cione (Orcagna) depicting Cecco in the flames of hell.

JACQUES LEFÈVRE D'ETAPLES (1450–1537)

French scholar, theologian, and reformer who lived near Paris and was a leader of biblical Humanism in pre-Reformation France. He wrote on arithmetic, Euclid's elements, Aristotle, and the planetary harmonies. After being condemned as a heretic for his religious and critical writings, he fled to Strasbourg, France (1525).

CORNELIUS HEINRICH AGRIPPA VON NETTESHEIM (1486–1535)

German physician, theologian, and scholar of natural and occult science who taught at Dôle, France, where he urged a return to the beliefs of the early Church. As a consequence, he was denounced as a heretic and forced to leave. He later taught at Pavia. He did not believe that natural disasters were caused by witches. As Syndic of Metz (present capital of the Moselle district, France), he tried to save a woman who was on trial for witchcraft before the Inquisition. The chief Inquisitor had him imprisoned (1520), then exiled. Though he was the historiographer to Emperor Charles V, he was persecuted for occult beliefs and was driven from one place to another by the Dominicans who ultimately placed a malevolent epitaph on his grave after his death on February 18, 1535, in Grenoble, France.

WILLIAM TYNDALE
(1492?–1536)

English translator of the New Testament and the Pentateuch from the Greek versions. Opposed by Cuthbert Tunstall, Bishop of London. After visiting Martin Luther at Wittenberg, he began the printing of the gospels of Matthew and Mark in Cologne, but was stopped by a warning from Henry VIII and Cardinal Wolsey. He then moved to Worms and printed three thousand copies of the New Testament in English, which were suppressed and mostly destroyed by English bishops. After being betrayed in Flanders by Henry Phillips, a Roman Catholic zealot, he was finally imprisoned at Vilvorde, Belgium, where he was partially strangled and, while still conscious, burned at the stake.

GERONIMO CARDANO
(1501–1576)

Italian mathematician and medical doctor, who was not admitted to the College of Physicians because he was illegitimate. He wrote books on mathematics, medicine, physics, cosmology, astrology, and made outstanding contributions to algebra. In one book, a best-seller in several languages, he included a dialogue between a Christian, a Jew, a Muslim, and a pagan. After being arrested for heresy and debt in 1570, he was jailed, lost his professorship at Bologna, was forced to leave the city, and forbidden to lecture or publish books.

BERNARD PALISSY
(1510?–1589)

French natural philosopher, glass-maker, and portrait painter. He was the inventor of pure white enamel and of Palissy ware, and he made figurines for Catherine de Medici. He also lectured on science and recognized fossils as formerly living animals, noting their similarity to creatures that were alive at his time. He followed Aristotle in denying the existence of a vacuum, and believed that a fire under the earth was responsible for earthquakes and volcanoes. After being arrested as a Huguenot and sent to the Bastille, he was hanged and burned for heresy.

MICHAEL SERVETUS
(1511–1553)

Spanish theologian and practicing physician who lectured on geometry, astronomy, and medicine, and discovered pulmonary circulation and blood purification by the lungs, partly anticipating Colombo (1588) and Harvey (1628). He opposed the doctrine of the Trinity and stated that Ptolemy had written that Judea was barren and inhospitable rather than "a land of milk and honey" (Exodus 2:8 and 33:3). After being arrested and brought to trial before the Inquisition at Vienne (France), he escaped but was captured at Geneva. Calvin acccused him of having "necessarily inculpated Moses and grievously outraged the Holy Ghost." At Calvin's request, he was imprisoned and burned at the stake as a heretic.

GERHARDUS MERCATOR (GERHARD KREMER)
(1512–1594)

Flemish mathematician, geographer, and map-maker who established a shop at Louvain for making maps, celestial globes, and scientific instruments. He was arrested in 1544 and prosecuted for heresy. Subsequently, he moved to Duisbourg, near Düsseldorf, where he published the first map on "Mercator's projection" (1568), and began an atlas which was completed by his son.

ANDREAS VESALIUS
(1514–1564)

Flemish anatomist and physician who performed many human dissections, almost the first since the school at Alexandria 1500 years earlier. He discovered that man does not have a missing rib from which Eve had emerged, and does not have an "imponderable bone," the nucleus for resurrection. He corrected many errors handed down from Galen, who had dissected animals. As physician to Emperor Charles V, he was protected from persecution but was in immediate danger from Charles's son and successor, Philip II, King of Spain (1558), who expanded and encouraged the Spanish Inquisition. He was then persecuted, and with no opportunity to continue his work, he made a pilgrimage to the Holy Land, believed to be in fulfillment of commutation of his death sentence. En route he was shipwrecked and drowned.

SEBASTIANUS CASTELLIO
(1515–1563)

French Protestant theologian and humanist. He was forced to relinquish the rectorship at Geneva (1545) because of deviation from Calvinist doctrine. He threw light on the Song of Songs. He was driven to starvation and death by Calvin and his successor, Theodorus Beza.

PETRUS RAMUS
(1515–1572)

French philosopher, mathematician, and reformer of sciences and humanities. He wrote a thesis (1536) entitled "All the things that Aristotle said are wrong." He somehow avoided trouble from this but later wrote many books on grammar and mathematics, which were censured by the Royal Council in 1544. Always critical of authority, he became a Protestant and was murdered on August 26, 1572, toward the end of the Massacre of St. Bartholomew's Day.

DIETRICH FLADE
(?–1589)

Rector of the University of Trèves and a revered scholar. As Chief Judge of the Electoral Court, he passed judgments on magicians and witches, but realizing the injustice of the sentences, he eventually refused to condemn them. After being arrested by authority of the Archbishop, he was forced to confess under torture that he had sold himself to Satan. He was strangled, and his body burned.

LUIS DE LEON
(1527–1591)

Spanish Augustinian monk, scholar, and lyric poet who threw new light on the Song of Solomon and other Old Testament books. He was kept in a dungeon by the Inquisition for five years until, a broken man, he agreed to rewrite his commentary on the Song.

CORNELIUS LOOS
(?–1595?)

Professor at the University of Trèves and a staunch Catholic, he wrote a book called *True and False Magic*, criticizing sorcery and questioning occult powers. Since this was opposed to certain biblical passages, publication was stopped and the manuscript destroyed. After being thrown into a dungeon, he was forced (1593) to recant on his knees before Church dignitaries. He was imprisoned, kept under surveillance, and died from the plague. The Jesuit Delrio declared that he would have been burned at the stake if the plague had spared him.

GIORDANO BRUNO
(1548–Feb. 17, 1600)

Italian philosopher who was a champion of Lucretius and of Copernicus's model of the solar system, and was critical of the logic of Aristotle. He was a follower of Raimundus Lullius in believing that all truth could be demonstrated by reason. He conjectured that stars could be centers of other planetary systems. His ideas marked the turning point from medieval to modern scientific thinking. Accused of heresy in Toulouse (1576), he fled and wandered throughout Europe. He was excommunicated by the Lutheran Church in Helmstedt, Germany, in 1589, and was arrested by the Inquisition in Venice, Italy, in 1593. He was then transferred to Rome, imprisoned, tried, and was finally burned at the stake in Campo dei Fiori, Rome.

MARCO ANTONIO DE DOMINIS
(1566–1624)

Archbishop of Spalato (present Split, Yugoslavia) who was involved in a quarrel between the papacy and Venice. He moved to England, converted to Anglicanism, and became Dean of Windsor. He studied the properties of light, saw color as a property of reflected light, used the principle of refraction to partially explain the rainbow, and attributed tides to the effects of the sun and moon. Returning to Italy in 1622, he was imprisoned by the Inquisition for heresy and sorcery, and died in a dungeon. His body and writings were publicly burned.

GALILEO GALILEI
(1564–1642)

Italian physicist and astronomer, who discovered isochronism of the pendulum and experimentally analyzed motions of projectiles and falling bodies. He also improved refracting telescopes and discovered the four largest moons of Jupiter, thereby offering strong support to the theory of Copernicus. Bishops and priests warned their congregations that this was heresy, as well as his discovery of mountains on the moon and spots on the sun. In 1615, he was summoned before the Inquisition at Rome. "All writings that affirm the motion of the earth" were condemned by papal bull and placed on the Index. Returning to Florence, he wrote his "Dialogue" (1632), was summoned again to the Inquisition in Rome, threatened with torture, imprisoned by command of the pope, and forced to recant publicly on his knees.

TOMMASO CAMPANELLA
(1568–1639)

Italian philosopher and opponent of scholasticism and Aristotelian logic. He defended Galileo's research and wondered if the inhabitants of other planets also thought themselves to be at the center of the universe. He asserted that self-consciousness is the basic fact of knowledge, anticipating Descartes; advocated the experimental method; and argued that all bodies would fall at the same speed in a vacuum. He was imprisoned in Naples for twenty-seven years (1599–1626) by the Spanish Inquisition, was tortured seven times, was released by Pope Urban VIII, re-arrested, and released again (1628). He finaly fled to Paris.

LUCITIO VANINI
(1585–1619)

Originator of a mechanistic philosophy of natural development, according to which there is a gradation from the lowest to the highest created beings. He was sentenced at Toulouse to have his tongue torn out, and to be burned alive.

URIEL ACOSTA
(1591?–1647)

Portuguese philosopher who left Portugal to escape the Inquisition. He converted from Roman Catholicism to Judaism and joined a synagogue in Amsterdam. Because he believed that the Old Testament did not imply the immortality of the soul, and that Moses's law was of human origin, he was arrested three times. He was forced to confess and recant to the Jewish congregation, and was stripped, scourged, and stepped on. He finally committed suicide. "All evils," he wrote, "come from not following Right Reason and the Law of Nature."

GIOVANNI ALFONSO BORELLI
(1608–1679)

Italian physicist and astronomer. He was one of the founders of Accademia del Cimento (1657), under Prince Leopold de'Medici, with "the obligation to investigate Nature by the pure light of experiment." Leopold was promoted to Cardinal, so that the Academy could be accused of being irreligious. Continuous attacks were successful, and the Academy was forced to disband after ten years. Borelli died a pauper.

ISAAC DE LA PEYRÈRE
(1594–1676)

French Protestant who wrote a book called *Preadamites*, about men who existed before the time of Adam, criticizing the biblical account and the forgeries of some early Christian zealots. He was thrown into prison by the Grand Vicar of the Archdiocese of Mechlin (presently Belgium), and was forced to retract his Protestantism and his ideas of human life before Adam. The Parliament of Paris ordered copies of his book to be burned by the executioner.

BARUCH SPINOZA
(1632–1677)

Dutch philosopher and follower of Descartes who showed that Moses could not have been the author of the Pentateuch in the form existing at that time. He argued that claims for inerrancy of biblical books cannot be maintained and that religious teaching should emphasize conduct, not creed. Consequently, he was excommunicated from his synagogue and barely escaped assassination by a fanatic. He backed up his arguments with detailed scholarship in his *Tractatus Theologico-Politicus* (1670), the second edition (1674) of which was banned. Because he was abhorred as a heretic also by Christians, and regarded as a forerunner of the Antichrist, his main works were published only after his death. A monument to him was erected in Amsterdam two hundred years after his death, but it was preached in churches and synagogues that this would bring the wrath of Heaven, and the statue had to be protected by the police.

BALTHASAR BEKKER
(1634–1698)

Dutch Protestant theologian, suspected of being a follower of Descartes. In his book (1691) *De Betoverde Wereld* (*The Bewitched World*), he questioned the devil's bodily existence, and condemned belief in witchcraft and magical powers. As a result, he was tried by the Synod of his Church, expelled from his pulpit, and condemned as a heretic.

JUANA INÉS DE LA CRUZ
(1651–1695)

Spanish nun, poet, and scholar, born in Mexico, lady-in-waiting to the wife of the viceroy of New Spain. She acquired a sizeable library for that time and place, and when she entered San Jerónimo Convent, she took her library with her. She was the first woman to write about the status of women and to recommend changes. She studied physical phenomena, including the motion of a top, and is reported to have communicated with Isaac Newton on the subject. For her activities, she was condemned by the bishop, had her library confiscated, and was forced to abandon her work and ask forgiveness (1691). She spent the rest of her life in the convent, saddened by having the things she loved prohibited by the church she loved.

THOMAS WOOLSTON
(1670–1733)

English deist who challenged the clergy in freethinking tracts and questioned the miracles of the New Testament in his *Discourses* (1727–1729). Arrested, fined (1730), and jailed until his death.

GEORGES LOUIS LECLERC, COMTE DE BUFFON
(1707–1788)

French naturalist and Director of Jardin du Roi and Royal Museum. He studied the variation of species and had early ideas about evolution. After being reminded by the Sorbonne that "in the beginning God made the heavens and the earth," he was forced by the faculty of theology to recant publicly and to print his recantation: "I abandon everything in my book respecting the formation of the earth, and generally all which may be contrary to the narrative of Moses."

ANTOINE LAURENT LAVOISIER
(1743–1794)

Wealthy French chemist, considered the founder of modern chemistry. He conducted quantitative measurements, disproved the theory of phlogiston, explained combustion, and determined the composition of various organic compounds. After being denounced by the terrorist and third-rate chemist Marat, "Friend of the People," he was arrested by order of the Convention of the French Revolution and was guillotined May 8, 1794. "It took only a moment to cut off his head, and a hundred years may not give us another like it." (Attributed to Joseph Louis Lagrange.)

JOSEPH PRIESTLEY
(1733–1804)

English clergyman and chemist who rejected miracles and believed that the soul dies with the body. He was the first to publish the discovery of oxygen (by focusing a burning lens on cinnabar), and he also investigated the properties of other gases, such as ammonia, sulphur dioxide, and carbon monoxide. His appointment to the scientific expedition of Captain Cook (1772) was successfully opposed by the Cambridge and Oxford clergy. His house, library, manuscripts, and laboratory in Birmingham were burned (1791) partly because of his sympathy with the French Revolution, and partly because the mob threatened "to kill all philosophers." He fled to London, where he was branded as a heretic by Edmund Burke. He then fled to the United States, but was threatened with deportation as an undesirable alien. He was saved by the election of Thomas Jefferson as president.

JOHN WILLIAM COLENSO
(1814–1883)

English Bishop of Natal, South Africa, who translated the New Testament into Zulu. He fought against the doctrine of eternal punishment, declared the Pentateuch a postexile forgery, and that in all of these books there is much that is mythical and legendary. He was deposed and excommunicated by the Bishop of Cape Town (1863) and the congregation was enjoined to treat him as "a heathen man and a publican." He was exonerated by the British courts but was socially ostracized, branded by the English populace as an "infidel" and "traitor."

JOHN THOMAS SCOPES
(1900–1970)

Science teacher and football coach at Rhea County High School, Dayton, Tennessee, U.S.A. He taught the theory of evolution in defiance of state law against teaching it in tax-supported schools. As a result, he was tried

(July 1925), convicted, and fined $100. Charles Darrow defended him against prosecutor William Jennings Bryan. His conviction was reversed by the Tennessee Supreme Court on technical grounds. The anti-evolution law was in force until 1967. Similar laws in other states led publishers to omit evolution from high school textbooks.

Helpful Despots

Fortunately, there is another side to the eternal conflict between free inquiry and oppression. Every now and then there have arisen kings and dictators who were sympathetic to the excitement of new scientific knowledge and, in some cases, of the arts and humanities as well. The men whose names are given later in this section acquired their power by inheritance, intrigue, or invasion. Despite ruthless extermination of their enemies, they had the foresight to use some of their power so acquired to provide havens where scholars could come together in relative peace. Strangely, three of these men were named Frederick II. We should be grateful to all of them for wanting to understand various aspects of nature, and for supporting many studies without the requirement of submission to an ecclesiastical authority.

ASHURBANIPAL OF ASSYRIA
(died 625 B.C.)

Sent scholars throughout Mesopotamia to collect ancient Sumerian and Akkadian cuneiform tablets for his great library in Nineveh. Copies and bilingual (Sumerian-Akkadian) editions of the ancient Mesopotamian literary corpus were made.

PERICLES OF ATHENS
(died 429 B.C.)

Made Athens a center for art, architecture, and literature, and supported Anaxagoras, Phidias, and other scholars and artists.

ALEXANDER III (THE GREAT), KING OF MACEDONIA (356–323 B.C.)

Sent back fauna and flora from his expeditions for his former teacher Aristotle and Theophrastos to catalogue and study. Even when on the march, he surrounded himself with scholars and scientists. He wrote to Aristotle, "I had rather surpass others in the knowledge of what is excellent, than in the extent of my power and dominion." In that spirit, he supported a number of scientific enquiries.

PTOLEMAIOS SOTER (PTOLEMY I) (367–283 B.C.)

King of Egypt and founder of the thirty-first dynasty. He served as a general in Alexander's army, and was awarded Egypt and Libya. He founded the library and museum of Alexandria and made it a haven for scholars.

PTOLEMAIOS PHILADELPHOS (PTOLEMY II) (309–246 B.C.)

He succeeded his father as King of Egypt and made Alexandria the center of Hellenistic culture, encouraging art, literature, and medicine. Ptolemy II is said to have arranged the translation of the Hebrew scriptures into Greek.

THEODOSIUS II (401–450)

Ruled the Eastern Roman Empire jointly with his sister Pulcheria. He established a university at Constantinople (425) with thirty-one faculty members, some in philosophy, and law, but mostly in Greek and Latin grammar, literature, history, and rhetoric. He set up a library and supported the copying of Greek manuscripts, thereby transmitting them to posterity. (Some were destroyed when Constantinople was pillaged by the Western Christians in 1204.) He also encouraged poetry, e.g., Musaeus's *Hero and Leander*. In 421, he married Eudocia, the daughter of the philosopher Leontius of Athens.

KHOSRAU I (died 579 A.D.)

King of Persia who provided a haven for refugees from Plato's Academy when it was closed by Justinian I. Pahlavi literature flourished under him.

HARUN AL-RASHID
(764?–809)

Caliph of Baghdad, described in "Arabian Nights," who made Baghdad the center of Arabic culture. He was a poet and scholar and supported medicine and the arts, offering his own criticism of the ideas and performances of those he invited to join him.

AL-MA'MUN
(786–833)

Caliph of Baghdad and son of Harun al-Rashid. He established an academy of science and a library, collected from other cities many Greek and Hindu manuscripts, and paid for their translation into Arabic. Al-Ma'mun offered limited freedom of thought and religion, including in his council representatives of all major faiths. He also established observatories at Baghdad and at Tadmor; supported the arts, sciences, philosophy, and letters; and invited scholars from all over the world, making Baghdad the premier center of learning.

CHARLEMAGNE
(742–814)

King of the Franks (768–814), Holy Roman Emperor of the West (800–814), and patron of literature, science, and the arts. He invited foreign scholars to restore the schools of France, and to the royal palace at Aachen. At the latter school he himself and his family became students, studying astronomy, Latin, and rhetoric.

CHARLES THE BALD
(833–877)

King of France (840–877) who supported the palace school started by Charlemagne, and improved it greatly by his encouragement and protection. He invited Irish and English scholars to participate, notably the Irish scholar Erigena, the first original thinker among the Scholastics.

ALFRED THE GREAT
(849–901)

King of the West Saxons (871–901) and King of England (886–901). He established a palace school at his capital, Reading, giving one-eighth of the income of the realm for educational and religious works in monasteries and churches. Alfred sent abroad for scholars, including Erigena, and supported them. He had others read and translate for him until he learned enough Latin to translate works of Boethius and Bede into English himself. In addition, he arranged for the translation of other Latin manuscripts. and helped to restore the libraries that had been destroyed by a series of invasions by the Danes during the previous hundred years.

SAMAN

Patriarch of the Samanid dynasty of eastern Persia, centered in Bokhara (874-999). He was a Zoroastrian noble from Balkh, Afghanistan, whose great-grandson was the first ruler of a dynasty that continuously supported literature, the arts, science, and medicine. Rhazes' ten-volume survey of medicine was dedicated to one of the Samanid princes. Protection and use of the fine library was given to the young Avicenna, who later became a great philosopher and physician. Saman made Bokhara and Samarkand rivals to Baghdad as centers of learning and art.

AL-HAKAM II
(913?–976)

Caliph of Cordoba, Spain (961–976) and a patron of learning. He made the University of Cordoba the greatest educational institution in the world at that time. He encouraged studies in astronomy, medicine, mathematics, and other areas; invited scholars from abroad; and made funds available for studies. He also developed a library of over 400,000 catalogued volumes, employing agents to collect books from distant countries. (Those books judged inconsistent with the Sunni Mohammedan sect were destroyed by Hakam's son and successor, Hisham II.)

MAHMUD OF GHAZNI
(971?–1030)

Dominant member of the Ghaznevid dynasty that ruled (977–1186) southwest Asia from his capital at Ghazni, Pakistan. After conquering the Samanids, he expanded his empire by multiple invasions of India. He supported a number of scholars, including al-Biruni; built a university and library; and kept the great Persian poet Firdausi on his staff until Firdausi fled after writing a satire about him. He then issued a command for Avicenna to join his staff, but Avicenna refused and went into hiding.

ABU-ALI-MANSUR AL-HAKIM
(985–1021)

Fatimid Caliph of Cairo who destroyed the Holy Sepulcher at Jerusalem, persecuted Christians and Jews, and declared himself an incarnation of the Almighty. As a result, he was referred to as "The Mad Caliph." Nonetheless, he sponsored scholars in medicine and astronomy, and the *Hakimite Astronomical Tables* were named for him.

FREDERICK II OF HOHENSTAUFEN
(1194–1250)

King of Sicily (1198–1212), Holy Roman Emperor (1215–1250), speaker of nine languages, patron of science and literature, and founder of the University of Naples (1224). He kept a large menagerie, and even traveled with it, in order to study animal behavior. He also performed experiments on artificial incubation, supported dissection of cadavers and regulated the medical profession. Frederick opposed trial by fire as not leading to the truth and doubted that the world was created in six days. He was excommunicated three times (1227, 1239, 1245).

ALFONSO X (THE WISE)
(1221–1284)

King of Castile and Léon. A founder of Spanish Science, he established Latin and Arabic Colleges in Seville, and the University of Salamanca, which helped communicate Arabic knowledge to Europe. He sponsored translations of the Old Testament by Jews in Toledo. He also prepared the best planetary tables of the Middle Ages (1252) and other books on astronomy. The moon crater Alphonsus is named in his honor.

ALFONSO V (THE MAGNANIMOUS)
(1385–1458)

King of Naples who supported Latin classical scholars financially, and protected them from the Inquisition. He spent much time with scholars, leaving his brother to rule most of the kingdom, and even had Latin classics read to him during meals. Among those he supported were Lorenzo Valla, referred to in chapter 14, and Poggio Bracciolini (1380–1459), who discovered in various monasteries some lost writings of Cicero, Quintillian, Plautus, and Tacitus. We are specially indebted to Poggio for finding a copy of Lucretius's *De Rerum Natura.*

GIOVANNI DE'MEDICI
(1475–1521)

He was appointed cardinal at the age of thirteen and became Pope Leo X in 1513. He united the College of the Holy Palace and City College to form the University of Rome, brought in and supported many scholars, expanded the medical school, introduced the teaching of Hebrew, supported a translation of the Old Testament into Latin, and encouraged the study of Greek and the writing of a Greek-Latin dictionary. In additon, he built up a library with many ancient manuscripts, supported archeology, encouraged scholarship throughout Italy, opened a theater on the Capitol, and supported music, poetry, and art.

WILLIAM IV OF HESSE-KASSEL
(1532–1592)

Called "The Wise." A pioneer astronomer, he built the first observatory with a revolving dome (1561) at Kassel, Germany, and supported astronomers. After an extended meeting with Tycho Brahe, he urged King Frederick II of Denmark and Norway to provide him with an observatory.

FREDERICK II
(1534–1588)

King of Denmark and Norway, and patron of astronomy, philosophy, and the arts. He gave to Tycho Brahe the island of Hveen (1576), on which to build a house and observatory, and also provided Brahe funds for construction and an excellent salary, on recommendation by William IV of Kassel. The work of Tycho Brahe at the observatory of Uraniborg, which was built at his direction, transformed observational astronomy.

FREDERICK II (FREDERICK THE GREAT)
(1712–1786)

King of Prussia (1740–1786). He decreed religious toleration and invited Voltaire and the materialistic philosopher La Mettrie to live at his court. He was a notable patron of philosophy and literature, especially in the earlier years of his reign, but, like other monarchs of the time, did not allow "academic" thinking to influence his imperial policies.

Select Bibliography

ABBREVIATIONS

ADW	Andrew D. White. *A History of the Warfare of Science with Theology in Christendom.*
CB	*Pictorial Cyclopaedia of Biography.* Edited by Francis L. Hawks.
daV	*The Notebooks of Leonardo da Vinci.* Edited by Jean Paul Richter.
Du	Will and Ariel Durant. *The Story of Civilization.*
GS	George Sarton. *A History of Science.*
HCl	*Harvard Classics.*
HFO	Henry Fairfield Osborn. *From the Greeks to Darwin.*
L	Lucretius. *On the Nature of Things.*
UA	*The Universal Anthology.* Richard Garnett. Edited by Leon Vallée and Alois Brandl.
We	Jesse L. Weston. *From Ritual to Romance.*
WP	*The World of Physics.* Jefferson Hane Weaver.

GENERAL REFERENCES

Some scholarly books have been extremely helpful during the writing of almost every chapter of this book:

Bartlett, John. *Familiar Quotations.* Boston: Little, Brown and Company, 1940.

Cane, Philip. *Giants of Science.* New York: Pyramid Books, 1966.

*Debus, A. G., ed. *World Who's Who in Science.* Chicago: Marquis, 1968.

*Durant, Will and Ariel. *The Story of Civilization.* 11 vols. New York: Simon and Schuster, 1935–1967.

*Eliot, Charles W., ed. *Harvard Classics.* 50 vols. New York: P. F. Collier and Son Corporation, 1938.

---

*References marked with an asterisk are those that were especially useful.

*Garnett, Richard. *The Universal Anthology.* Edited by Leon Vallée and Alois Brandl. 33 vols. London: Clarke Co., 1899.
*Garraty, John A., and Peter Gay, eds. *The Columbia History of the World.* New York: Harper and Row, 1972.
*Hawks, Francis L., ed. *Pictorial Cyclopaedia of Biography.* New York: D. Appleton and Company, 1856.
Mackay, Alan L. *The Harvest of a Quiet Eye.* London: Institute of Physics, 1977.
Sambursky, S. *Physical Thought from the Pre-socratics to the Quantum Physicists.* New York: Pica, 1974.
* *Webster's Biographical Dictionary.* Springfield, Mass.: G. and C. Merriam Co., 1943 and 1983.
Webster's New Collegiate Dictionary. Springfield, Mass.: G. and C. Merriam Co., 1953.
* *Webster's New International Dictionary of the English Language.* Springfield, Mass: G. and C. Merriam Co., 1925.
*Wells, H. G. *The Outline of History,* Garden City, N.Y.: Doubleday, 1956.
Westfall R. S., and V. E. Thoren, eds. *Steps in the Scientific Tradition: Readings in the History of Science.* New York: J. Wiley & Sons, 1968.

I have also used the King James Bible, the Revised Version, the new Oxford Bible, *The Bible as Literature*, and *Holy Bible: From the Ancient Eastern Text* by G. M. Lamsa (New York: Harper and Row, 1986). The following magazines and other periodicals have also supplied material: *American Journal of Physics, Associated Press, Free Inquiry, Horizon, Mississippi Press, Mobile Press Register, Nature, Physics Today, Science Digest, Science News, Secular Humanist Bulletin, Skeptical Inquirer, Sun-Herald, Time, USA Today, U.S. News and World Report, The Washington Spectator,* and *World Press Review.*

In addition, many books that I have consulted refer to particular periods or disciplines. The following lists the works consulted by chapter:

CHAPTERS 1–6

Allen, Reginald E., ed. *Greek Philosophy: Thales to Aristotle.* New York: Free Press, 1966.
Archimedes. *The Works of Archimedes.* Edited by W. D. Ross. Oxford: Oxford University Press, 1922.
Baumgartner, Anne S. *A Comprehensive Dictionary of the Gods.* New York: Carol Communications, 1984. (Especially helpful for chapter 1.)
Bulfinch, Thomas. *Mythology.* New York: Modern Library.
Clagett, Marshall. *Greek Science in Antiquity.* London: Collier Books, 1962.
Cotterell, Arthur. *The Macmillan Illustrated Encyclopedia of Myths and Legends.* New York: Macmillan Publishing Company, 1989.
Dreyer, J. L. E. *A History of Astronomy from Thales to Kepler.* Revised by W. H. Stahl. New York: Dover, 1953.
Gibbon, Edward. *The Decline and Fall of the Roman Empire.* William Benton,

publisher, 1952.
Goodrich, Norma Lorre. *Ancient Myths*. New York: Mentor Book, New American Library, Times-Mirror, 1960.
Gray, Louis Herbert, ed. *The Mythology of All Races*. 13 vols. New York: Cooper Square Publishers Inc., 1964.
Hamilton, Edith. *The Greek Way*. New York: Mentor, New American Library, 1962.
Hawkes, Jacquetta. *The Atlas of Early Man*. New York: St. Martin's Press, 1976.
Kaplan, Justin D., ed. *The Pocket Aristotle*. New York: The Pocket Library, 1956.
Lindeman, Eduard C., ed. *Plutarch's Lives*. New York: Mentor, New American Library of World Literature, 1957.
*Lucretius. *On the Nature of Things*. Translated by Palmer Bovie. New York: Times Mirror, 1974.
Neugebauer, O. *The Exact Sciences in Antiquity*. 2nd ed. New York: Dover, 1969.
Plato. *Great Dialogues of Plato*. Edited by Eric H. Warmington and Philip G. Rouse. Translated by W. H. D. Rouse. New York: Mentor, New American Library of World Literature, 1963.
*Sarton, George. *A History of Science*. New York: W. Norton and Co., 1959.
*Smith, Sir William. *Smaller Classical Dictionary*. Revised by E. L. Blakeney and John Warrington. New York: E. P. Dutton and Co., 1958.
Time (Oct. 30, 1989): p. 80. A description of the 1989 findings at Nimrud.
Weston, Jessie L. *From Ritual to Romance*. New York: Doubleday Anchor Books, 1957.

CHAPTER 7

Haggard, Howard W. *The Doctor in History*. New York: Dorset Press, 1989.
———. *Devils, Drugs, and Doctors*. New York: Harper and Row, 1929.
Zinsser, Hans. *Rats, Lice and History*. New York: Bantam Books, 1960.

CHAPTER 8

A Collection from Mysteries of the Unknown. New York: Time-Life Books, 1989.
Frazier, Kendrick, ed. *Skeptical Inquirer*. Buffalo, N.Y.: CSICOP.

CHAPTERS 9 AND 11

For these chapters the literature is enormous. The most useful scholarly work, one that I use often, is: White, Andrew D. *A History of the Warfare of Science with Theology in Christendom*, 1896 (Dover edition, 1960). It was also used for chapters 13 and 14).

Also:

Cross, F. L., and E. A. Livingstone, eds. *The Oxford Dictionary of the Christian Church*. 2nd ed. Oxford: Oxford University Press, 1974.

Delaney, John I. *Dictionary of Saints.* Garden City, N.Y.: Doubleday and Company Inc., 1980.
Drake, Durant. *Problems of Religion.* New York: Houghton Mifflin Company, 1916.
Draper, John William. *History of the Conflict between Religion and Science.* New York: D. Appleton and Company, 1897.
Eliade, Mircea, ed. *The Encyclopedia of Religion.* 16 vols. New York: Macmillan Publishing Company, 1987.
Gaer, Joseph. *How the Great Religions Began.* New York: American Library, 1929.
Grunwald, Henry Anatole. "André Malraux: The Gods in Art." *Horizon* 1 (2) (November 1958): p. 4.
Harden, John A. *Modern Catholic Dictionary.* Garden City, N.Y.: Doubleday and Company Inc., 1980.
Kent, Charles Foster. *The Life and Teachings of Jesus.* New York: Charles Scribner's Sons, 1913.
Stein, Gordon, ed. *The Encyclopedia of Unbelief.* Buffalo, N.Y.: Prometheus Books, 1985.
Wells, Donald A. *God, Man, and the Thinker.* New York: Random House, 1962.

CHAPTER 10

I found Durant's *The Story of Civilization* very helpful, along with most of the books listed at the beginning of this bibliography under general references. Of special value was A. G. Debus's *World Who's Who in Science*, and some of the books listed under chapters 7 and 14. See also: Duggan, Alfred. "Richard and Saladin." *Horizon* 1 (3) (January 1959): p. 87.

CHAPTER 12

Galilei, Galileo. *Dialogue concerning the Two Chief World Systems—Ptolemaic and Copernican.* Translated by Stillman Drake. 2nd ed. Berkeley: University of California Press, 1967.
Kearney, Hugh. *Science and Change (1500–1700).* New York: World University Library, McGraw-Hill, 1971.
*Koestler, Arthur. *The Sleepwalkers.* London: Penguin Books, 1964.
Plumb, J. H. *The Penguin Book of the Renaissance.* London: Penguin Books, 1964.
Richter, Jean Paul, ed. *The Notebooks of Leonardo da Vinci.* New York: Dover Publications, 1970.
Sarton, George. *Six Wings: Men of Science in the Renaissance.* Cleveland: World Publishing Company, 1966.
Wilson, Derek. *The Circumnavigators.* New York: M. Evans and Company, 1989.

CHAPTER 13

*Chicago, Judy. *The Dinner Party*. New York: Anchor Books, Doubleday, 1979.
Trevor-Roper, H. R. "The Persecution of Witches." *Horizon* 2 (2) (November 1959): p. 57.

CHAPTER 14

Greenblatt, Robert B. *Search the Scriptures.* Totowa, N.J.: Barnes and Noble, 1985.
Mathison, Richard R. *The Eternal Search.* New York: G. P. Putnam's Sons, 1958.
Pachter, Henry M. *Magic Into Science: The Story of Paracelsus.* New York: Henry Schuman, 1951.
Rapport, Samuel, and Helen Wright. *Great Adventures in Medicine.* New York: Dial Press, 1961.
Singer, Charles, and E. Ashworth Underwood. *A Short History of Medicine.* Oxford: Oxford University Press, 1962.
Williams, Guy. *The Age of Miracles: Medicine and Surgery in the Nineteenth Century.* Chicago: Academy Chicago Publishers, 1987.

CHAPTER 15

Cooper, Lane. *Aristotle, Galileo and the Tower of Pisa.* Ithaca, N.Y.: Cornell University Press, 1935.
Kuhn, Thomas S. *The Structure of Scientific Revolutions.* 2nd ed. Chicago: University of Chicago Press, 1990.
McGuire, J. F., and P. M. Rattensi. *Notes and Records of the Royal Society* (December 1966). London: Royal Society.
Thorndyke, Lynn. *A History of Magic and Experimental Science.* 8 vols. New York: Columbia University Press, 1958.
Thorpe, Sir Edward. *History of Chemistry.* Vol. 2: 1850–1910. London: Franklin Watts and Co., 1910.
Weaver, Jefferson H. *The World of Physics.* New York: Simon and Schuster, 1987.

CHAPTER 16

Bronowski, J. *The Ascent of Man.* Boston: Little, Brown and Company, 1973.
Lee, Mrs. R. (formerly Mrs. T. Ed Bowdich). *Memoirs of Baron Cuvier.* New York: J. and J. Harper, 1833.
*Osborn, Henry Fairfield. *From the Greeks to Darwin.* New York: Charles Scribner's Sons, 1929.
Rimmer, Harry. *The Theory of Evolution and the Facts of Science.* Grand Rapids, Mich.: William B. Eerdmans Publishing Co., 1946.

CHAPTER 17

For excellent discussions of modern physics, astronomy, and religion, see:

Davies, Paul. *God and the New Physics.* New York: Touchstone, Simon and Schuster, 1983.

Hawking, Stephen W. *A Brief History of Time.* New York: Bantam Books, 1988.

Herbert, Nick. *Quantum Reality.* Garden City, N.Y.: Anchor Books, Doubleday, 1987.

Kaku, Michio, and Jennifer Trainer. *Beyond Einstein.* 2nd ed. New York: Bantam Books, 1987.

EXTRA READING

(I cannot claim to have read all of these.)

Andersson, T. H., N. F. Garsoïan, H. L. Kessler, J. Leyerle, and A. L. Udovitch, eds. *Dictionary of the Middle Ages.* 12 vols. New York: Charles Scribner and Sons, 1982.

Boorstin, Daniel. *The Discoverers.* New York: Vintage Books, Random House, 1985.

Browne, E. Martin, ed. *Religious Drama.* New York: Living Age Books, Meridian Books, 1958.

Budge, Sir E. A. Wallis. *The Gods of the Egyptians.* New York: Dover reprint, 1990.

Campbell, Joseph. *Myths to Live By.* New York: Bantam Books, 1973.

———. *The Masks of God* and *The Hero with a Thousand Faces.* New York: Barnes and Noble reprints, 1990.

Davies, P. C. W., and R. Brown. *The Ghost in the Atom.* Cambridge: Cambridge University Press, 1986.

Ecker, Ronald L. *Dictionary of Science and Creationism.* Buffalo, N.Y.: Prometheus, 1990.

Eldridge, Niles. *Time Frames: The Rethinking of Darwinian Evolution and the Theory of Punctuated Equilibria.* New York: Simon and Schuster, 1985.

Gould, Stephen Jay. *Wonderful Life.* New York: W. W. Norton & Co., 1989.

Hofstadter, Richard. *Anti-Intellectualism in American Life.* New York: Alfred A. Knopf, 1963.

Huxley, Francis. *The Way of the Sacred.* New York: Barnes and Noble reprint, 1990.

Kealey, E. J. *Medieval Medicus.* Baltimore: Johns Hopkins University Press, 1981.

Pagels, Elaine. *The Gnostic Gospels.* New York: Vintage Books, Random House, 1981.

Penrose, Roger. *The Emperor's New Mind.* Oxford: Oxford University Press, 1989.

Revel, Jean-François. *Without Marx or Jesus.* New York: Doubleday and Company, 1970.

Stephen, Sir Leslie, and Sir Sidney Lee, eds. *Dictionary of National Biography.* 22 vols. Oxford: Oxford University Press, 1921.

Acknowledgments

I wish to thank my family to whom this book is dedicated for helpful comments and support, to thank Beverly in particular for the story of Xa, Yo and Zoda, to thank Mr. R. Schenk for introducing me to some of this literature, and to thank Mrs. Helen White for turning my handwritten scribble into readable prose. I am also grateful to the Centro de Investigación y Museo de Altamira for information about the Altamira caves, and to Dr. Mark Hall for his suggested improvements, particularly in chapters 2 and 3.

Index 1

Gods, Goddesses, and Their Associates

Index 2

People—Real and Legendary*

*See also Select Bibliography.

Index 3

General*

*Includes titles of books, essays, etc. See also Select Bibliography.